Process Plant Machinery

Process Plant Machinery

Edited by
Heinz P. Bloch, P.E.

Butterworths
Boston London Singapore Sydney Toronto Wellington

Library of Congress Cataloging-in-Publication Data

Process plant machinery / edited by Heinz P. Bloch.
    p.   cm.
  Includes bibliographies and index.
  ISBN 0-409-90087-7
    1. Chemical plants—Equipment and supplies. I. Bloch, Heinz P.
TP157.P73   1989   88-28731
660.2'83—dc19   CIP

British Library Cataloguing in Publication Data

Process plant machinery.
  1. Chemical engineering plants
  I. Bloch, Heinz P.
  660.2'8

  ISBN 0-409-90087-7

Butterworth Publishers
80 Montvale Avenue
Stoneham, MA 02180

10 9 8 7 6 5 4 3 2 1

Printed in the United States of America

To Bill Tillman, with thanks for getting me started

Contents

Preface

When I graduated from the New Jersey Institute of Technology in 1962, I had a degree in Mechanical Engineering and a fairly good knowledge of thermodynamics, fluid mechanics, and machine design. Little did I know, however, that application of these concepts to the real world of process machinery would be an altogether different learning process.

My early suspicion that the recent engineering graduate would greatly benefit from a down-to-earth, application-oriented text on process plant machinery was further reinforced when I accepted employment with the Exxon Corporation. There, I dealt with chemical engineers who had an equal and sometimes greater need to understand the basic operating concepts and application criteria of the machines that both of us encountered in modern process plants. It was then that I searched unsuccessfully for such a text and perhaps also thought I might, some day, assemble the material for this book.

The opportunity to synthesize the best available information into a cohesive overview text presented itself in 1986 and 1987 when I elected to retire from an interesting career in machinery reliability and maintenance-related engineering work with Exxon.

I then asked some of the best companies in the process machinery field for the material to be included in this text, and many of them responded to the challenge. The guidelines were straighforward: We wanted to present the younger engineer or technician with an overview of the machinery categories he or she was likely to encounter in most process plants. Furthermore, even the experienced individual should be able to benefit from a reference text that didn't dwell on theory but went quickly into a thorough explanation of how the equipment was designed and what made it work. Stated differently, this text is an attempt to close the gap between the machinery engineer and the chemical engineer.

Together with my many contributors, I express the hope that we have accomplished this goal.

Heinz P. Bloch
Baytown, Texas

Acknowledgments

It would have been impossible to assemble this text without the help and cooperation of many experienced equipment manufacturers and key individuals at each of the companies who devoted time and effort to the task. When we explained our intention to combine the best available information into a relevant text book, we received support from professionals who shared our goal of providing a practical, up-to-date text on the subject of process plant machinery. The following companies and individuals have earned our gratitude by allowing us to reprint, adapt, or otherwise incorporate their work in this book: Reliance Electric Co., Cleveland, OH; The Marley Cooling Tower Co., Mission, KS (electric motors and control); Prager Inc. Gear and Machine Products, New Orleans, LA; Westech Gear Corporation, Lynnwood, CA; Maag Gear Co., Zurich, Switzerland (gear speed transmission equipment); General Electric Co., Schenectady, NY (gas turbines); Cooper-Bessemer Reciprocating and Mr. J. Helmich, Grove City, PA (gas engines); Elliott Co., A United Technologies Subsidiary, Jeannette, PA; Siemens Energy and Automation, Bradentown, FL (steam turbines); Mafi-Trench Corp. and Mr. Jens E. Hanson, Santa Maria, CA (turboexpanders); Goulds Pumps, Inc., Seneca Falls, NY; Kontro Co., Inc., Orange, MA; Sundstrand Fluid Handling and Mr. John Alexander, Arvada, CO; Girdlestone Pumps Limited, Woodbridge, England (centrifugal pumps); Industrial Press, Inc., New York, NY; Dresser-Worthington, Harrison, NJ; Foster Pump Works, Inc., Westerley, RI; Milton Roy Co., St. Petersburg, FL; Pulsafeeder Co., Rochester, NY; Wallace & Tiernan, Belleville, NJ; The Hydraulic Institute, Cleveland, OH; McGraw-Hill Book Co., New York, NY; National Mastr Pump, Inc., Houston, TX; Netzsch Pumps, Philadelphia, PA; Leistritz Pump Corp., Allendale, NJ; American Lewa, Inc., Holliston, MA (positive displacement pumps); Kinney Vacuum Co., Boston, MA; SIHI Pumps, Inc., Grand Island, NY; Stokes Division of Pennwalt Corp., Philadelphia, PA; Sulzer-Burckhardt, Basel and Zurich, Switzerland (vacuum pumps); The Marley Cooling Tower Co. and Mr. John C. Hensley, Mission, KS (cooling water supply systems); Mannesmann-Demag and Mr. Otto Robert Hoesel, Houston, TX and Duisburg, Germany; Sundstrand Fluid Handling and Mr. Larry Glassburn, Arvada, CO (centrifugal compressors); Dresser-Rand Co. and Ms. Kathie H. Bulson, Phillipsburg, NJ (axial compressors); Garden City Fan Co. and Mr. Eugene Folkers, Niles, MI; Cemax Fans, Boston, MA; ACME Engineering and Manufacturing Corp., Muskogee, OK (propeller, axial, and centrifugal fans); Sulzer-Burckhardt and Messrs. H.R. Klaey and C. Matile, Basel and Zurich, Switzerland; Transamerica-De Laval, Trenton, NJ (reciprocating compressors); Aerzen USA Corp. and Mr. Pierre Noack, Exton, PA (rotating positive displacement compressors); Mixing Equipment Co., Inc. and Dr. James Y. Oldshue, Rochester, NY; Burgmann Seals America, Houston, TX (mixers and agitators); Werner & Pfleiderer Corp. and Mr. Adam Dreiblatt, Ramsey, NJ (internal mixers, single and twin-screw extruders); Sandvik Process Systems, Inc. and Mr. Jack Kahrs, Totowa, NJ (conveyor-based processing systems).

Three well-qualified machinery engineers and long-term associates have assisted me in compiling several chapters of this book. Hurlel Elliott, Gampa Bhat, and Perry Monroe deserve credit for their contributions to the chapters on fans, gas turbines, and electric motors, respectively.

Last, but not least, I'm indebted to my editor, Greg Franklin, who was compelled to show more than the usual patience in bringing it all together.

Introduction

Modern process plants would be unthinkable without the machinery to transport, modify, mix, compress, or otherwise manipulate the gases, liquids, and solids that move through the plant at any given time.

There are literally hundreds of types of plant machines, and it would not be at all unusual to find 20 or 30 different types at a single process plant site. Moreover, these could be further subdivided into numerous modified versions, depending on desired throughput, pressure, processing temperature, product characteristics, design life, and a host of other parameters.

In determining the scope of coverage it quickly became evident that we had to be highly selective in our choice of machinery to be included in this text. Dealing with every conceivable machine type would be an encyclopedic task and could result in superficial descriptions. Fortunately for the reader, and also the writer, the many process plant machines can be assigned to a few major classifications, or functional categories. Once the operating principles and typical configurations of important machine components are understood, the reader will find it considerably easier to think through the design variations or derivatives of a given machine.

Making this our basic premise, we decided that the majority of chemical engineers working in the process industries will probably encounter machinery that fits into one of the four primary classifications:

- prime movers and power transmission machinery
- pumping equipment
- gas compression machinery
- mixing, conveying, and separation equipment

We describe the essentials and, if necessary, important details of a number of process machines and drivers that fit into these four primary classifications. Starting with electric motors, the reader will find substantial information on turbines, pumps, and compressors, but may be disappointed to realize that thin-film evaporators, rotary filters, centrifuges, and other machines are not represented.

We also had to grapple with the question of whether the United Customary System of units (USCI) or International System of units (SI) should be used. From a purely practical view, it was realized that—like it or not—the reader will continue to (1) encounter both systems for the foreseeable future, and (2) have to be able to convert from one system to the other. Accordingly, we opted to generally leave the decision to the individual contributor; thus the reader will find himself or herself immersed in both USCI and SI units.

Finally, to avoid duplication, we decided for instance to explain compressor sealing components only in the chapter on centrifugal compressors, although other compressor types may well utilize the same styles of seals as well. The reader is thus encouraged to make ample use of the cross-reference index at the end of this text.

Process Plant Machinery

Chapter 1

Electric Motors and Controls*

Electric motors are by far the most prevalent ''machines'' in use in process plants around the world. The engineer, technician, and operator will benefit from an overview-type knowledge of electric motors; accordingly, we have elected first to introduce the reader to this machine category. Basic motor types, their major component parts, selection criteria, and other topics will be reviewed. To assist the reader, we have included a ''motor glossary'' at the end of this chapter.

Squirrel-cage induction motors are the most widely used type in the size range up to 200 horsepower. An induction motor is an alternating current device in which the primary winding on one member (usually the stator) is connected to the power source and the secondary winding or a squirrel-cage secondary winding on the other member (usually the rotor) carries the induced current. There is no physical electrical connection to the secondary winding; its current is induced. Induction motors are simple, rugged, and reliable because they have no rotating windings, slip rings, or commutators. They have good efficiency and high starting torque, but a lagging power factor; the induction motor operates below the synchronous speed that is set by the power cycles and the number of poles in the stator. For example, a two-pole, 60-Hertz (cycle-per-second) motor has a synchronous speed of 3600 RPM, and a four-pole motor operates at 1800 RPM. Normal operating speeds for induction motors are 3550 RPM and 1750 RPM. This deviation from synchronous speed is called *slip,* and it varies with load. Full load (maximum) slip varies from 1 percent in large motors to as high as 5 percent in small units. Most motors have an average slip of 3 percent.

Synchronous motors are used in cases where a fixed speed is required or to correct a lagging power factor in the power distribution system. A synchronous motor operates at constant speed up to full load, with a rotor speed equal to the speed of the rotating magnetic field of the stator, i.e., there is no slip. There are two major types of magnets on the synchronous motor: reluctance and permanent. Because of a relatively low inrush current, the synchronous motor has low starting torque and will trip out on overload when started under a load.

Electric motor configurations can be divided into two major categories: horizontal and vertical. The output shafts of horizontal motors are constructed parallel (horizontal) to the ground or mounting base, while the output shaft of a vertical motor will be perpendicular (or at a right angle) to the conventional mounting base. There is a growing interest in the use of vertical motors as pump drivers in the petroleum, petrochemical,

---

*Source: Reliance Electric Company, Cleveland, OH. Adapted by permission.

and chemical industries, because the entire assembly requires less space than a horizontal pump.

Both horizontal and vertical motors are available in a variety of designs and enclosures. They are classified as standard duty, large heavy-duty, special industry designs, and special application designs. Each of these classifications is available in different enclosures to meet a multitude of environmental requirements.* The two basic types of motor enclosures are *open* and *totally enclosed.* The open motor circulates external air inside the enclosure for cooling, whereas a totally enclosed motor prevents outside air from entering the enclosure. Both types of motors are available in either fractional or integral horsepowers.

Open motors are further classified as *drip-proof, splash-proof, guarded,* and *weather protected,* the distinction between them being the degree of protection provided against falling or airborne water gaining access to live and rotating parts. Drip-proof motors, as defined by the National Electrical Manufacturers Association (NEMA), are never used on outside installations. Integral-horsepower open drip-proof motors now marketed (sometimes called "protected") usually meet NEMA requirements for weather-protected type I enclosures, except they will not prevent the passage of a ¾" diameter rod. These enclosures are used on a wide variety of process machinery in outside locations. NEMA-defined weather-protected type II enclosures require oversized housings for special air passages to remove airborne particles. This type of enclosure is not available in the smaller motor sizes.

Totally enclosed motors used on such process equipment as cooling towers are classified as non-ventilated (TENV), fan-cooled (TEFC), air-over (TEAO), and explosion-proof. Whether a motor is TENV or TEFC is dependent on the need for an internally mounted fan to keep the operating temperature of the motor within the rating of its insulation. Air-over motors are TEFC motors without the fan and must have an outside cooling source. Totally enclosed motors are recommended and used for locations where fumes, dust, sand, snow, and high humidity conditions are prevalent, and they can provide a high quality installation either in or out of the air stream provided the typical problems of mounting, sealing, and servicing are properly addressed. In all cases, totally enclosed motors should be equipped with drain holes, and explosion-proof motors should be equipped with an approved drain fitting.

Explosion-proof motors are manufactured and sold for operation in hazardous atmospheres, as defined by the National Electrical Code. The motor enclosure must withstand an explosion of the specified gas, vapor, or dust within it, and prevent the internal explosion from igniting any gas, vapor, or dust surrounding it. The motors are Underwriters Laboratory (UL) approved and marked to show the class, group, and operating temperature (based on a 40°C ambient temperature) for which they are approved. In applying these motors, no external surface of the operating motor can have a temperature greater than 80 percent of the ignition temperature of the gas, vapor, or dust involved.

The National Electrical Code defines hazardous locations by class, group, and division. Class I locations contain flammable gases or vapors; class II locations contain combustible dust; and class III locations contain ignitible fibers or flyings. *Group* defines the specific gas, vapor, dust, fiber, or flying. *Division* defines whether the explosive atmosphere exists continuously (division 1) or only in case of an accident (division 2).

Motors for division 1 applications must be explosion-proof. Standard open or totally enclosed motors that do not have brushes, switching mechanisms, or other arc-producing devices can be used in class I, division 2 applications. In some cases, they can also be used in class II, division 2 and class III, division 2 applications.

---

*This segment courtesy of the Marley Cooling Tower Corporation. Adapted by permission.

MOTOR DESIGN

Three-Phase Motors

Three-phase squirrel-cage induction motors have become the standard on the over-whelming majority of process plant machines. They do not have the switches, brushes, or capacitors of other designs and therefore require somewhat less maintenance. Where three-phase power is not available, single phase capacitor-start motors may be used, usually not exceeding 7.5 horsepower. Concerned process machinery manufacturers will supply motors that are a few steps beyond "off-the-shelf" quality. These motors are usually purchased from specifications developed after comprehensive, rigorous testing under simulated operating conditions.

Two-Speed Motors

Two-speed motors are of a variable torque design, in which the torque varies directly with the speed, with 1800/900 RPM being the most common speeds. Single-winding design motors enjoy greatest utilization, since they are smaller in size and less expensive than those of a two-winding design.

Occasionally, an installation deserves consideration of the use of two-speed motors. Cooling tower fans are a case in point. Whether operated seasonally or year-round, there will be periods when a reduced load and/or a reduced ambient temperature will permit satisfactory cold water temperatures with the fans operating at half-speed. The benefits accrued from this mode of operation will usually offset the additional cost of two-speed motors in a relatively short time.

Additionally, since nighttime operation is normally accompanied by a reduced ambient temperature, some operators utilize two-speed motors to preclude a potential noise complaint.

High-Efficiency Motors

Several motor manufacturers provide high-efficiency designs that are suitable for use on numerous types of process plant machinery. These motors are in the same frame sizes as standard motors, but they utilize more efficient materials. While the efficiency will vary with the manufacturer and the size of the motor, the efficiency will always be higher than that manufacturer's standard motor. Naturally, there is a price premium for high-efficiency motors, which must be evaluated against their potential for energy savings.

Motor Insulation

One of the most important factors contributing to long service life in an electric motor is the quality of the insulation. It must withstand thermal aging, contaminated air, moisture, fumes, expansion and contraction stresses, mechanical vibration and shock, as well as electrical stress.

Insulation is categorized by classes, which establish the limit for the maximum operating temperature of the motor. Classes A, B, F, and H are used in the United States, with class A carrying the lowest temperature rating and class H the highest. Standard integral horsepower motors have class B insulation and are designed for a maximum altitude of 3300 feet and a maximum ambient temperature of 40 °C.

Class F insulation is used for higher altitudes, as well as higher ambient temperatures, and it is gaining increased use as a means of improving the service factor of a motor of given frame size.

Motor Service Factor

The service factor of a motor is an indication of its maximum allowable continuous power output, as compared with its nameplate rating. A 1.0 service factor motor should not be operated beyond its rated horsepower at design ambient conditions, whereas a 1.15 service factor motor will accept a load 15 percent in excess of its nameplate rating. Usually, motor manufacturers will apply the same electrical design to both motors but will use class B insulation on 1.0 service factor motors and class F insulation on 1.15 service factor motors. Class B insulation is rated at a total temperature of 130 °C and class F is rated at 155 °C.

More important, a 1.15 service factor motor operates at a temperature from 15 °C to 25 °C lower (compared with the temperature rating of its insulation) than does a 1.0 service factor motor operating at the same load. This, of course, results in longer insulation life and, therefore, longer service life for the motor. For this reason, many equipment manufacturers will recommend the use of 1.15 service factor motors for loads at or near nominal horsepower ratings.

Since increased air density increases the load on air movers, an added attraction for using 1.15 service factor motors is that there is less chance of properly sized overloads tripping out this equipment category during periods of reduced heat load and low ambient temperatures.

Motor Heaters

Although the insulation used in quality electric motors is considered to be nonhygroscopic, it does slowly absorb water and, to the degree that it does, its insulation value is reduced. Also, condensed moisture on insulation surfaces can result in current leakage between pin holes in the insulation varnish. Because of this, it is advisable on installations exposed to high humidity to keep the inside of the motor dry.

This can be done by keeping the temperature inside the motor 5 °C to 10 °C higher than the temperature outside the motor. Motors in continuous service will be heated by the losses in the motor, but idle motors require the addition of heat to maintain this desired temperature difference.

One recommended method of adding heat is by the use of electric space heaters, sized and installed by the motor manufacturer. Another method is single-phase heating, which is simply the application of reduced voltage (approximately 5 to 7.5 percent of normal) to two leads of the motor winding. Both of these methods require controls to energize the heating system when the motor is idle. If low voltage dynamic braking is used to prevent an inoperative motor from rotating, it will add sufficient heat to the motor windings to prevent condensation. A typical application would be in cooling tower fans.

Motor Torques

High starting torque motors are neither required nor recommended for most process machines. Normal torque motors perform satisfactorily for pumps, fans, blowers, etc., and cause far less stress on the driven components. Normal torque motors should be specified for the bulk of single-speed applications, and variable torque in the case of two-speed.

There are five points along a motor speed-torque curve that are important to the operation of many machines: (1) locked-rotor torque, (2) pull-up torque (minimum torque during acceleration), (3) breakdown torque (maximum torque during acceleration), (4) full-load torque, and (5) maximum plugging torque (torque applied in reversing an operating motor). Compared with full-load torque, the average percentage values of the other torques are as follows: locked-rotor torque = 200%; pull-up torque = 100%; breakdown torque = 300%; and plugging torque = 250%.

MOTOR CONTROLS

Control devices and wiring, the responsibility for which usually falls to the purchaser, can also be subjected to demanding service situations. Controls serve to start and stop the motor and to protect it from overload or power supply failure, thereby helping assure continuous reliable equipment operation. They are not routinely supplied as a part of a machinery procurement contract, but because of their importance to the system, the need for adequate consideration in the selection and wiring of these components cannot be overstressed.

The various protective devices, controls, and enclosures required by most electrical codes are described in the following paragraphs. *In all cases, motors and control boxes must be grounded.*

1. Fusible Safety Switch or Circuit Breaker: This device provides the means to disconnect the controller and motor from the power circuit. It also serves to protect the motor-branch-circuit conductors, the motor control apparatus, and the motors against overcurrent due to short circuits or grounds. It must open all ungrounded conductors and be visible (not more than 50 feet distant) from the controller or be designed to lock in the open position. The design must indicate whether the switch is open or closed, and there must be one fuse or circuit breaker in each ungrounded conductor. A disconnect switch must be horsepower rated or must carry 115 percent of full-load current and be capable of interrupting stalled-rotor current. A circuit breaker must also carry 115 percent of full-load current and be capable of interrupting stalled-rotor current.
2. Nonfused Disconnect Switch: This switch is generally only required if the fusible safety switch or circuit breaker either cannot be locked in the open position or cannot be located in sight of the motor.
3. Manual and Magnetic Starters: These controls start and stop the motor. They also protect the motor, motor control apparatus, and the branch-circuit conductors against excessive heating caused by low or unbalanced voltage, overload, stalled rotor, and too frequent cycling. Starter requirements are determined by the basic horsepower and voltage of the motor. Overloads in a starter are sized to trip at not more than 125 percent of full-load current for motors having a 1.15 or higher service factor, or 115 percent of full-load current in the case of 1.0 service factor motors. Single phase starters must have an overload in one ungrounded line. A three-phase starter must have overloads in all lines. If a magnetic controller is used, it may be actuated by devices sensing certain process fluid parameters. Temperature sensors sensing cooling water temperature would be a typical example.
4. Control Enclosures: NEMA has established standard types of enclosures for control equipment. The types most commonly used in conjunction with process plant machinery are as follows:
    a. NEMA Type 1—General Purpose: Intended primarily to prevent accidental contact with control apparatus. It is suitable for general purpose applications indoors, under normal atmospheric conditions. Although it serves as a protection against dust, it is not dust-proof.

b. NEMA Type 3—Dust-tight, Rain-tight, and Sleet-resistant: Intended for outdoor use and for protection against wind-blown dust and water. This sheet metal enclosure is usually adequate for use outdoors on a cooling tower. It has a watertight conduit entrance, mounting means external to the box, and provision for locking. Although it is sleet-resistant, it is not sleet-proof.

c. NEMA Type 3R: This is similar to type 3, except it also meets UL requirements for being rainproof. When properly installed, rain cannot enter at a level higher than the lowest live part.

d. NEMA Type 4—Watertight and Dust-tight: Enclosure is designed to exclude water. It must pass a hose test for water and a 24-hour salt spray test for corrosion. This enclosure may be used outdoors on a cooling tower. It is usually a gasketed enclosure of cast iron or stainless steel.

e. NEMA Type 4X: Similar to type 4, except it must pass a 200-hour salt spray test for corrosion. It is usually a gasketed enclosure of fiber-reinforced polyester.

f. NEMA Type 6—Submersible, Watertight, Dust-tight and Sleet-resistant: Intended for use where occasional submersion may be encountered. Must protect equipment against a static head of water of 6 feet for 30 minutes.

g. NEMA Type 12—Dust-tight and Drip-tight: Enclosure intended for indoor use. It provides protection against fibers, flyings, lint, dust, dirt, and light splashing.

h. NEMA Type 7—Hazardous Locations—Class I Air-Break: This enclosure is intended for use indoors in locations defined by the National Electrical Code for class I, division 1, groups A, B, C, or D hazardous locations.

i. NEMA Type 9—Hazardous Locations—Class II Air-Break: Intended for use indoors in areas defined as class II, division 1, groups E, F, or G hazardous locations.

WIRING SYSTEM DESIGN

The design of the wiring system for the numerous process machines, fans, compressors, pumps, and controls is the responsibility of the owner's engineer. Although the average installation presents no particular problem, there are some systems that require special consideration if satisfactory operation is to result. Conductors to motors must be sized both for 125 percent of the motor full-load current and for voltage drop. If the voltage drop is excessive at full load, the resultant increased current can cause overload protection to trip. (Although motors should be operated at nameplate voltage, they *can* be operated at plus or minus 10 percent of nameplate voltage.)

In a normal system with standard components, even the larger machines will often attain operating speed in less than 15 seconds. During this starting cycle, although the motor current is approximately 600 percent of full-load current, the time delays in the overload protective devices prevent them from breaking the circuit.

Because of the high starting current, the voltage at the motor terminals is reduced by line losses. Within certain limits, the output torque of a motor varies as the square of the voltage. Thus, under starting conditions, the current increases, the voltage decreases, and the torque decreases, with the result that the starting time is increased. Long conductors that increase voltage drop, low initial voltage, and high-inertia fans can all contribute to increased starting time, which may cause the protective devices to actuate. In extreme cases, the starting voltage may be insufficient to allow acceleration of the fan to full speed regardless of time.

The wiring system design must consider pertinent data on the available voltage (its actual value, as well as its stability), length of lines from the power supply to the motor, and the motor horsepower requirements. If this study indicates any question as to the startup time of the motor, the inertia of the load as well as that of the motor should be determined. This is the commonly known "flywheel" effect (WK^2 factor). Once the WK^2 of the load (referred to the motor shaft) is obtained, the acceleration time can be determined using the motor speed-torque and speed-current curves, compared with the speed-torque curves for the fan. If the calculated time and current is greater than allowed by the standard overload protection of the motor, the condition may be corrected by increasing voltage, by increasing conductor size, or by providing special overload relays. Given no solution to the base problem, special motors or low-inertia fans may be necessary.

CYCLING OF MOTORS

The high inrush current that occurs at motor startup causes heat to build up in the windings and insulation. For this reason, the number of start-stop or speed-change cycles should be limited in order to allow time for excessive heat to be dissipated. As a general rule, 30 seconds of acceleration time per hour should not be exceeded. A fan-motor system that requires 15 seconds to achieve full speed, therefore, would be limited to two full starts per hour. Smaller or lighter fans, of lesser inertia, permit greater frequency of cycling.

STANDARD INDUCTION MOTORS *

Standard induction motors are typically used for industrial applications such as machine tools, material handling equipment, processing lines, pumps, fans, blowers, and countless others.

The protected motor shown in Figure 1–1 (sometimes called the drip-proof or open) is typical of general purpose AC motors. This enclosure is suited for most

FIGURE 1–1   *Energy-efficient electric motor typical of general purpose AC motors. (Source: Reliance Electric, Cleveland, OH.)*

---

*Source: Reliance Electric Company, Cleveland, OH. Adapted by permission.

FIGURE 1–2   *Totally enclosed fan-cooled motor. The externally mounted fan pushes cooling air over the fin-equipped frame. (Source: Reliance Electric, Cleveland, OH.)*

industrial environments when temperatures are 40 °C maximum with ambient air relatively clean and dry.

The totally enclosed fan-cooled (TEFC) motor (Figure 1–2) is the enclosed motor most often selected for indoor or outdoor industrial environments containing dust, dirt, water, etc., in modest amounts that are best kept out of the interior of motors. The external fan is used to cool the motor, since there is not a free exchange of air.

Explosion-proof motors have a dual purpose: to withstand an explosion from within and to prevent the explosion of gases in the atmosphere. Both requirements place special emphasis on motor design. Explosion-proof motors should meet the rigid requirements of UL for most National Electrical Code class, group, and temperature code restrictions.

Motor sizes and dimensions have been standardized by the NEMA. NEMA frames 48, 56, and 140 T encompass single-phase capacitor start and polyphase motors. These motors are designed for continuous duty operations in a 40 °C maximum ambient environment—obviously not typical plants. Similarly, NEMA frames 180 T through 449 T include standard AC polyphase induction motors and also certain single-phase motors designed for a maximum environmental temperature of 40 °C. They are, however, available in explosion-proof executions for use in class I, group D, and class II, groups F and G hazardous locations.

LARGE HEAVY-DUTY ALTERNATING CURRENT MOTORS

Large heavy-duty AC motors cover the range from 250 to 5000 horsepower and are available in a variety of frames or enclosures. Totally enclosed motors, typically of cast iron frame design, range to roughly 500 horsepower. Weather-protected designs and tube or water-cooled enclosures can be supplied with these motors. They range through 1500 horsepower with cast iron, and 5000 horsepower with fabricated steel construction (Figure 1–3).

FIGURE 1–3   *Weather-protected motor with cooling facilities enclosure and fabricated frame design. (Source: Reliance Electric, Cleveland, OH.)*

SPECIAL INDUSTRY AND APPLICATION DESIGNS

Capable motor manufacturers can offer suitably modified motors for specific industries. Typical are motors designed for corrosive atmospheres such as those found in the paper, chemical, petroleum, and metals industries. These motors would be available from fractional to 500 horsepower.

Motors for the food and dairy industries must be designed for easy cleaning and hose washdowns to meet the rigid sanitary codes of all government agencies, the Baking Industry Sanitation Standards Committee, and Dairy Standards.

Figure 1–4 depicts an important special application design, a brake motor that combines a motor and an integrally mounted disc brake into one unit. These direct action brakes are spring set, electrically released, and designed for stopping and holding a load.

Energy-saving motor designs are available and have reduced full load motor losses through the use of optimum electrical designs and increased active material. They are also designed to operate at low noise levels and are available up to 300 horsepower.

Multispeed motors are motors with special electrical characteristics for a wide variety of two-speed applications requiring constant or variable torque.

Vertical motors (Figure 1–5) are flanged, footless designs used in direct coupled vertical applications. Vertical motors are available with normal and medium thrust

FIGURE 1–4   *Brake motor combining a motor and integrally mounted disc brake. (Source: Reliance Electric, Cleveland, OH.)*

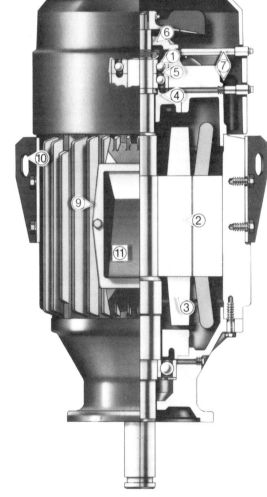

FIGURE 1–5   *Cross section of NEMA frame, vertical AC motor. 1 = top-mounted thrust bearing; 2 = steel laminations; 3 = rotor fins; 4 = bearing cap; 5 = lower thrust bearing; 6 = grease reservoir; 7 = grease relief ports; 8 = drip cover; 9 = conduit box; 10 = lifting plates; 11 = cover, with grounding provision. (Source: Reliance Electric, Cleveland, OH.)*

FIGURE 1-6 *Submersible motor for use in hazardous areas. (Source: Reliance Electric, Cleveland, OH.)*

capabilities on many different frames. They are widely applied in pumps and mixers, and sizes can exceed 1000 horsepower.

A submersible motor is shown in Figure 1-6. These motors are designed for continuous pumping duty submerged in liquids containing a maximum solid content of 10% by weight and 90% liquid.

For use in hazardous environments, these motors are UL listed for use in class I, group D, division I hazardous locations in air or submersible in water or sewage.

MAJOR COMPONENTS OF INDUCTION MOTORS

The major components of an AC induction motor are illustrated in Figure 1-7. They consist of the following:

FIGURE 1–7   *What to look for in major components of an AC motor. (Source: Reliance Electric, Cleveland, OH.)*

- frame
- stator or windings
- rotor
- bearings
- end shield, end bell, or end bracket
- cooling fan
- conduit box

Frame

Cast iron is the most popular frame material. The frame must be made of a rigid material that will absorb noise as well as vibration because it provides the structural support for all the other motor components. The stator or windings are mechanically attached to the frame prior to the rotor installation. Tight machining tolerances are required on the mating surfaces of the frame and end bells or end shields. These tight tolerances ensure accurate bearing locations that allow the rotor to run at the proper air gap. All of this helps to increase motor efficiency.

Stator and Rotor

The stator is made up of copper wire that is formed into coils that are insulated. Stator coil construction procedures are sometimes complicated and must be entrusted to capable manufacturers.

The rotor is the rotating member of an induction motor that is made up of stacked laminations. Thin, high silicone steel stampings are aligned on a keyed mandrel and welded in the axial direction in three or four places to form a rigid cylinder. This cylinder is removed from the mandrel and pressed on the motor shaft. Slots in each lamination of the cylinder are filled with molten aluminum to form the squirrel cage. The cast aluminum bars of the squirrel cage act as conductors for the induced magnetic fields. Some motors are made with copper bars instead of cast aluminum.

The shaft portion of the rotor is precision machined on both ends to fit the bearings. Motors in the 5- to 200-horsepower sizes are usually fitted with antifriction (ball) bearings. Roller or journal bearings are used for very large motors.

Bearings and End Shields

Bearings are used to reduce friction and wear while supporting the rotating element (rotor). The bearing acts as the connection point between the rotating (rotor) and stationary (end bell or end shield) elements of a motor. There are various types, such as roller, ball, sleeve (journal), and needle bearings.

The ball bearing is used in virtually all types and sizes of electric motors. It exhibits low friction loss, is suited for high-speed operation, and is compatible in a wide range of temperatures. There are various types of ball bearings such as open, single-shielded, or sealed. Although not mandatory, some manufacturers offer a special bearing design for oil-mist lubrication.

End shields or end bells cover the ends of the motor frame. They protect the internal electrical and mechanical parts from moisture and dirt and provide support for the bearings.

Cooling Fan

The cooling fan is a small but important part of an electric motor that is often overlooked. It provides the cooling air across the TEFC motor and is made of a non-sparking, corrosion-resistant material. Most motors come with a bidirectional fan so as not to be sensitive to the direction of motor rotation. Some motors have low noise or high-efficiency fan designs and will pump air in only one direction of rotation. This type of fan generally has a direction of rotation arrow cast on the fan hub.

Conduit Box

The conduit box is the metal container on the side of the motor that houses the electrical connections. It must provide a waterproof environment where the stator (winding) leads are connected to the incoming power leads. If oil mist is used to lubricate the motor bearings, a gas-tight seal must be provided around the stator leads at the penetration of the conduit box.

A MOTOR AS PART OF A SYSTEM

Selecting a Motor

Motor selection is a complicated process containing numerous trade-offs. Efficiency is only one of several important considerations. The objective of informed motor selection is to arrive at the best possible installation consistent with minimum cost, horsepower, and frame size for the specified life expectancy, load torque, load inertia, and duty cycle of the specified application.

To satisfy the torque, horsepower, and speed requirements of a large variety of motor applications, polyphase AC motors are designed and manufactured in four groups classified design A, B, C, and D by NEMA. Each classification of motors has its own distinctive speed-torque relationship (Figure 1–8) and inherent expectations regarding motor efficiency.

Motors intended for loads that are relatively constant and run for long periods of time are of low slip design (less than 5 percent) and are inherently more efficient than design D motors, which are used in applications where heavy loads are suddenly applied, such as hoists, cranes, and heavy metal presses. Design D motors deliver high starting torque and are designed with high slip (more than 5 percent) so that motor

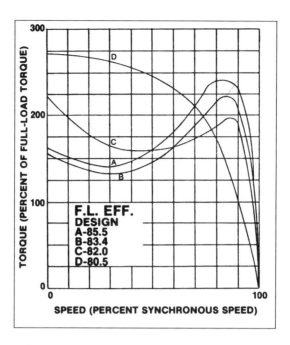

FIGURE 1–8   *Speed-torque curves for a 5-HP motor, NEMA design A and D, and full-load efficiencies. (Source: Reliance Electric, Cleveland, OH.)*

speed can drop when fluctuating loads are encountered. Although design D motor efficiency can be less than other NEMA designs, it is not possible to replace a design D motor with a more efficient design B motor, because it would not meet the performance demands of the load.

The motor with the highest operating efficiency does not always provide the lowest energy choice. Figure 1–9 compares the watts loss of a NEMA design D and a design B motor, in a duty cycle that accelerates a load inertia of 27 lb. ft.^{2} to full speed and runs at full load for 60 seconds. During acceleration, the lower curve represents the performance of the design D motor, while the upper curve reflects the NEMA design B motor. The shaded area between the curves represents the total energy difference during acceleration. In this example, this area is approximately 6.0 watt-hours, the energy saved accelerating this load with a design D motor instead of a design B. During the run portion of this duty cycle, the energy loss differential favors the NEMA design B, because it has a higher operating efficiency. In this example, the energy saved operating this load with a design B motor instead of a design D motor is approximately 2.8 watt-hours.

The bar chart shown in Figure 1–10 summarizes acceleration and running loss/ cycle on both the NEMA design B and design D. Comparison of the total combined acceleration and running portions of this duty cycle indicates a total energy savings of 3.2 watt-hours favoring the use of the design D motor, even though the design B motor has an improved operating efficiency. The key is the improved ability of the design D motor to accelerate a load inertia at minimum energy cost.

Other Components Affecting Efficiency

Because a motor buyer selects the most efficient motor of a given size and type does not mean that energy savings are being optimized. Every motor is connected to some

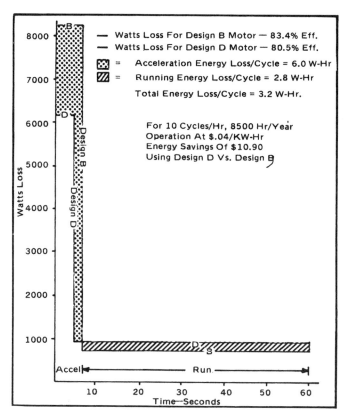

FIGURE 1-9 *Energy usage on duty cycle application 5-HP, 4-pole, TEFC accelerating 27 lb-ft² inertia. (Source: Reliance Electric, Cleveland, OH.)*

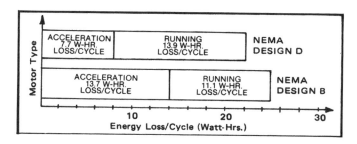

FIGURE 1-10 *Acceleration and running loss per cycle on NEMA D and NEMA B motors. (Source: Reliance Electric, Cleveland, OH.)*

form of driven equipment: a crane, a machine tool, a pump, etc., and motors are often connected to their loads through gears, belts, or slip couplings. By examining the total system efficiency, the component that offers the greatest potential improvements can be identified and money allocated to the component offering the greatest payback.

In the case of new equipment installations, a careful application analysis, including load and duty cycle requirements, might reveal that a 7 ½-horsepower pump, for example, could be utilized in place of a 10-horsepower pump, thereby reducing motor horsepower requirements by one third. By reducing the mass of the moving parts, the energy required to accelerate the parts is also proportionately reduced. Or in the instance of an air compressor application, the selection as to the size and type of compressor relative to load and duty cycle will affect system efficiency and energy usage. Of course, the most efficient equipment should be selected whenever possible.

Reduced system efficiency and increased energy consumption are also possible with existing motor drive systems due to additional friction that can gradually develop within the driven machine. This additional friction could be caused by a build-up of dust on a fan, the wearing of parts causing misalignment of gears or belts, or insufficient lubrication in the driven machine. All of these conditions cause the driven machine to become less efficient, which causes the motor to work harder. Rather than replace the existing motor with a higher efficiency model, replacing either critical machine components or the machine itself may result in greater system efficiency and energy savings.

Choosing the Best Applications

Energy-efficient motors may be the most cost-effective answer for certain applications. Simple guidelines are listed below:

- Choose applications where motor running time exceeds idle time.
- Review applications involving larger horsepower motors, where energy usage is greatest and the potential for cost savings can be significant.
- Select applications where loads are fairly constant, and where load operation is at or near the full load point of the motor for the majority of the time.
- Consider energy-efficient motors in areas where power costs are high. In some areas, power rates can run as much as $.12 per kilowatt-hour. In these cases, the use of an energy-efficient motor might be justified in spite of long idle times or reduced load operations.

Using these simple guidelines, followed by an analysis and cost justification based on various techniques, can yield results that will influence motor choice beyond just-in cost consideration.

Other Determinants of Operating Cost

Voltage Unbalance

Although efficiency is a commonly used indicator of energy usage and operating costs, there are several important factors affecting motor operating costs. Rated performance as well as selection and application considerations of polyphase motors requires a balanced power supply at the motor terminals. Unbalanced voltage affects the motor's current, speed, torques, temperature rise, and efficiency. NEMA Standard MG 1-14.34 recommends derating the motor where the voltage unbalance exceeds 1 percent and recommends against motor operation where voltage unbalance exceeds 5 percent. Voltage unbalance is defined as follows:

$$\text{Voltage Unbalance (\%)} = 100 \times \frac{\text{Maximum Voltage Deviation From Average Voltage}}{\text{Average Voltage}}$$

Voltage imbalance is not directly proportional to the increase in motor losses, as a relatively small unbalance in percent will increase motor losses significantly and decrease motor efficiency as Figure 1–11 shows. An effort to reduce losses with the purchase of premium priced, premium efficiency motors that reduce losses by 20 percent can easily be offset by a voltage unbalance of 3.5 percent that increases motor losses by 20 percent.

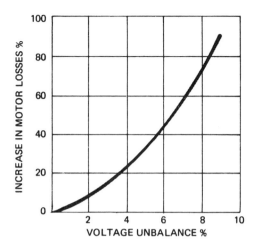

FIGURE 1–11 *Motor loss percentage as a function of voltage unbalance. (Source: Reliance Electric, Cleveland, OH.)*

Energy cost can be minimized in many industrial applications by reducing the additional motor watts loss due to voltage unbalance. Uniform application of single-phase loads can assure proper voltage balance in a plant's electrical distribution system used to supply polyphase motors.

Motor Loading

One of the most common sources of motor watts loss is the result of a motor not being properly matched to its load. In general, for standard NEMA frame motors, motor efficiency reaches its maximum at a point below its full-load rating, as indicated in Figure 1–12. This efficiency peaking below full load is a result of the interaction of the fixed and variable motor losses resulting in meeting the design limits of the NEMA standard motor performance values, specifically locked rotor torque and current limits.

Power factor is load variable and increases as the motor is loaded, as Figure 1–12 shows. At increased loads, normally in the region beyond full load, this process reverses as the motor's resistance to reactive ratio begins to decrease and power factor begins to decline.

In some applications where motors run for an extended period of time at no load, energy could be saved by shutting down the motor and restarting it at the next load period.

Maintenance

Proper care of the motor will prolong its life. A basic motor maintenance program requires periodic inspection and, when encountered, the correction of unsatisfactory conditions. Among the items to be checked during inspection are lubrication, ventilation, and presence of dirt or other contaminants; alignment of motor and load; possible changing load conditions; belts, sheaves, and couplings; and tightness of hold-down bolts.

Total Energy Costs

There are three basic components of industrial power cost: cost of real power used, power factor penalties, and demand charges. To understand these three charges and

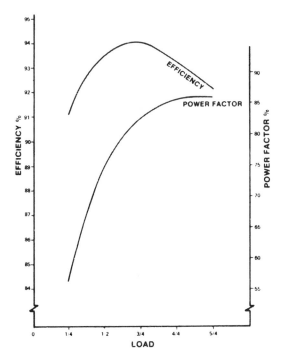

FIGURE 1–12   *Power factor and efficiency changes as a function of motor load. (Source: Reliance Electric, Cleveland, OH.)*

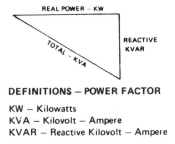

DEFINITIONS – POWER FACTOR

KW – Kilowatts
KVA – Kilovolt – Ampere
KVAR – Reactive Kilovolt – Ampere

FIGURE 1–13   *Electrical power vector diagram. (Source: Reliance Electric, Cleveland, OH.)*

how they are determined, a review of the power vector diagram (Figure 1–13) identifies each component of electrical energy and its corresponding energy charge.

Real Power

The real power-kilowatt (kw) is the energy consumed by the load. Real power-kw is measured by a watt-hour meter and is billed at a given rate ($/kw-hr). It is the real power component that performs the useful work and is affected by motor efficiency.

Power Factor

Power factor is the ratio of real power-kw to total KVA. Total KVA is the vector sum of the real power and reactive KVAR. Although reactive KVAR performs no actual work, an electric utility must maintain an electrical distribution system (i.e., power

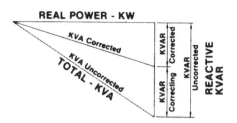

FIGURE 1–14 *Effect of corrective capacitance on total KVA vector. (Source: Reliance Electric, Cleveland, OH.)*

transformers, transmission lines, etc.) to accommodate this additional electrical energy. To recoup this cost burden, utilities may pass this cost on to industrial customers in the form of a power factor penalty for power factor below a certain value.

Power factors in industrial plants are usually low due to the inductive or reactive nature of induction motors, transformers, lighting, and certain other industrial process equipment. Low power factor is costly and requires an electric utility to transmit more total KVA than would be required with an improved power factor. Low power factor also reduces the amount of real power that a plant's electrical distribution system can handle, and increased line currents will increase losses in a plant's distribution system.

A method to improve power factor, which is typically expensive, is to use a unity or leading power factor synchronous motor or generator in the power system. A less expensive method is to connect properly sized capacitors to the motor supply line. In most cases, the use of capacitors with induction motors provides lower first cost and reduced maintenance expense. Figure 1–14 graphically shows how the total KVA vector approaches the size of the real power vector as reactive KVAR is reduced by corrective capacitors. Because of power factor correction, less power need be generated and distributed to deliver the same amount of useful energy to the motor.

Just as the efficiency of an induction motor may be reduced as its load decreases, the same is true for the power factor, only at a faster rate of decline. A typical 10-horsepower, 1800 RPM, three-phase, design B motor with a full-load power factor of about 80 percent decreases to about 65 percent at half-load. Therefore, it is important not to overmotor. Select the right size motor for the right job. Figure 1–15 shows that the correction of power factor by the addition of capacitors not only improves the overall power factor but also minimizes the fall-off in power factor with reduced load.

Demand Charges

The third energy component affecting cost is demand charge, which is based on the peak or maximum power consumed or demanded by an industrial customer during a specific time interval. Because peak power demands may require an electric utility to increase generating equipment capacity, a penalty is assessed when demand exceeds a certain level. This energy demand is measured by a demand meter, and a multiplier is applied to the real power-kw consumed.

Industrial plants with varying load requirements may be able to affect demand charges by (1) load cycling, which entails staggering the starting and use of all electrical equipment and discontinuing use during peak power intervals, and (2) using either electrical or mechanical "soft start" hardware, which limits power inrush and permits a gradual increase in power demand.

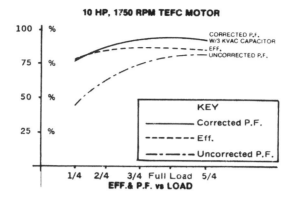

FIGURE 1-15   *Effect of capacitors on fall-off in power factor with reduced load. (Source: Reliance Electric, Cleveland, OH.)*

ADJUSTABLE FREQUENCY MOTOR CONSIDERATIONS

Speed control by way of adjusting power frequency is becoming more and more important for economical throughput or pressure capacity variation of modern process machinery. Several key parameters that must be considered when applying induction motors to adjustable frequency controllers include the load torque requirements, current requirements of the motor and the controller current rating, the effect of the controller wave-shape on the motor temperature rise, and the required speed range for the application.

In order to properly size a controller for a given application, it is necessary to define the starting torque requirements, the peak torque requirements, and the full-load torque requirements. These basic application factors require reexamination because the speed-torque characteristics of an induction motor/controller combination are different from the speed-torque characteristics of an induction motor operated on sine-wave power.

The motor current requirements should be defined for various load points at various speeds in order to ensure that the controller can provide the current required to drive the load. The current requirements are related to the torque requirements, but there are also additional considerations due to the harmonics of adjustable frequency control power that must be taken into account.

Temperature rise and speed range must be considered when applying induction motors to adjustable frequency controllers because this nonsinusoidal power results in additional motor losses, which increase temperature rise and reduce motor insulation life.

Before dicussing the speed-torque characteristics of a motor/controller combination, it is useful to review the speed-torque characteristics of an induction motor started at full voltage and operated on utility power (Figure 1-16). Here we see the speed-torque curve for a 100-horsepower, 1800 RPM, high-efficiency motor. When this motor is started across the line, the motor develops approximately 150 percent of full-load torque for starting and then accelerates along the speed-torque curve through the pull-up torque point, through the breakdown torque point, and, finally, operates at the full-load torque point, which is determined by the intersection of the load line and the motor speed-torque curve.

In this case, we have shown an application such as a conveyor where the load-torque requirement is constant from 0 RPM to approximately 1800 RPM. The difference between the motor speed-torque curve and the load line is the accelerating torque and is indicated by the cross-hatched area.

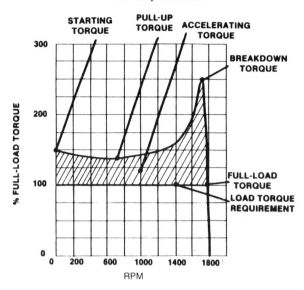

FIGURE 1–16   *Speed-torque characteristics of induction motors started at full voltage. (Source: Reliance Electric, Cleveland, OH.)*

If the load-torque requirement ever exceeded the maximum torque capability of the induction motor, the motor would not have enough torque to accelerate the load and would stall. For instance, if the load line required more torque than the motor could produce at the pull-up torque point, i.e., 170 percent load torque versus 140 percent pull-up torque, the motor would not increase in speed past the pull-up torque speed and would not be able to accelerate the load. This would cause the motor to overheat. It is, therefore, important to ensure that the motor has adequate accelerating torque to reach full speed.

Normally, the motor accelerates the load and operates at the point of intersection of the load line and the motor speed-torque curve. The motor then always operates between the breakdown torque point and the synchronous speed point that corresponds to the 1800 RPM location on the horizontal axis. If additional load torque is required, the motor slows down and develops more torque by moving up toward the breakdown torque point. Conversely, if less torque is required, the motor would speed up slightly toward the 1800 RPM point. Again, if the breakdown torque requirements were exceeded, the motor would stall.

Figure 1–17 depicts the same motor speed-torque curve, but now the motor current has been shown for full voltage starting.

Typically, when a NEMA design B induction motor is started across the line, an inrush current of 600 percent to 700 percent occurs corresponding to the starting torque point. As the load is accelerated to the full-load torque point, the current decreases to 100 percent full-load current at 100 percent full-load torque. High currents, however, are drawn during the acceleration time.

The amount of time that the motor takes to accelerate the load will depend on the average available accelerating torque, which is the difference between the motor speed-torque curve and the load speed-torque curve, and the load inertia.

Figure 1–18 illustrates a blown-up view of the region between the breakdown torque point and the synchronous speed point, which is where the motor would operate. This is of particular interest because the current for various torque requirements can easily be seen. This would directly affect the size of the controller required to produce a given torque because controllers are current-rated.

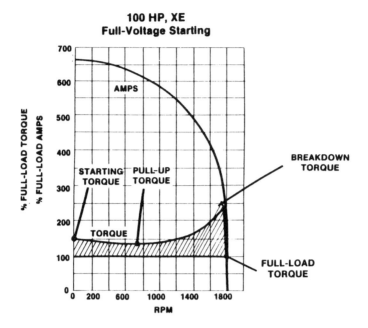

FIGURE 1–17   *Motor current of induction motor started at full voltage. (Source: Reliance Electric, Cleveland, OH.)*

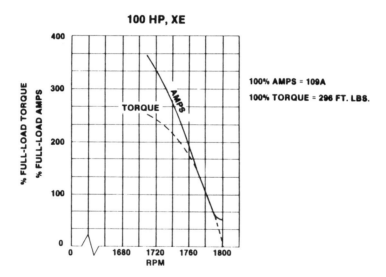

FIGURE 1–18   *Motor current and torque as full operating speed is approached. (Source: Reliance Electric, Cleveland, OH.)*

At 100 percent full-load torque, 100 percent full-load nameplate current is required. At 150 percent torque, 150 percent full-load nameplate current is required. Beyond the 150 percent full-load torque point, however, the torque-per-amp ratio is no longer proportional. For this case, 251 percent breakdown torque would require 330 percent current.

Adjustable frequency controllers are typically rated for a maximum of 100 percent continuous or 150 percent for one minute of the controller full-load current. This would generally provide a maximum of 100 percent or 150 percent of motor full-load torque. This would not, however, provide the same amount of torque as the motor could potentially develop if it were operated from utility power, which could normally provide as much current as the motor required.

It would generally be uneconomical to oversize a controller to obtain the same amount of current (torque), since the controller size would actually triple for this example in order to provide 251 percent torque.

Two basic concepts that can explain adjustable speed operation of induction motors can be summarized as follows:

$$\text{Speed} \propto \frac{\text{Frequency}}{\text{Poles}}$$

$$\text{Torque} \propto \text{Magnetic Flux} \propto \frac{\text{Volts}}{\text{Hertz}}$$

The speed of an induction motor is directly proportional to the applied frequency divided by the number of poles. The number of poles is a function of how the motor is wound. For example, for 60 Hertz power, a two-pole motor would operate at 3600 RPM, a four-pole motor at 1800 RPM, and a six-pole motor at 1200 RPM.

The torque developed by the motor is directly proportional to the magnetic flux or magnetic field strength, which is proportional to the applied voltage divided by the applied frequency or Hertz. Thus, in order to change speed, all that must be done is to change the frequency applied to the motor. If the voltage is varied along with the frequency, the available torque would remain constant. It is necessary to vary the voltage with the frequency in order to avoid saturation of the motor, which would result in excessive currents at lower frequencies, and to avoid underexcitation of the motor, which would result in excessive currents, both of which would cause excessive motor heating.

In order to vary the speed of an induction motor, an adjustable frequency controller would have an output characteristic as shown in Figure 1–19. The voltage is varied

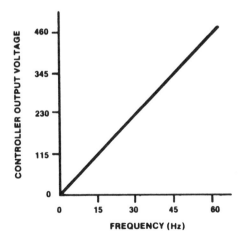

FIGURE 1–19　*Controller output voltage versus frequency relationship for adjustable speed reduction motors. (Source: Reliance Electric, Cleveland, OH.)*

directly with the frequency. For instance, a 460-volt controller would normally be adjusted to provide 460 volts output at 60 Hertz and 230 volts at 30 Hertz.

A controller would typically start an induction motor by starting at low voltage and low frequency and increasing the voltage and frequency to the desired operating point. This would contrast with the conventional way of starting induction motors of applying full voltage, 460 volts at 60 Hertz, immediately to the motor. By starting the motor with low voltage and low frequency, the inrush current associated with across-the-line starting is completely eliminated. This results in a soft start for the motor. In addition, the motor operates between the breakdown torque point and synchronous speed point as soon as it is started, as compared with starting across the line, in which case the motor accelerates to a point between the synchronous speed and breakdown torque point.

Summary

- The maximum torque for an induction motor is limited by the adjustable frequency controller current rating. In order to determine the maximum torque that would be available from an induction motor, it would be necessary to define the motor torque at the controller's maximum current rating.
- The starting torque equals the maximum torque for a motor/controller combination.
- The starting torque current is substantially less for an adjustable frequency controller/motor combination than the locked rotor current for an induction motor started across the line. This results in a soft start for the controller/motor combination.
- The motor load inertia capability for a controller/motor is much higher, since the controller can limit the motor current to 100 percent or less. This would result, however, in longer acceleration times than starting the motor across the line. Harmonics cause additional motor temperature rise over the temperature rise that occurs for sine-wave power operation. As a rule of thumb, for every 10°C rise in temperature, the motor insulation life is cut in half. This explains why it is important to consider the additional temperature rises associated with adjustable frequency control power and to follow the suggested rating curves provided by capable motor manufacturers.
- NEMA design C and D motors are not recommended for use on adjustable frequency control power because these motors have high watts loss due to higher rotor watts loss over design B motors and resulting high temperature rises when operated on adjustable frequency control power.
- Key application points must be defined in order to properly apply an induction motor to a solid-state adjustable frequency controller torque, speed range, motor description, and environment. In order to ensure that adequate torque is available to drive the load and adequate current is available to produce the required torque, the starting torque, the peak running torque, and the continuous torque requirements must be defined. The continuous torque is usually defined, but the peak and starting torques are more difficult to define. For the case of retrofit applications, the speed-torque curve of the existing motor might be used as a reference to define the starting and peak-load torque. Sizing the controller for these points, however, would frequently result in a larger controller than necessary.
- The speed range affects the motor thermal rating. The controllers will typically provide a 10 to 1 speed range below 60 Hertz.
- The motor description will permit selection of a controller size for the motor horsepower, voltage, and current rating. The motor insulation class and design type will permit the motor to be rated properly to ensure that its thermal limitations are not exceeded.

• It is necessary to consider the environment to choose the proper motor enclosure. Explosion-proof motors usually have a UL label certifying that they are suitable for the defined classified area. The UL label, however, is suitable only for 60 Hertz sine-wave power. When an explosion-proof motor is operated on adjustable frequency control power, the 60 Hertz sine-wave UL label is voided. In addition, induction motors are normally rated for 40 °C (104 °F) ambient temperature. Use in a higher ambient temperature may require additional cooling or overframing.

Chapter 2

Gear Speed Transmission Equipment*

In numerous types of process machinery, gears are employed to transmit motion and power from one revolving shaft to another or from a revolving shaft to a reciprocating element. When used between revolving shafts, the shafts can take one of only three positions: they may be parallel, may intersect at an angle, or may cross without intersecting. If the shafts are parallel, the basic friction wheels and the gears developed from them assume the shape of cylinders (Figures 2-1 and 2-2). When the shafts are not parallel, the shapes of the friction wheels and gears will be different. For example, on intersecting shafts the wheels become cones and the gears developed on these conical surfaces are called bevel gears (Figure 2-3). Where motion is transmitted from a shaft to a reciprocating element, a cylindrical friction wheel may engage a flat surface, and the gears assume a similar form (Figure 2-4). All of these methods of transmitting power between cylinders, cones, and flat surfaces involve rolling motion. When shafts cross (one above the other), the friction wheels may be cylindrical or may be of hyperbolic cross section. The gears developed on these surfaces are helical (Figure 2-5) or hypoid (Figure 2-26). In neither case will there be pure rolling action, because when shafts are crossed there is unavoidable side-slip between the surfaces of the wheels.

It is important to appreciate that the type of contact that occurs between the surfaces of friction wheels will also be the type of contact that will occur between the meshing teeth of the corresponding gears. Thus the contact between two cylinders on parallel shafts takes place along a line, and the contact between the teeth developed on those surfaces (Figure 2-2) also occurs along a line. Likewise, the contact between two cones and the teeth developed on those cones (Figure 2-3) will occur along lines. The same condition exists on hypoid gears. However, the contact between friction surfaces is not always a line. Where cylindrical surfaces are on crossed, nonintersecting shafts (Figure 2-5), contact occurs at a point instead of along a line. When this is the case, the teeth developed on these surfaces also make point contact.

TYPES OF GEARS

Spur Gears

When the shafts are parallel, the teeth of the meshing gears may be cut straight across the faces of the gear blanks (Figure 2-1). Gears of this kind are called spur gears. There

---

*Source: Prager Incorporated, New Orleans, LA, unless otherwise noted. Adapted by permission.

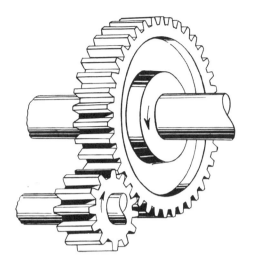

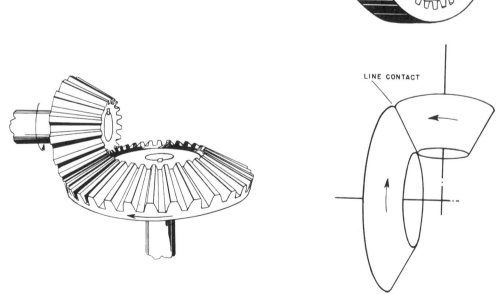

FIGURE 2–1   *Typical external spur gear set. (Source: Prager, Inc., New Orleans, LA.)*

FIGURE 2–2   *Internal spur gears. The pitch surfaces on friction wheels contact along a line (upper view) to transmit motion between parallel shafts. The teeth are developed on internal and external cylindrical pitch surfaces as shown in the lower view. (Source: Prager, Inc., New Orleans, LA.)*

FIGURE 2–3   *Bevel gears. These shafts intersect at a right angle, although bevel gears may also be used between shafts that intersect at larger or smaller angles. (Source: Prager, Inc., New Orleans, LA.)*

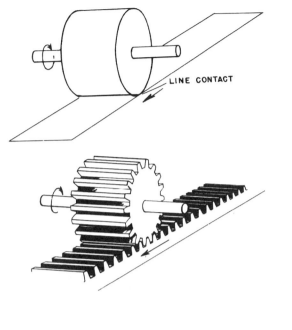

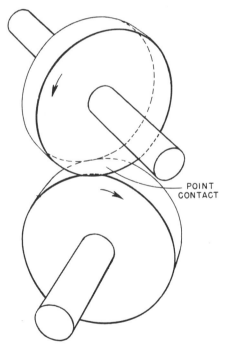

FIGURE 2–4 *Rack-and-pinion gear set. (Source: Prager, Inc., New Orleans, LA.)*

FIGURE 2–5 *Principle of helical gearing. (Source: Prager, Inc., New Orleans, LA.)*

are many special kinds of spur gears, some of which are not commonly encountered. Although none of these special forms differ materially in tooth action from the usual spur gears, it may be beneficial to mention some of them that are sometimes found in special applications.

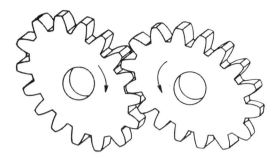

FIGURE 2-6   *Elliptical gears as occasionally used in special machinery. (Source: Prager, Inc., New Orleans, LA.)*

A spur gear meshing with a straight element (Figure 2-4) is known as a *rack and pinion*. The rack may be visualized as a short section of an infinitely large spur gear; a gear so large in diameter that the teeth lie on a straight line. *Elliptical gears* (Figure 2-6) are used to convert the uniform rotary motion of a driving shaft to a rhythmic, pulsating rotation of the driven shaft. An *equalizer gear* and eccentric pinion (Figure 2-7) are sometimes used to drive large chain conveyors in order to prevent the changes of conveyor speed that would occur when the long bar links pass around the sprockets at the driving end. The eccentric driving pinion revolves at a constant speed, but it imparts an irregular motion to the equalizer gear that is calculated to provide a smooth, unvarying speed to the chains. This type of drive reduces what otherwise might be excessive shock on the chains. Another special type of spur gear is sometimes encountered in what are known as *stop motion* or *intermittent* gears (Figure 2-8). In this modification, the driving gear rotates continuously but actuates the driven gear only when the teeth of both gears are engaged. At other times, the driven gear is locked in a fixed position.

Ordinary gears are sometimes used in special groupings; for example, the *epicyclic* or *planetary* arrangement of gears in Figure 2-9 makes possible a very compact reduction gear set. Such a unit usually consists of a central pinion, several planetary gears mounted on a spider, and an internal gear (ring gear) encircling the entire unit. One of the three elements—pinion, spider, or ring gear—must be held stationary, the other two then being used as power input and output elements. Usually it is the ring gear that is rigidly fixed in the housing of the gear set. The pinion is usually the driving element with the spider the power take off, that is, the driven element. In this arrangement, the driving and driven shafts are in line with each other and rotate in the same direction but at different speeds.

Epicyclic gearing is often the ideal solution for transmitting high horsepower at high speeds where a compact, in-line, and lightweight drive unit is required. These epicyclic drives are available in both planetary and star planetary configurations. The free-floating sun design results in balanced, equal load sharing to maximize reliability and life. Whether mounted integrally with a turbine or free standing, epicyclic drives have a low inertia for quick start-up in standby power generation systems. Installation versatility makes these drives convenient for use in fixed, mobile, and marine applications. Capable of handling high ratios, epicyclic drives have an operating efficiency that may reach up to 99 percent.

Epicyclic gears are available in different variations and configurations. Figure 2-10 shows the longitudinal section of a planetary gear with rotating planet carrier. The sun pinion and the planet carrier with the three planets rotate in the bearings. The outer ring with internal teeth (annulus) is fixed firmly to the casing.

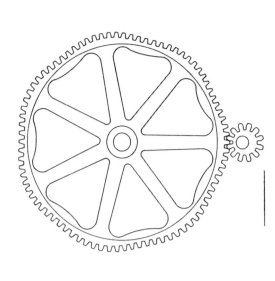

FIGURE 2–7 *Equalizer gear with eccentric pinion. (Source: Prager, Inc., New Orleans, LA.)*

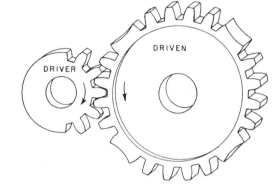

FIGURE 2–8 *Stop-motion or intermittent gearing. (Source: Prager, Inc., New Orleans, LA.)*

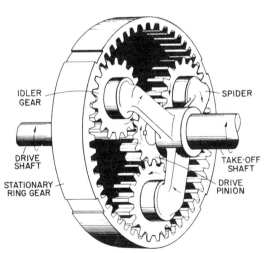

FIGURE 2–9 *Epicyclic or planetary gearing. (Source: Prager, Inc., New Orleans, LA.)*

The permissible speed of the slow-running shaft is limited principally by the centrifugal force of the planets and the resulting bearing pressure on the bearing pins. The sense of rotation of input and output shafts is the same.

Planetary gears with rotating annuli are shown in Figure 2–11. Here, the stationary planet carrier with the three planets is fixed to the gear casing, and the internal tooth annulus connected with the low-speed shaft rotates. Input and output shafts rotate in opposite directions.

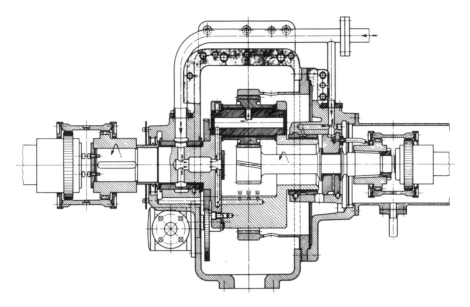

FIGURE 2–10. *Longitudinal section of planetary gear with rotating planet carrier. (Source: Maag Gear Company, Zurich, Switzerland.)*

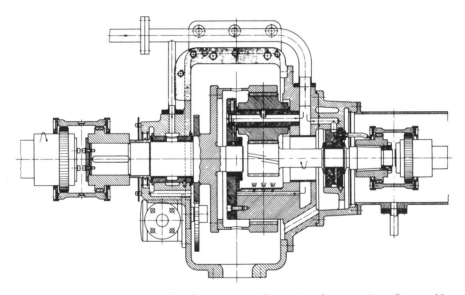

FIGURE 2–11 *Planetary speed-reducing gear with rotating planet carrier. (Source: Maag Gear Company, Zurich, Switzerland.)*

It should be noted that Figures 2–10 and 2–11 show the gearing with helical teeth, whereas Figure 2–9 illustrates spur gears incorporated in the epicyclic package. The reason for frequently using helical gears is perhaps best explained by reviewing their precursor, *stepped gears.*

Since only a few teeth of a pair of spur gears are in contact at the same time, meshing of these gears may be accompanied by a slight impact as the load shifts from tooth to tooth. This is because the teeth undergo slight deformation as the load moves

FIGURE 2-12 *Stepped gearing predates helical gears. (Source: Prager, Inc., New Orleans, LA.)*

over the tooth surface. At low speeds, this is not a serious factor, but as speeds become higher and higher, this deformation makes it more difficult to mesh spur gear teeth without noise and shock. Although accurate machining of spur gear teeth is a major factor in smoothing out any slight irregularities of transmitted motion and torque, these irregularities are minimized only when the gears are designed to distribute the load on several engaging teeth. An early method of accomplishing this result was by the use of stepped gears. In this construction, an assembly of two or more narrow gears are mounted on the shaft in such a way that the teeth are staggered (Figure 2-12). Sudden transfer of load from one tooth to another is minimized as each adjacent stagger tooth absorbs part of the load before the preceding tooth leaves the mesh. This increases the number of teeth in contact at any one time. In rare cases, stepped gears are used to permit operation with zero backlash. When used for this purpose, one of the gears is an ordinary spur gear. The meshing gear consists of two narrow gears bolted together in such a position that all play between the meshing gear teeth is eliminated. In general, the stepped type of gear is seldom used for ordinary transmission of power because of the difficulty of equalizing the load among the various tooth faces. For this use, however, the modern development of the stepped gear—the helical gear—is widely employed.

Helical Gears, Parallel Shafts

If, instead of two or three narrow gears, a stepped gear were composed of innumerable staggered laminations—each lamination so thin that it no longer appeared as an individual unit—the result would be a gear with smoothly twisted teeth. In actual practice, twisted-tooth gears are machined from solid gear blanks with the twist in the same direction. Such a uniform twist is a true helix, and the resulting gears are called helical gears (Figure 2-13). The angle of twist (helix angle) may range from about 20 degrees to 45 degrees. The helix angle is selected so that several teeth will be in mesh at the same time. Even if only two of these helical teeth are in mesh, a very smooth transfer of power results. As the helix angle is increased, the number of teeth in simultaneous contact and the smoothness of tooth engagement are correspondingly increased. Single helical gears are used in speed reducers as well as speed increasers. The speed reducer shown in Figure 2-14 connects a steam turbine to a reciprocating compressor. It accomplishes the speed reduction in two steps, hence the designation *double reduction* gear.

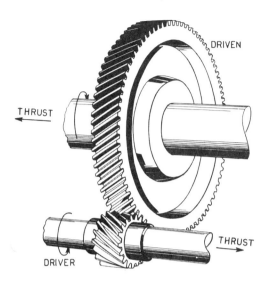

FIGURE 2–13 *Helical gears. (Source: Prager, Inc., New Orleans, LA.)*

FIGURE 2–14 *Double reduction, single helical gear speed reducer for reciprocating compressor drive. (Source: Maag Gear Company, Zurich Switzerland.)*

Another single helical gear is shown in Figure 2–15. Used as a speed increaser between a gas turbine operating at 4860 RPM and a compressor operating at 12507 RPM, this speed increaser transmits 23,800 horsepower and has a pitch-line velocity of 540 feet per second (fps).

Due to the angularity of their teeth, the operation of helical gears produces axial thrusts that must be absorbed by thrust bearings. In most cases, properly selected rolling element bearings will take care of this thrust. However, by using two pairs of opposed helical gears (Figure 2–16), the thrust of one set of gears balances that of the other. This practice was developed before adequate thrust bearings were available. Later developments in the economical cutting of gear teeth made it possible to machine

FIGURE 2-15 *Single reduction, single helical gear increaser for centrifugal compressor drive. (Source: Maag Gear Company, Zurich, Switzerland.)*

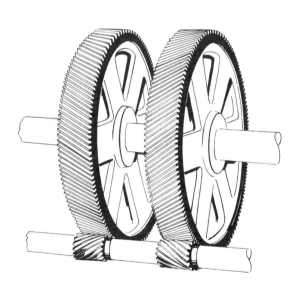

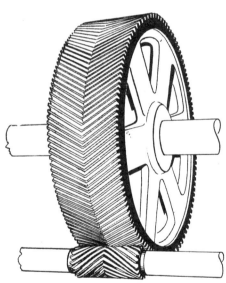

FIGURE 2-16 *Two single helical gears mounted mirror-style to equalize axial thrust generation. (Source: Prager, Inc., New Orleans, LA.)*

FIGURE 2-17 *Double-helical or herringbone gear. (Source: Prager, Inc., New Orleans, LA.)*

two opposed helical gears on a single gear blank. Such a gear is commonly known as a double-helical or herringbone gear (Figure 2-17). High-speed herringbone gears often have a continuous groove machined between the sets of teeth to assist the escape of oil as the gears pass through mesh. Alternatively, the two sets of teeth may be staggered for the same purpose. Either double- or single-helical gear units are widely used

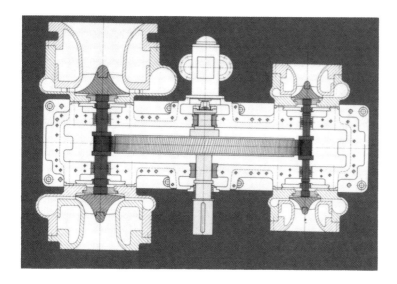

FIGURE 2–18   *Single helical speed increaser set driving high-speed compressor wheels. (Source: Bayerische Huetten-werke Sonthofen, F.R. Germany.)*

in demanding applications where reliability and low maintenance are a must. For example, in the hydrocarbon processing industry, they are specified for process compressors, pump pipeline compressors, fan drives, generator drives, and blower drives. Figure 2–18 illustrates a typically demanding application. The centrally located input gear drives two pinions that are fitted with compressor impellers at each end.

Helical Gears, Nonparallel Shafts

When shafts are not parallel but cross one another, the provision of slanted helical teeth will allow a limited amount of power to be transmitted irrespective of the angle between the shafts. The gears are true helical gears (Figure 2–19) but are sometimes called spiral gears.

Worm Gears

Where the driving gear of a helical right angle drive is much smaller in diameter than the driven gear, the combination could be called a nonthroated worm gear set (Figure 2–20), the smaller gear being the worm. Worm length, as compared with diameter, permits the helical teeth to encircle the shaft more than once, thus giving the teeth the appearance of threads and giving the worm the appearance of a screw. When the worm has only one thread (tooth), it is commonly called a single-thread worm. If there is more than one thread, it is known as a double-thread, triple-thread, etc., worm. The relative number of teeth on the worm and wheel determines the ratio of speed reduction.

When the teeth of both worm and worm gear are of true helical form, the contacts concentrate on a series of points. This limits the power that can be transmitted by such gears. Although the gears transmit motion very smoothly, excessive wear occurs if much power is involved. For this reason, nonthroated worm gears and helical gears on crossed shafts are not very extensively used.

Since commercial worm gear sets must transmit considerable power, it is usual to machine the worm gear so that a considerably increased area of tooth surface will make contact (Figure 2–21). This is done by changing the shape of the teeth of

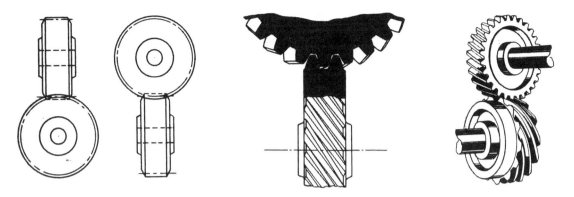

FIGURE 2–19 *Spiral gearing used in certain right-angle drives. (Source: Prager, Inc., New Orleans, LA.)*

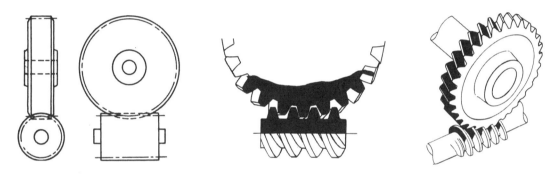

FIGURE 2–20 *Nonthroated worm gear set. The teeth are straight and do not envelop the worm. (Source: Prager, Inc., New Orleans, LA.)*

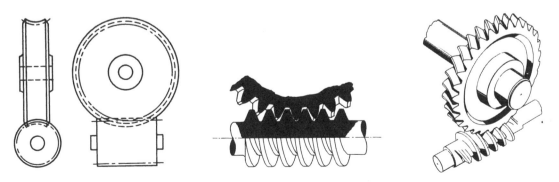

FIGURE 2–21 *Single-throated worm and gear. This is the type of gear used most commonly in industrial worm gear sets. The teeth of the gear are throated (curved) to partially envelop the worm. (Source: Prager, Inc., New Orleans, LA.)*

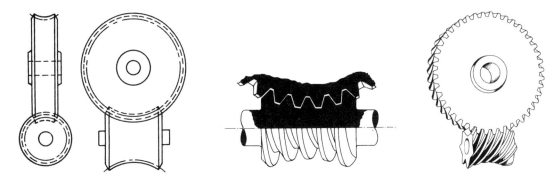

FIGURE 2–22   *Double-throated worm and gear. Both the worm and gear are throated. As a result, they partially envelop each other. The increased contact area permits heavy loads to be carried. (Source: Prager, Inc., New Orleans, LA.)*

the driven gear so that these teeth partly encircle the worm. Such a type is called a throated, or single-enveloping, worm gear and is the type most commonly used in worm gear sets.

Another type of worm gear is the double-throated, or double-enveloping, gear set, employing a throated worm and a throated gar (Figure 2–22). Not only does the gear partly envelop the worm, but the worm also partly envelops the gear, thus further increasing the area of the contacting surfaces. When properly designed and manufactured, these gears are able to carry very heavy loads, although not generally at high speeds.

Bevel Gears

When shafts intersect, the teeth of meshing gears may be cut straight across the faces of conical gear blanks. Such gears are called bevel gears (Figure 2–3). Bevel gears are widely used where a right angle change in direction of shafting is required, although occasionally the shafts may intersect at acute or obtuse angles. When of equal size and mounted on shafts at right angles, they are sometimes referred to as miter gears. Usually, however, the driving gear is smaller than the driven gear, because in the majority of cases gear sets are employed to obtain a reduction of operating speed.

Bevel gears may be assembled in a special grouping known as a differential gear set (Figure 2–23) such as is used in automotive vehicles. This arrangement of gears is intended to divide power between two variable speed shafts, e.g., to permit the wheels of motor vehicles to rotate at different speeds when the vehicle is turning corners. Observing the left side of Figure 2–23, we note a large ring gear that is rigidly attached to the differential gear case. The planetary idler pinions, meshing with the side gears (right view) are pivoted on and revolve with the case, thereby turning the takeoff shafts. When the load is equal on the takeoff shafts, the entire unit revolves as a solid block. Unbalanced load, however, causes the more heavily loaded shaft and its side gear to slow down. Since the ring gear and case are driven at a constant speed, the planetary idler pinions are forced to turn on their pins as they revolve around the slower side gear. This turning of the idler pinions causes an increase in the speed of the less heavily loaded shaft.

Bevel gears may also be arranged to form a strong and compact planetary reduction gear set (Figure 2–24) similar in principle to that shown in Figure 2–9. The drive pinion engages several idler (planetary) gears, causing them to rotate. These gears then roll

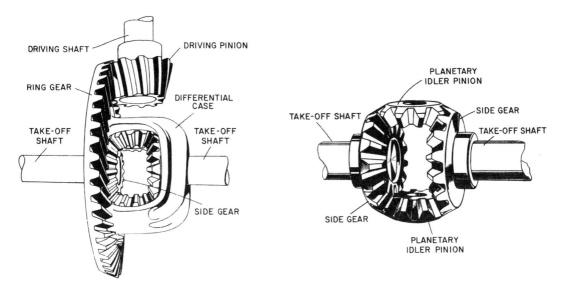

FIGURE 2–23 *Differential gear set. Turning of the idler pinions causes an increase in the speed of the less heavily loaded shaft. (Source: Prager, Inc., New Orleans, LA.)*

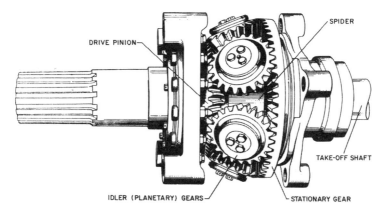

FIGURE 2–24 *Bevel gears arranged to form a strong and compact planetary reduction gear set. (Source: Prager, Inc., New Orleans, LA.)*

around the stationary gear, dragging the spider with them and causing the takeoff shaft to rotate. Speed reduction depends on the ratio of the number of teeth on the drive pinion to the number of teeth on the stationary gear.

Spiral Bevel Gears

In the same way that the teeth of a spur gear can be twisted to make a helical gear, the teeth of an ordinary bevel gear can be twisted to form a spiral bevel gear (Figure 2–25). Because the teeth of a bevel gear are developed on the surface of a cone, these twisted teeth will take the form of a spiral; thus the gears are called spiral bevel gears.

FIGURE 2–25   *Spiral bevel gears. (Source: Prager, Inc., New Orleans, LA.)*

The angle of the spiral is selected so that one end of each tooth enters mesh before the other end of the preceding tooth has disengaged. As with helical gears, this results in very smooth transfer of power.

Hypoid Gears

Hypoid gears (Figure 2–26) are used where shafts cross, one below the other, and the design of the machine precludes the use of a worm and gear. This may result where space limitations require that one of the shafts be moved aside, where several pinions on a single shaft drive several cross shafts, where a small pinion must transmit high power, or where rigidity requires a supporting bearing on each side of each gear. Where any of these conditions exist, hypoid gears provide a strong, smooth, and quiet drive. The shafts of practically all hypoid gear sets cross at right angles.

Although the ordinary hypoid gear is similar in appearance to a spiral bevel gear, it is not developed on the same type of pitch surfaces. Two conical pitch surfaces on intersecting shafts roll on each other with line contact and without side-slip, but if the

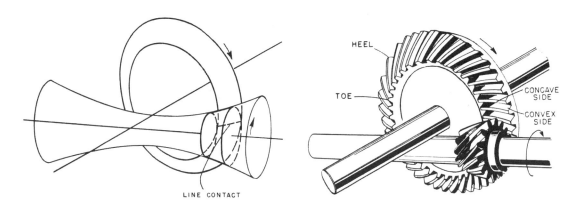

FIGURE 2–26   *Hypoid gear and pinion. These gears transmit motion between nonintersecting shafts crossing at a right angle. The pitch surfaces are hyperbolic in form. (Source: Prager, Inc., New Orleans, LA.)*

FIGURE 2–27 *Hypoid gear with extreme offset will require pinion to resemble a worm. (Source: Prager, Inc., New Orleans, LA.)*

shafts do not intersect, i.e., if they cross one below the other, the cones do not make contact along a line and do not roll without sliding. Instead, they meet at a point and roll with more or less side-slip, depending on the positions of the shafts. Gear teeth developed on such surfaces on crossed shafts would also make point contact and would give poor service.

To increase contact area and improve gear service, it is necessary to replace the cones with surfaces that will bear on each other along a line of contact. This line contact is obtained by employing curved pitch surfaces of hyperbolic contour. The teeth developed on these hyperbolic pitch surfaces also meet in line contact, thus distributing the load over considerable tooth surface. Unit loading on the metal is reduced, and the ability to transmit power is increased. The working surfaces of the meshing teeth, however, are always subject to side-slip and consequent friction. Such gears are called hypoid gears. Although most hypoid gears look like spiral bevel gears, this is not always the case. In rare cases, extreme offset of the shafts and high ratios of reduction will require gears such as shown in Figure 2–27. In this case, the pinion has only one tooth, which causes it to resemble a worm.

Special Gear Sets for Process Plant Applications

Process plant gear sets often incorporate one or more of the gear types, configurations, and arrangements that we have discussed here. Figures 2–28 through 2–31 show high-power-density parallel shaft double-reduction gears (Figure 2–28), special or customized process machine drives (Figure 2–29), cooling tower fan drives with double-reduction and right-angle output (Figure 2–30), and completely packaged drive systems that include driver, speed change, driven machine, and support systems such as lube oil supplies (Figure 2–31).

FIGURE 2–29   *Special process machine drives can be built in almost limitless variations of geometry and arrangements. (Source: WesTech Gear Corporation, Los Angeles, CA.)*

FIGURE 2–28   *This double-reduction helical gear set employs the gear geometries shown earlier in Figures 2–16 and 2–17. (Source: WesTech Gear Corporation, Los Angeles, CA.)*

FIGURE 2–30   *Right-angle drives such as this large cooling tower fan gear often use a combination of gear types to accomplish both speed change and changes in drive direction. (Source: WesTech Gear Corporation, Los Angeles, CA.)*

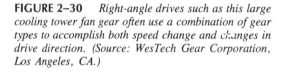

FIGURE 2–31   *Complete process systems often include speed-up gears for motor-driven blowers and high-speed pumps. This entire unit is being applied in a solvents recovery process. (Source: WesTech Gear Corporation, Los Angeles, CA.)*

Chapter 3

Gas Turbines*

Within the context of process plant machinery, the engineer is likely to encounter gas turbines as drivers for electric power generators and as mechanical drive turbines for large compressor trains.

Although gas turbines have been on the industrial scene since the late 1920s, large-scale applications had to wait until the 1950s when rapid advancements in aircraft jet engines brought significant improvement and vastly enhanced acceptance of industrial gas turbines. These improvements touched virtually every requirement cited for modern process plants: low initial cost, good efficiency, maintainability, reliability, operational ease, process flexibility, and environmental acceptability.

From a thermodynamic point of view, a gas turbine—or gas turbine engine—is a machine that accepts and rejects heat at different energy levels and, in the process, produces work. While this work is converted to pressure and velocity energy in the aircraft jet engine, the commercial or industrial gas turbine is arranged to convert this work into shaft rotation or, more correctly, torque.

The gas turbine (Figure 3–1) consists of an air compressor and gas combustion, gas expansion, and exhaust sections. The gas turbines cycle is composed of four energy exchange processes: an adiabatic compressor, a constant-pressure heat addition, an adiabatic expansion, and a constant-pressure heat rejection. The four thermodynamic processes can be accomplished either in an open-cycle or a closed-cycle system. The open-cycle gas turbine takes ambient air into the compressor as the working substance that, after compression, is passed through a combustion chamber where the temperature is raised to a suitable level by the combustion of fuel. It is then expanded inside the turbine and exhausted back to the atmosphere. Most industrial-type gas turbines work on this principle, and Figures 3–2 and 3–3 illustrate simple, open-cycle gas turbines. The use of two or more hot gas expansion stages makes it possible to produce the two-shaft turbine of Figure 3–3. This configuration has greater speed flexibility than single-shaft machines.

The closed-cycle gas turbine uses any gas as the working substance. The gas passes through the compressor, then through a heat exchanger where energy is added from a source, then expanded through the turbine and finally back to the compressor through a precooler where some energy may be rejected from the cycle.

Perhaps the most important reasons why process plants use gas turbines are summarized as high system reliability and high combined energy system and process efficiency. Where the forced outages of a single driver can shut down an entire complex, highest reliability is a must. For projects involving process system modifications of a new process design, choosing the most reliable turbine or energy system rather than maintaining an already existing process design can result in significantly higher reliability and reduced financial loss due to excessive process shutdowns.

---

*Source: General Electric Company, Schenectady, NY. Adapted by permission.

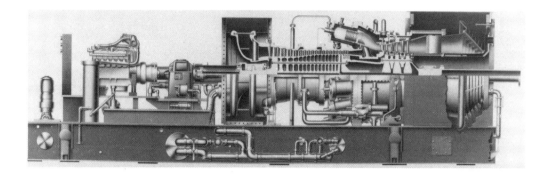

FIGURE 3–1   *Typical industrial gas turbine. (Source: General Electric Company, Schenectady, NY.)*

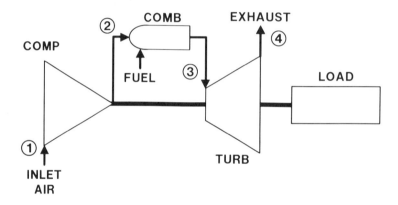

FIGURE 3–2   *Single-shaft, simple open-cycle industrial gas turbine. (Source: General Electric Company, Schenectady, NY.)*

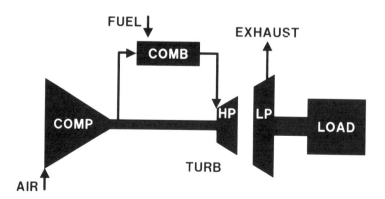

FIGURE 3–3   *Two-shaft, simple open-cycle industrial gas turbine. (Source: General Electric Company, Schenectady, NY.)*

With regard to the second point, high efficiency, the potential user may be confronted with an apparent mismatch between project needs and available machine sizes. In that case, it may save considerable money initially and over the life of the plant to revise the process design to match the best equipment and energy system available.

SIMPLE CYCLE GAS TURBINES

Most gas turbines in the process industries are operating in base load, or continuous, service. Fuel costs are an important consideration in determining the type of prime mover in these applications. However, there are hundreds of simple-cycle gas turbines installed in many areas of the world where fuel is relatively low cost, in underdeveloped areas, or in remote or harsh environments. Examples of simple-cycle gas turbine installations are in Indonesia, the North Sea, the Sahara Desert, and the Alaskan North Slope. The advantages of simple-cycle gas turbines include:

- Low capital cost
- Minimum installation cost
- No external power or cooling water required
- Minimum operating labor
- Low maintenance costs
- High reliability
- High availability

The disadvantage of the simple-cycle gas turbine is its relatively low system efficiency and higher fuel costs, compared with gas turbine systems with exhaust heat recovery.

HEAT RECOVERY CYCLES

Usually the economics of gas turbines in the process industries depend on effective use of the gas turbine exhaust energy. The most common use of this energy is for steam generation in heat recovery steam generators (HRSG), unfired, as well as fired designs. However, the gas turbine exhaust gases can also be used as a source of energy for unfired and fired process fluid heaters and direct drying applications, as well as for combustion air for power boilers, reformers, or other process equipment.

One of the more common gas turbine/heat recovery cycles is one where the exhaust energy is used to generate steam at conditions suitable for the process steam header (Figure 3–4). The HRSG may be unfired or have supplementary firing to increase steam output. Power generation capability for these gas turbine-HRSG cycles, per unit of heat delivered to process, ranges from approximately 150 to 250 kw per million British thermal units (Btu)-net heat to process (NHP).

Generation of steam at higher initial steam conditions than those required for process heat will allow use of a steam turbine in the cycle in addition to the gas turbine, as shown in Figure 3–5. This combined cycle will result in a higher power generation-to-process heat ratio than the gas turbine-HRSG cycle shown in Figure 3–4, with power generation in the range of 200 to 400 kw per million Btu NHP.

A typical upper limit for steam conditions of unfired HRSGs is 1315 pounds per square inch gauge (psig)/950 °F. The HRSG steam temperature is usually 75 to 100 °F or more below the gas turbine exhaust gas temperature. Fired HRSGs have been applied with steam generation pressure and temperature as high as 1525 psig/955 °F.

A multiple-pressure-level HRSG combined-cycle system is shown in Figure 3–6. This arrangement is common for unfired and moderately fired (up to approximately

TYPICAL INDUSTRIAL
GAS TURBINE CYCLES

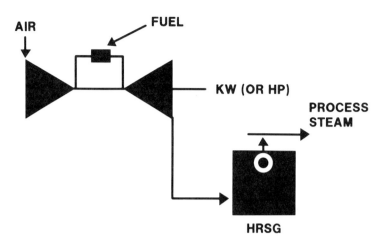

FIGURE 3–4   *Typical industrial gas turbine cycle employing heat recovery steam generator (HRSG). (Source: General Electric Company, Schenectady, NY.)*

TYPICAL INDUSTRIAL
GAS TURBINE CYCLES

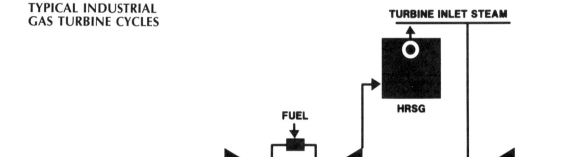

FIGURE 3–5   *Combined cycles employ both gas and steam turbines. (Source: General Electric Company, Schenectady, NY.)*

1200 °F) HRSGs. The multipressure-level HRSG results in increased recovery of the gas turbine exhaust energy compared with an unfired, single-pressure-level HRSG system, thus increasing the cycle thermal efficiency.

The steam turbine in a combined cycle may be a noncondensing or a condensing design, depending primarily on process heat requirements. The steam turbine design shown schematically in Figure 3–6 provides considerable cycle flexibility in industrial process applications. The condenser provides a heat sink for HRSG steam-generating capability in excess of that extracted from the turbine for process use. Furthermore, the optional admission capability permits the introduction of lower pressure process steam into the turbine for expansion to the condenser during times of excess low-pressure steam.

**TYPICAL INDUSTRIAL
GAS TURBINE CYCLES**

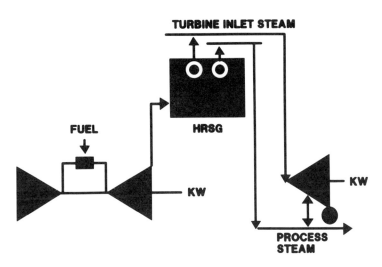

FIGURE 3–6 *Cycle flexibility is provided through utilization of a two-level heat recovery steam generator and admission-extraction steam turbine. (Source: General Electric Company, Schenectady, NY.)*

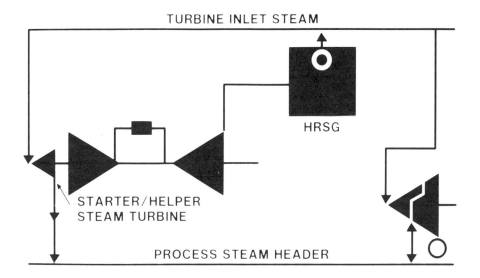

FIGURE 3–7 *Starter/helper steam turbine arrangement in a combined gas turbine cycle. (Source: General Electric Company, Schenectady, NY.)*

Even though gas turbines are not available in an infinite number of ratings, the application of a helper steam turbine may permit utilization of the capability of standard, proven gas turbine units in certain mechanical drive or generator drive applications. In this cycle, shown schematically in Figure 3–7, the helper steam turbine can augment gas turbine power generation as load requirements vary. In most instances, depending on horsepower and steam conditions, the helper steam turbine can be mounted on the gas turbine base and shipped as an integral driver unit to minimize installation costs. For instance, base-mounted, noncondensing helper steam

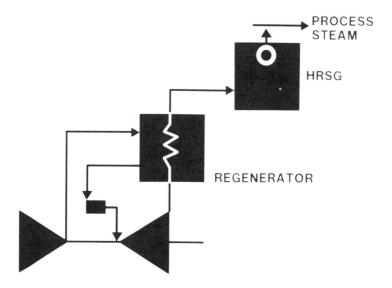

FIGURE 3–8 *Regenerative-cycle gas turbine followed by a low-pressure process HRSG. (Source: General Electric Company, Schenectady, NY.)*

turbines for a medium-to-large gas turbine may range from small, single-stage turbines to multistage, multivalue turbines rated up to 8000 horsepower (HP). Control of the steam turbine helper and the gas turbine is integrated into a single governing system.

In the regenerative-cycle gas turbine, exhaust heat is recovered by heating the turbine's combustion air after compression, but before it enters the combustion chambers to reduce gas turbine fuel. The Figure 3–8 schematic diagram shows a regenerative-cycle gas turbine followed by a low-pressure process HRSG. One of the consequences of the low fuel consumption of the regenerative-cycle gas turbine is a reduction of the gas turbine exhaust gas temperature to approximately 600 °F. This cycle arrangement can be an option when a relatively small amount of process steam is required.

TYPE SELECTION

The choice of gas turbine type or model to be favored at a given plant site depends on the economic evaluation of the system. Heat rate, site fuel, water and steam requirements, part-load efficiency of the gas turbine and heat recovery system, unused installed capacity, and capital investment are some of the considerations.

In petroleum refineries, power generator drives predominate, as illustrated in Figure 3–9. Petroleum production facilities, both onshore and offshore, make use of gas turbines for gas reinjection compressor, waterflood pumping, and power generator drive applications.

In modern ethylene production facilities, three and sometimes four major compressor strings require variable speed drivers. The charge gas compressor driver is typically an extraction-condensing steam turbine that supplies 600 psig steam to the cracking process. Additional process steam can be furnished by the extraction-condensing propylene compressor driver or sometimes by a noncondensing ethylene driver. However, typical energy balances require large quantities of power boiler fuel and large blocks of condensing power. To reduce overall fuel consumption and cooling water

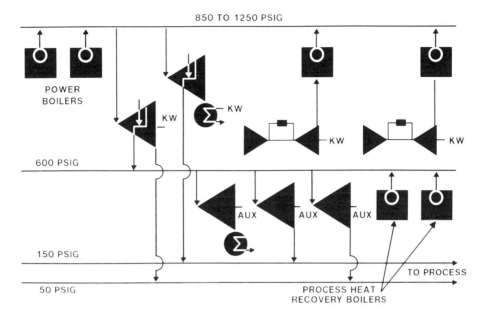

FIGURE 3-9 *Typical refinery power plant schematic shows power generator drives predominating. (Source: General Electric Company, Schenectady, NY.)*

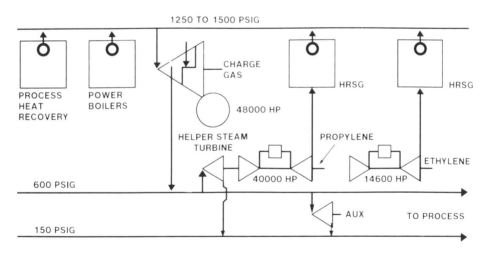

FIGURE 3-10 *Ethylene plant with combined cycle drives. (Source: General Electric Company, Schenectady, NY.)*

requirements, gas turbines have been selected as the propylene and ethylene compressor drivers for a number of large plants. A typical layout is shown in Figure 3-10.

One major plastics plant uses the gas turbine exhaust to first heat the oil used in a high-temperature distillation process. The exhaust gases then pass to a supplementary fired heat recovery steam generator where plant process steam is produced. Figure 3-11 depicts this sequence.

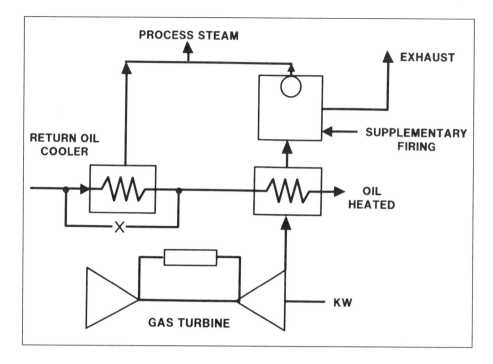

FIGURE 3–11   *Special process requirements may employ gas turbines in a variety of arrangements. (Source: General Electric Company, Schenectady, NY.)*

DESIGN

The design of a gas turbine largely depends on its flow path layout, which, in turn, is based on aerodynamic requirements. However, mechanical considerations also have significant influence on the design. In aircraft jet engines, which must fit within a given diameter, the compressor, combustor, turbine blades, and nozzles are arranged in-line, and both air and gas flow axially through them. This usually results in long rotors requiring three or more bearings. In automotive and industrial gas turbines, axial length is at a premium, while diameter is not so important, and the combustor is a single can-type placed out of line with the compressor and turbine. This results in a shorter rotor, and it is possible to use only two or three bearings for full and adequate rotor support.

As mentioned earlier and as shown in Figure 3–12, gas turbines can be viewed as being made up of different sections. The first of these sections comprises the axial compressor. This compressor is no different from the axial process gas compressors dealt with independently in Chapter 12 of this text. Hence, the design of the axial compressor section of a gas turbine is dictated by air flow, pressure ratio, and turbine inlet temperature. The compressor section is designed in such a way that the work done on the air per compressor stage is much less than the work extracted from the gas in a single turbine stage. This reduces the air loading on the compressor blades, and the blades can be very light and small, limited mainly by the susceptibility of the blading to erosion and foreign body damage. In an industrial gas turbine, as many as 19 rows of compressor blades are not uncommon. The compressor section includes a rotor and a stator cas-

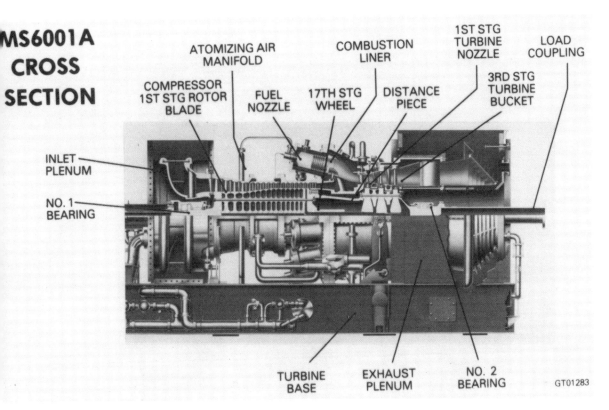

MS6001A CROSS SECTION

ATOMIZING AIR MANIFOLD

COMBUSTION LINER

1ST STG TURBINE NOZZLE

LOAD COUPLING

COMPRESSOR 1ST STG ROTOR BLADE

FUEL NOZZLE

17TH STG WHEEL

DISTANCE PIECE

3RD STG TURBINE BUCKET

INLET PLENUM

NO. 1 BEARING

TURBINE BASE

EXHAUST PLENUM

NO. 2 BEARING

GT01283

FIGURE 3–12 *Major components of an industrial gas turbine. (Source: General Electric Company, Schenectady, NY.)*

ing. Typically, the compressor rotor is built up of separately forged wheels in which the blades are mounted. Some rotor configurations allow blades to be changed without the rotor being lifted out of the casing. In certain types of rotors, the individual wheels are held together with pilot fits and multiple through-bolts.

The stator casing is normally constructed with angular grooves that hold the stator blading. The casing is generally horizontally split for ease of assembly, maintenance, and inspection. The stator consists of inlet and discharge casings. A cast bellmouth inlet provides for even air flow to the inlet guide vanes. The casing incorporates an aft diffuser section that is supported by the turbine shell and the inner barrel. In most industrial gas turbines, the compressor blades are made of stainless steel with special coatings to resist corrosion.

Turbine Buckets and Wheels

Most industrial-type gas turbine buckets (hot path blades) and wheels (blade anchoring disks) are designed to withstand temperatures in excess of 2000 °F. The buckets are high-quality castings of a corrosion-resistant super alloy. The buckets (Figure 3–13) feature long shanks that isolate the wheel rim from the gas path to reduce temperatures in the wheel rim and dovetail region. The increased length also provides greater bending flexibility, particularly at the vane platform. Convective cooling of the first and second stage buckets (Figure 3–14) is afforded by holes through the vanes, which main-

FIGURE 3-13   *Gas turbine blades, or "buckets." (Source: General Electric Company, Schenectady, NY.)*

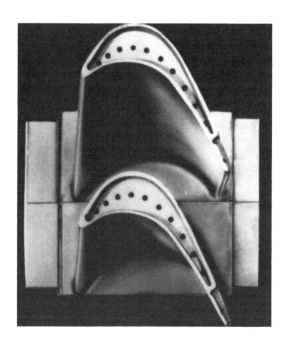

FIGURE 3-14   *Cooling air passages in first-stage gas turbine buckets. (Source: General Electric Company, Schenectady, NY.)*

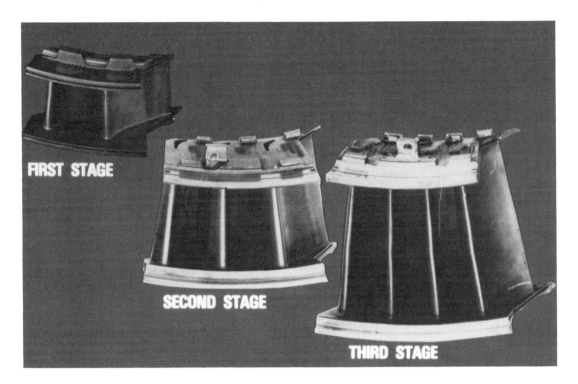

FIGURE 3–15   *Gas turbine nozzle segments. (Source: General Electric Company, Schenectady, NY.)*

tain acceptably low metal temperatures. Usually, the second- and third-stage buckets have integral tip shrouds that eliminate the need for tie wires that are sometimes used for damping vibration.

The turbine stages incorporate segmented nozzles (Figure 3–15), which are made from corrosion-resistant alloys. Usually, the first- and second-stage nozzles are air-cooled by a combination of internal impingement and external film cooling. The vanes are hollow with an internal sheet metal core plug. The compressor discharge air feeds into the inside of the core plug and then discharges through small holes in the core plug wall, impinging against the inside of the vane wall. From this space, the air flows around the core plug and exits to the gas path through holes in the vane wall, which provides external film cooling.

Combustion Section

The combustion section (Figure 3–16) includes multiple small combustors or an externally mounted single large combustor. In multiple combustor design, individual combustor chambers are connected by cross-fire tubes to ensure that the flame in one chamber will ignite the fuel in all of the other chambers. Each combustor is provided with a fuel nozzle that can utilize either oil or gas fuel. The combustors are provided with slot-cooled liners. The cooling protects the liner metal walls from the combustion

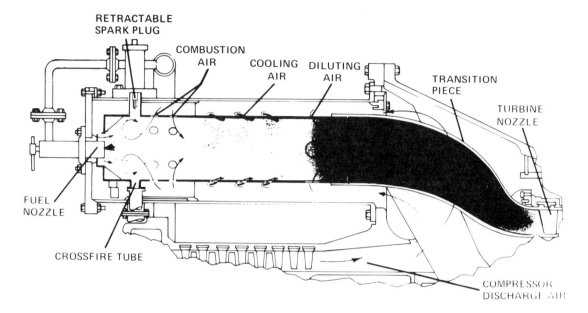

FIGURE 3–16   *Combustion section of modern industrial gas turbine. (Source: General Electric Company, Schenectady, NY.)*

zone gas temperatures, which can exceed 3000 °F. The transition pieces connect the combustors to the first-stage nozzles.

Bearing Systems

The bearing arrangement may vary with the type of gas turbine. One turbine manufacturer uses a three-bearing support system to maintain the concentricity between the rotating and stationary parts. The three-bearing system rotors will enable the manufacturer to opt for smaller shaft diameters, which in turn reduces rotor inertia, starting time, and power requirements and increases the rotor stability. The bearings are pressure-lubricated. Normally, the lube oil system is mounted on the base of the gas turbines. The rotors are held in position during operation by thrust and radial bearings that are quite similar in construction to the bearings described in Chapter 11 (see Centrifugal Compressors) of this text.

Size Ranges

As discussed earlier, heavy-duty industrial gas turbines are available as both simple cycle and regenerative cycle machines. These can be single- or two-shaft configurations for both mechanical and generator drive applications. Mechanical drive gas turbines cover a range from 14,000 HP to 45,000 HP at International Standards Organization (ISO) conditions burning gaseous fuels. One leading U.S. manufacturer has five different models that can be used as mechanical- or power generation-type turbines. Tables 3.1 and 3.2 indicate cycle, fuel used, output, heat rate, and shaft speed of both mechanical drive and generator drive gas turbines available from this manufacturer. Of course, other manufacturers may produce machines in different or overlapping size categories.

Table 3.1 Performance Specifications for Major Mechanical Drive Gas Turbines: 1988 Data

Model[a]	LM5000-PC	LM1600	LM2500-PH STIG[b]
Year[c]	1986	1988	1986
ISO rating continuous (HP)[d]	46,210	16,500	36,000
Heat rate (Btu/HP-hr)[e]	7,040	7,120	6,321
Air flow (lb/sec)[f]	265	96	151
Turbine speed (rpm)[g]	3,600	7,000–9,000	3,600
Pressure ratio[h]	30.0	21.5	20.0
Turbine inlet temp. (°F)[i]	—	1,344[j]	1,475[j]
Exhaust temp. (°F)[i]	834	880	952
Dry weight (~lb)	89,000	7,000	10,500
Length × width × height (~ft)	58 × 11 × 12	17 × 6 × 6.5	21 × 8 × 7

All ratings without losses and zero humidity, except for STIG units.
[a] Series designation
[b] All ratings on gas fuel, 4″/10″ H_2O inlet/exhaust losses
[c] First year unit was available
[d] 6,000 hours per year or more
[e] Lower heating value
[f] At base load
[g] Output shaft speed
[h] Overall compressor
[i] At continuous rating
[j] Power turbine inlet temperature

Table 3.2 Performance Specifications for Major Electric Power Generator Drive Gas Turbines: 1988 Data

Model[a]	PG5371 (PA)	PG6541 (B)	PG7111 (EA)	PG9161 (E)	PGLM5000-PC
Year[b]	1987	1978	1976	1987	1986
ISO base rating (kw)[c]	26,300	38,340	81,700	116,900	33,090
Heat rate (Btu/kw-hr)[d]	11,820	10,860	10,610	10,310	9,860
ISO peak rating (kw)[e]	28,150	41,400	89,200	126,100	—
Heat rate (Btu/kw-hr)[d]	11,730	10,780	10,580	10,280	—
Pressure ratio[f]	10.2	11.8	12.4	12.1	30.0
Air flow (lb/sec)[g]	270	301	641	889	278
Turbine speed (rpm)[h]	5,100	5,100	3,600	3,000	3,000
Turbine inlet temp. (°F)[i]	1,755	2,020	2,020	2,020	—
Exhaust temp. (°F)[i]	901	1,003	992	985	834
Dry weight (~lb)	570,000	700,000	1,070,000	1,900,000	314,200
Length × width × height (~ft)	115 × 19 × 34	123 × 24 × 34	132 × 71 × 31	115 × 77 × 39	119 × 20 × 31

All units are package power plants; all ratings are on gas fuel.
[a] Series designation
[b] First year unit was available
[c] 6,000 hours per year or more
[d] Lower heating value
[e] Good for up to 2,000 hours per year
[f] Overall compressor
[g] At base load
[h] Output shaft speed
[i] At base rating

MAINTENANCE

The combustion gas turbine, as does any rotating power equipment, requires a program of planned periodic inspection, with repair and replacement of parts to achieve optimum availability and reliability. The major structural components of the heavy-duty combustion gas turbine are designed according to long-established standards derived from steam turbine design and manufacture. Major differences occur between the steam turbine and the combustion gas turbine due to the fact that the combustion gas turbine is a complete, self-contained, prime mover. This combustion process to develop energy does not require a boiler with its associated limitations; therefore, the cycle temperatures are considerably higher. The parts that are unique to the gas turbine because of this feature are combustion caps, liners, and transition pieces. These, along with the turbine nozzles and buckets, are referred to as the "hot-gas path" parts.

The inspection and repair requirements of the gas turbine lend themselves to establishing a pattern of inspections, starting with very minor work and increasing in magnitude to a major overhaul, and then repeating the cycle. These inspections can be optimized to reduce unit outages and maintenance cost for the user's specific mode of operation, while maintaining maximum availability and reliability. Inspections can be classified as operational or shut down. The operational inspections are used as indicators of the general condition of the equipment and as guides for planning the disassembly maintenance program. The entire scope of inspections can be described as standby, running, combustion, hot-gas path, and major.

Standby Inspection

Standby inspections pertain particularly to gas turbines used in intermittent service, such as peaking and emergency duty. Starting reliability is of prime concern, as a delay in starting usually means that the demand for the unit has passed. This includes routine servicing of the battery system, lubrication, changing of filters, checking oil and water levels, cleaning relays, checking device calibrations, and other general preventive maintenance. This servicing can be performed in off-peak hours without interrupting the availability of the turbine. A periodic test run is an essential part of the standby inspection.

Running Inspection

Running inspections consist of the observations made while a unit is in service. The turbine should be observed on a programmed schedule, which should be established as part of the unit maintenance program consistent with the operator's requirements.

Operating data should be recorded to permit an evaluation of equipment performance and maintenance requirements. Typical running inspections (Table 3.3) include load versus exhaust temperature; vibration; fuel flow and pressure; exhaust temperature control and variation; and startup time.

The general relationship between load and exhaust temperature should be observed and compared with previous data. Ambient temperature and barometric pressure will have some effect on the absolute temperature level. High exhaust temperature can be an indicator of deterioration of internal parts, excessive leaks, axial-flow compressor fouling, or improper control settings. Initial startup data should be used as the reference point for checking.

Power loss resulting from deteriorated parts or leaks may require disassembly of the turbine to restore power. This can be done with on-site labor and equipment. Loss due to dirt fouling of the axial flow compressor can usually be restored by cleaning the compressor while in service. This is accomplished by injecting 10 to 20

Table 3.3 Typical Running Inspections Recommended for
Gas Turbines

- Load versus exhaust temperature
- Vibration
- Fuel flow and pressure
- Exhaust temperature control
- Exhaust temperature variation
- Start-up time

---

pounds of mild abrasives such as hard rice or screened crushed nut shells into the compressor inlet. A successful cleaning will reduce the exhaust temperature for a given load and will increase the compressor discharge pressure. If the need to clean the compressor is frequent, the causes of the fouling condition should be determined and corrected.

The vibration level of the unit should be observed and recorded. Minor changes will occur with changes in operating conditions. However, major changes, or a continuous trend to increase, indicate that corrective action is required.

The fuel system should be observed for general fuel flow versus load relationship. Fuel pressures through the system should be observed. Changes in fuel pressure can indicate that fuel nozzle passages are plugged or fuel metering elements are damaged or out of calibration.

Probably the most important control function to be observed is the exhaust temperature-fuel override system and its backup overtemperature trip system. Routine verification of the operation and calibration of these devices will minimize wear on the hot-gas path parts.

The variation in turbine exhaust temperature should be measured. An increase in temperature spread indicates combustion deterioration or fuel distribution problems. If not corrected, reduced life of downstream parts can be expected.

Startup time (when the gas turbine is new) is an excellent reference against which subsequent operating parameters can be compared and evaluated. A curve of starting parameters of speed, fuel signal, exhaust temperature, and critical sequence benchmarks versus time from the initial start signal will give a good indication of the condition of the control system. Deviations from normal conditions help pinpoint impending trouble, changes in calibration, or damaged components.

Combustion Inspection

This is a shutdown inspection to inspect combustion liners and fuel nozzles; these are recognized as the first parts requiring replacement and repair for a good maintenance program. Proper attention to these items will optimize the life cycle of downstream parts, such as turbine nozzles and buckets.

Figure 3–17 illustrates the section of a typical unit that is disassembled for this inspection. The combustion liners and fuel nozzles should be removed and replaced with new or repaired liners and new or clean fuel nozzles. This method of inspecting allows for minimum unit downtime and maximum utilization of manpower. A visual inspection of the transition pieces and nozzles (first-stage) at this time optimizes the scheduling of the hot-gas path inspection. This visual inspection is accomplished by an optical instrument called a ''borescope,'' which is inserted through the combustion liner area to allow examination of the transition pieces and first-stage nozzle. The typical intervals for combustion inspections are shown in Tables 3.4 and 3.5.

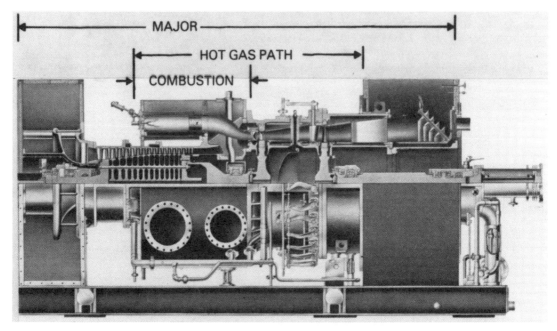

FIGURE 3–17   *Major areas of gas turbine inspection. (Source: General Electric Company, Schenectady, NY.)*

Table 3.4   Typical Inspection Intervals for Gas Turbines in Continuous Duty*

	Combustion	*Hot Gas*	*Major*
A. Gas	—	—	30,000–40,000
Distillate	10,000–14,000	—	20,000–28,000
B. Gas	8,000–10,000	16,000–20,000	30,000–40,000
Distillate	5,000– 7,000	10,000–14,000	20,000–28,000

*One start per 1,000 fired hours.

A. Continuous process—Gas turbine outage results in process shutdowns. Loss in production exceeds savings from optimum maintenance.

B. Interruptable process—Scheduled outages coincide with other equipment inspections (i.e., underwriter requirement). Maintenance costs optimized.

Table 3.5   Typical Inspection Intervals Recommended for Gas Turbines in Peaking Service

Starts per Fired Hour	*Base Temperature Control*	*Peak Temperature Control*
1–3	2,100	700
1–5	3,000	1,000
1–10	3,450	1,150

Fuel: Natural gas/light distillate

Hot-Gas Path Inspection

This inspection includes the work necessary for a combustion inspection plus the removal of the upper half turbine shell, and on applicable turbines, removal of the upper half combustion chamber wrapper.

The inspection involves all hot-gas path parts. These include the turbine buckets, shrouds, nozzles, transition pieces, and exhaust hood turning vanes. The typical recommended intervals are given in Table 3.6.

Major Inspection

A major inspection includes the work items outlined for combustion inspections and hot-gas path inspections; it also includes "laying the turbine on the half shell," and completely inspecting the axial flow compressor stator and rotor parts, turbine buckets and shrouds, bearings, and seals. Table 3.7 gives the recommended work scope for major inspections and Table 3.8 gives estimated man-hours and work shifts for popular gas turbines.

Table 3.6 Recommended Inspection Intervals Linked to Fired Hours, Starts, and Elapsed Time

	Whichever Comes First		
	Fired Hours	*Starts*	*Time*
Combustion inspection			
Natural gas	8,000–10,000	300–400	Annually
Distillate	5,000– 8,000	300–400	
Hot-gas path			
Natural gas	20,000–24,000	600–800	—
Distillate	20,000–24,000	600–800	—
Major inspection			
Natural gas	42,000–48,000	1,600–2,400	6 Years
Distillate	42,000–48,000		

Table 3.7 Recommended Work Scope for Major Inspections*

Part	*Action*	*Inspection For*
Bearings, seals	Clean	Wear, fouling, leaks, wiping, scoring, deterioration of babbitt
Blading	Clean manually, loose parts check	Foreign object damage, erosion, corrosion, cracks, fouling
Buckets	Remove from rotor-grit blast— loose parts check	Foreign object damage, cracks, erosion, corrosion
Turbine wheel	Clean, loose parts check in dovetail area	Cracks in dovetail area
Journals and seal fits	—	Wear, scoring, wear on seal fits
Inlet system and exhaust system	Inspect, repair, paint	Corrosion, cracks, loose parts

*Step 1: Same as for combustion and hot-gas-path inspections;
 Step 2: Remove remaining upper half casings and bearing covers;
 Step 3: Remove rotors.

Table 3.8 Estimated Time to Perform Recommended Gas
Turbine Inspections on Popular GE Sizes

Inspection	Model	Hours	Work Shifts
Combustion	5001	160	5
	6001	240	6
	9001	480	10
Hot-gas path	5001	480	10
	6001	672	12
	9001	1,120	20
Major	5001	1,280	20
	6001	1,600	25
	9001	2,560	40

Estimated Hot-Gas Path Parts Lives (Peaking Duty)

The application, cyclic or continuous duty, starting frequency and time, internal temperatures as a result of loading duty, and type of fuel used all determine parts life and maintenance cost. A peaking plant has many thermal cycles, resulting in a requirement to inspect the unit on a shorter fired-hour basis than is required with a continuous-duty unit. The normal variance of peaking units will be one start per one fired hour to one start per six fired hours. Within this range, the planned hot-gas path inspection should take place each 6,000 to 10,000 fired hours of operation, depending on the evaluations made at combustion inspections, factoring in the effects of fuel and metal temperature. This is approximately one third of the fired hours expected to be attained on a continuous-duty unit before an inspection will be routinely scheduled. At this inspection, the affected parts may be replaced for minimum downtime or may be repaired and reinstalled in the unit with a longer outage. The parts under consideration for this inspection interval are transition pieces, first-stage nozzles, and second-stage nozzles. The combustion system parts will be repaired, using these criteria at approximately 1000- to 1500-hour intervals, or once per year. These parts include the combustion caps, liners, and cross-fire tubes.

Turbine buckets should require little repair except for foreign object damage caused by ingesting external material or for restoration of bucket tip clearances for continued efficiency.

The inspection interval hours stated previously for the hot-gas path parts will be maximized by optimization of the combustion system. It is important that the maintenance program be used to maintain proper control settings and that the combustion parts be kept in proper working order. The fuel nozzles, for example, will have a direct effect on the liners, transition pieces, and nozzles. Balanced firing temperatures will maintain minimum temperature differentials and assure that one combustion chamber and nozzle segment will not experience excessive temperatures. This occurs because the transition pieces and first-stage nozzles are exposed to the direct discharge from the combustion process.

First- and second-stage turbine nozzles can be repaired several times with a resultant extension of the total life. The economic determination of repair versus replacement will govern the feasibility and number of times the nozzles are repaired.

Operating Factors Affecting Maintenance

The factors having the greatest influence on the life of parts for any given machine are type of fuel, starting frequency, load duty, environment, and maintenance practices.

Fuel

The effect of the type of fuel on parts life is associated with the radiant energy in the combustion process and the ability to atomize various liquid fuels. Therefore, natural gas, which does not require atomization, has the lowest level of radiant energy and will produce the longest life of parts. Diesel fuels will produce the next highest life, and the crude oils and residual oils, with higher radiant energy and more difficult atomization, will produce shorter life of parts, as shown in Figure 3–18.

Contaminants in the fuel also affect the maintenance interval. This is particularly true for liquid fuels where dirt results in accelerated replacement of pumps, metering elements, and fuel nozzles. Contaminants in fuel gas systems can erode or corrode control valves and fuel nozzles. Filters must be observed and changed when practical to assure against the carrying of these contaminants through the fuel system. Clean fuels will invariably result in reduced maintenance and extended parts life.

Starting Frequency and Time

Each start, stop, and load change of a combustion gas turbine subjects its hot-gas path parts to thermal cycling. Control systems are designed, programmed, and adjusted to apply temperatures that are compatible with material properties to minimize required maintenance from this cycling effect. However, a unit in a peaking application will demonstrate parts lives (Figures 3–19 and 3–20) that are shorter than a similar unit in base-load continuous duty service, as with any equipment subject to cycling conditions.

The normal programmed starting time for a peaking unit is designed to minimize transient thermal stresses and maximize parts life. Fast start/load programs are available that compromise these objectives and are therefore used primarily in emergencies or to periodically demonstrate fast starting capability. These effects are shown in Figure 3–19 as a function of starting frequency. The maintenance penalty for fast start/load occurs mainly from the load application, since a fast start from standstill to rated speed occurs in approximately 2.25 minutes, with a temperature change of 800°F maximum; but a fast load application is accomplished in 30 seconds, with a resulting temperature change of 1000°F. The differences in rate of temperature change are obvious and explain the increased maintenance cost.

Load Duty

Utility units are usually supplied with a designated peak and peak reserve rating higher than the normal base rating. These ratings will affect the life of hot-gas path parts due

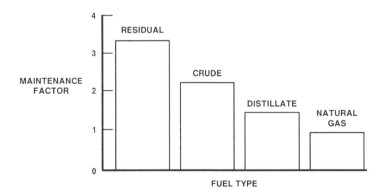

FIGURE 3–18 *Effect of fuel on gas turbine maintenance. (Source: General Electric Company, Schenectady, NY.)*

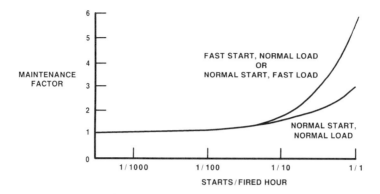

FIGURE 3–19   *Effect of number of starts on gas turbine maintenance. (Source: General Electric Company, Schenectady, NY.)*

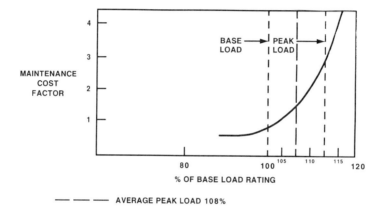

FIGURE 3–20   *Effect of load duty on gas turbine maintenance. (Source: General Electric Company, Schenectady, NY.)*

to the higher firing temperatures that exist in the unit (Figure 3-20). These ratings are used to allow the operator flexibility in the use of this equipment for the system needs. Usual daily peaking applications justify loading the units only to the base rating, with peak and peak reserve capabilities assigned for additional flexibility in emergency conditions.

Maintenance requirements are affected by the assignment of loading temperatures, and the economics of use must be balanced to arrive at overall use factors.

Environment

The condition of inlet air to the gas turbine can have a significant effect on maintenance if it is either abrasive or corrosive. In the case of abrasives in the inlet air, such as ash particles, careful attention should be paid to inlet filtering to minimize or eliminate this effect.

In the case of corrosive atmospheres, careful attention should be paid to air inlet arrangement and the application of correct materials and protective coatings.

Maintenance Practices

Parts condition information is based on estimates only and will vary with machines and specific operating conditions. However, estimates can be very useful in planning a maintenance program. Making the necessary adjustments, as actual operating data are accumulated on a specific application, should be the next step in a well-planned program.

The initial inspection program should be derived by identifying the turbine with the following parameters:

1. Intended load duty (peaking or continuous)
2. Fuel
3. Starting frequency

Knowing these parameters, the applicable time intervals may be extracted from Tables 3.4 through 3.7.

Standby and running inspections will be dictated by the physical location of the turbine, i.e., remote unattended pipeline stations versus the attended turbine driving process equipment in a petrochemical complex, or a remotely controlled turbine in utility peaking service that is observed by an operator on a daily or weekly basis.

Chapter 4

Gas Engines*

Gas engines are internal combustion engines and incorporate many of the operating principles of modern automobile engines. They are reciprocating machines and come in a variety of sizes and configurations.

Gas engines, as the term implies, operate on gaseous fuel that is ignited by an electric spark. Smaller gas engines, typically in the 200- to 400-kilowatt (kw) range, are frequently used to drive emergency generators or fire water pumps. Integral gas engines, with compressor cylinders mounted on the engine crankcase (Figure 4–1), are often found in the larger size ranges, up to several thousand kilowatts of power output.

Again analogous to automotive engines, reciprocating gas engines are manufactured either as two-stroke-per-cycle or four-stroke-per-cycle machines. With twice as many power strokes per revolution, the two-stroke-per-cycle engines tend to be smaller than four-stroke-per-cycle versions built for the source power output. Moreover, the two-stroke machine is generally less complex because it can dispense with valves and their associated mechanisms.

Reciprocating gas engines display typical efficiencies in the 28 to 42 percent range. The upper portion of this range belongs to turbocharged engines, whereas the lower range is populated by naturally aspirated machines.

Turbocharged equipment uses exhaust gas to drive a blower that forces combustion air through a suitable heat exchanger into the intake manifold. This cooled, compressed air is used for combustion and, in certain engine types, "scavenging." Scavenging air purges exhaust gases from engine cylinders before combustion air is admitted.

Reciprocating gas engines suffer from a few disadvantages that must be considered when selecting process plant equipment:

- They require a fair amount of competent surveillance and routine maintenance. Minor overhauls are typically needed after about 2500 operating hours. Major overhauls should be anticipated every three to five years.

- Their low speed, typically in the 180 to 900 RPM range, requires step-up gears for such process duties as pump drives.

- Fluctuations in the heating value of the fuel may require constant adjustment of spark timing and could also require derating of the power output capability.

- The sulfur content of gas engine fuels may adversely influence the extent and frequency of maintenance required.

Gas engines typically consume between 6500 and 8000 British thermal units (Btu)/brake horsepower (BHP)-hour.

---

*Source: Cooper-Bessemer Reciprocating, Grove City, PA. Adapted by permission.

FIGURE 4–1   *Large gas engine (ten cylinders) with built-on reciprocating compressor cylinders. (Source: Cooper-Bessemer Reciprocating, Grove City, PA.)*

TWO-STROKE GAS ENGINES

It could be stated that gas engines are basically blown-up versions of the conventional automotive engine with just a few important modifications: they are slow-running, they use a gaseous fuel/air mixture instead of the liquid fuel/air mixture typically found in most automotive engines, and they are often integrally arranged (or combined) with the process gas compressor cylinders that they are driving.

Since the combustion process takes place inside the cylinder, gas engines belong to the family of internal combustion engines. Like their cousins the automotive gasoline and automotive diesel engines, gas engines are either of the two-stroke or of the four-stroke per cycle variety. Two-stroke engines have one power stroke for every full revolution of the crankshaft, whereas in four-stroke engines, only every second revolution is accompanied by a power stroke. Four-stroke engines have inlet and exhaust valves; two-stroke engines have inlet and exhaust ports. Each type of engine has a spark plug; the two-stroke engine also incorporates a fuel admission valve.

With two-stroke engines considerably more prevalent in the process industries, we will confine our considerations to this type of engine.

Modern units are typically configured as shown in Figure 4–2. This cutaway view illustrates a Cooper-Bessemer model GMVH, essentially a combination V-type gas

FIGURE 4–2   *Cutaway view of a modern gas engine reciprocating compressor combination. (Source: Cooper-Bessemer Reciprocating, Grove City, PA.)*

engine and horizontal compressor built into one compact unit. The typical two-stroke engine is built in units of six, eight, ten, and twelve cylinders; the number of compressor cylinders varies according to requirements and arrangements yield any combination of volume and pressure within the rating of the engine.

The GMVH is a two-stroke-cycle loop-scavenged V-type engine, designed to use natural gas as a fuel. Scavenging air is supplied by an exhaust-driven turbocharger and a highly developed control system maintains optimum combustion under varying conditions of load and speed. Perhaps the best way to study this engine is to review its principal components identified in Figure 4–3.

The engine base (Figure 4–3, item 17) is a complex iron casting that forms the backbone of the entire structure. The main journal bearings are vertically split, leaving one side of the base open for easy access to the bearings and the crankshaft.

Once the alignment and fit of the main bearings have been established, the caps and outer shells are removed and the crankshaft (Figure 4–4, item 24) is installed. The bearing halves and caps are then installed and properly tightened once more. The crankshaft is a high-quality steel forging, machined with great precision and having a very fine finish on all bearing surfaces and fillets. Oil holes are drilled from the main journals to the adjacent crank pins to transmit large quantities of lubricating oil from the pressure-fed main bearings for lubrication of the connecting rod bearings and cooling of the power pistons.

The remaining connecting rod assembly depicted in Figure 4–4 is installed next. The main connecting rod, item 7, is used to drive the compressor piston. The power connecting rods, item 11, are articulated to the main connecting rod in a manner similar to that used in radial aircraft engines. In an engine of this type, the load on the power connecting rod is always compressive.

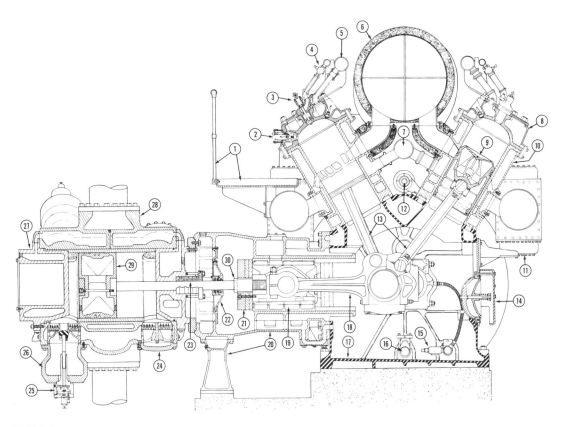

FIGURE 4–3   *Cross-section of a Cooper-Bessemer-type GMVH gas engine compressor. 1—platform and railing; 2—jet cell igniter; 3—gas injection valve; 4—load balancing valve; 5—jacket water outlet header; 6—insulated exhaust manifold; 7—jacket water inlet header; 8—power cylinder head; 9—power piston; 10—power cylinder; 11—air inlet manifold and intercooler; 12—layshaft; 13—articulated power rods; 14—crankcase relief valve; 15—lube oil pressure regulator; 16—lube oil suction header; 17—engine base (crankcase); 18—master rod; 19—crosshead and shoe; 20—crosshead guide housing and support; 21—crosshead balance weights; 22—crosshead diaphragm and packing; 23—compressor cylinder rod packing; 24—valve cap; 25—plug-type suction valve unloader; 26—unloader volume bottle; 27—compressor cylinder head; 28—compressor cylinder body; 29—compressor piston; 30—piston rod and nut. (Source: Cooper-Bessemer Reciprocating, Grove City, P.A.)*

The power cylinders (Figure 4–3, item 10) are next mounted on top of the base. The cylinder is a high-strength iron casting of some complexity, and Figure 4–3, item 10 shows clearly the air induction ports as well as the higher exhaust ports and the passages leading to each. The self-contained jacket, providing a flow of cooling water, is also shown. The bore of the power cylinder is chrome plated and then honed to a high degree of precision. It is common for such cylinders to be in continuous operation for several years without significant wear of the cylinder bore. The power piston (Figure 4–3, item 9) is an oil-cooled trunk-type piston using four compression rings and two oil control rings. Figure 4–4 shows these ring grooves as items 9 and 8, respectively. The piston pin housing, Figure 4–4, item 13, is a separate casting, bolted into the piston, containing the bronze bushing for the pin at the upper end of the power connecting rod. The space between the pin housing and the piston crown receives a continuous flow of lubricating oil for cooling. This oil comes through a longitudinal drilled hole in one flange of the power connecting rod and up through one of the vertical tubes in the pin housing. Oil is continually drained through the other tube and down through the other side of the connecting rod. This cooling prevents excessive thermal strains

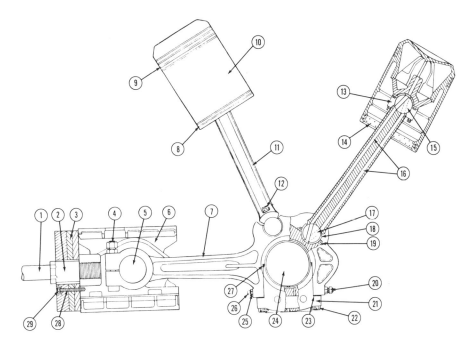

FIGURE 4–4   *Gas engine reciprocating compressor crankshaft assembly showing power pistons and compressor crosshead. (Source: Cooper-Bessemer Reciprocating, Grove City, PA.)*

in the piston even though the engine is operating at a high level of output. The cylinder head, (Figure 4–3, item 8) is an open-style iron casting with a cover plate to form a complete water jacket. The fuel gas injection valve is located in the center of the head, and there are two spark plugs, one on either side of the gas valve.

The crosshead guide, (Figure 4–3, item 19), is now installed. This serves both as a mount for a compressor cylinder and as a stationary slide for the crosshead at the outer end of the main connecting rod. The crosshead, Figure 4–4, item 6, has separate top and bottom shoes made either of aluminum or of cast iron with a babbitt overlay. These shoes are adjustable to fit with the proper clearance within the bore of the crosshead guide. The crosshead pin, item 5, is used to connect the crosshead to the eye in the outer end of the main connecting rod. The opposite side of the crosshead receives the end of the piston rod that drives the compressor piston. The piston rod packing will be discussed in more detail later.

The last major component to be added to the basic mechanical structure is the flywheel. The flywheel is bolted and doweled to a flange on the end of the crankshaft. Being of generous size, the flywheel serves to maintain the engine speed essentially constant, in spite of the variable turning effort of the power and compressor cylinders.

Gas Engine Compressor Support Systems

Pressure lubrication is supplied to practically all lubrication points of the engine except the power and compressor cylinders, which are lubricated by a force-feed lubricator system. The external lubricating oil system will vary according to installation requirements. Figure 4–5 is a schematic diagram of a typical system.

The wet-sump-type engine base serves as a reservoir for the lubricating oil. A sight gauge, located in the forward end cover, indicates the oil level at all times. The pump

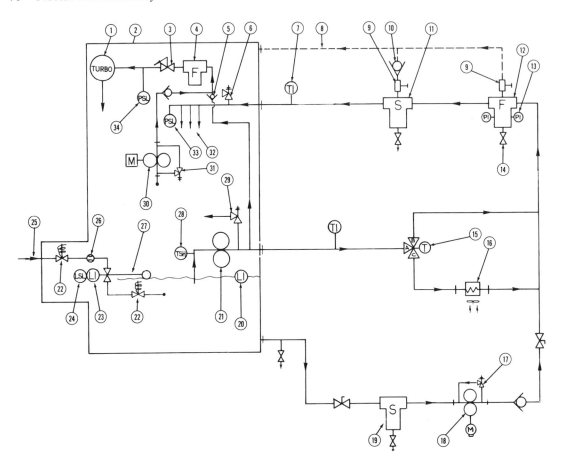

FIGURE 4–5   *Schematic representation of a gas engine pressure lube system. 1—turbocharger; 2—engine outline; 3—turbo oil variable pressure regulator; 4—turbo oil filter; 5—check valve; 6—engine oil pressure relief valve; 7—oil in temperature indicator; 8—filter and strainer vent line; 9—needle valve; 10—check valve; 11—oil strainer; 12—oil filter; 13—differential pressure gauge; 14—drain valve; 15—three-way thermostatic valve; 16—oil cooler; 17—pump relief valve; 18—engine prepost-lube pump; 19—oil strainer; 20—engine oil sight glass; 21—engine main oil pump; 22—fire safe shut-off valve; 23—oil level indicator; 24—engine low oil level alarm; 25—oil supply to engine; 26—flow meter (optional); 27—oil level regulator; 28—engine high oil temperature shutdown; 29—pump relief valve; 30—turbo prepost-lube pump; 31—pump relief valve; 32—oil to engine bearings; 33—engine low oil pressure shutdown; 34—turbo low oil pressure shutdown. (Source: Cooper-Bessemer Reciprocating, Grove City, PA.)*

suction header, in the bottom of the base, is perforated to form a strainer that prevents foreign matter from entering the pump. The oil pump (21) discharges the oil through an oil cooler (16), full flow filter (12), and strainer (11) and delivers it to the main oil header (32) in the engine base.

A thermostatic-operated three-way valve (15) is located in the system upstream of the cooler to direct oil through the cooler or around it to maintain the proper operating temperature. Thermometers should be installed in the line ahead of and after the oil cooler to give a constant reading of oil temperature.

From the lube oil distribution header, connections supply oil to all main bearings. From the main bearings, oil is delivered through drilled passages in the crankshaft to the master connecting rod bearings. The crankshaft (Figure 4-4, item 24) is drilled so that each master connecting rod receives oil from the two adjacent main bearings. From the master connecting rod, oil flows to the piston pin through a drilled passage in the articulated connecting rod. The crown of the power piston is jacketed, and oil from the piston pin is circulated through the jacket to cool the piston. Oil returns from the piston jacket through a second drilled passage in the articulated connecting rod, which connects with passages in the master connecting rod and is discharged to the base through holes in the master connecting rod cap. Oil also flows through a drilled passage in the master connecting rod to lubricate the crosshead pin and guide.

Vertical oil lines at the flywheel end carry oil from the main header to all the chain sprockets and bearings. Oil is returned to the base by gravity. At the other end, lines carry oil under pressure to all auxiliary drive shaft bearings, gears, chains, etc.

A pressure relief valve, located at the flywheel end of the oil header inside the base, protects the system against excessive pressure. A pressure gauge on the control panel indicates the oil pressure in the system. To maintain the required oil level in the base, an automatic control valve is sometimes installed in an oil makeup line.

The filter consists of a housing or shell containing replaceable, yarn-wrapped elements operating in parallel to give the desired capacity. The frequency of replacing the elements will vary with operating conditions. As the elements become contaminated, the pressure drop across the filter will increase.

Modern machines have the main lube-oil header installed in the engine base. High-pressure flexible lines supply oil from the header to the main bearing caps, from where it is carried to the rest of the running gear as previously described.

The engine is equipped with either multiple pumps or a block distribution-type force-feed lubrication system. All force-feed lubricator systems are divided into two separate sections. One section supplies lubricant for the power cylinders while the other section supplies lubricant to the compressor cylinders. This arrangement permits the use of different oils to lubricate the compressor cylinders when required. For certain types of compressor service, this is unnecessary and the same oil may be used for both power and compressor cylinders.

Fuel System

The fuel piping system (Figure 4–6) consists of a variable fuel gas pressure regulator, gas receiver, manual gas cock, safety shut-off and vent valve, gas accelerating valve, governor-operated gas regulating valve, gas injection valves, and isolating valves. The variable fuel gas pressure regulator regulates the gas supply pressure according to the governor speed signal. The receiver (located as close to the engine as possible) absorbs pulsations in the gas flow and ensures a more uniform gas pressure at the engine. The safety shut-off and vent valve will shut off the gas supply and vent the line to the engine if an abnormal operating condition occurs. The gas accelerating valve controls the amount of gas supplied to the engine by the governor-operated gas valve during starting.

The variable fuel gas pressure regulator is installed in the main fuel gas supply line upstream of the receiver. The gas pressure required at the engine will vary with the number of power cylinders and the heat content of the gas. In every case, the pressure should only be high enough to enable the engine to carry about 10 percent overload.

The gas regulating valve (Figure 4–7) is located in the gas inlet on the operating end. It is controlled by the speed-regulating governor to regulate the amount of gas according to load requirements. Gas enters the valve body from the supply line connected to the bottom of the body, passes through the valve port, and enters the inlet header of the engine. The valve is of the ported type and is designed to give a very fine regulation of flow for minimum travel of the valve and governor. A balancing piston on the valve stem will equalize the gas pressure in both directions of travel.

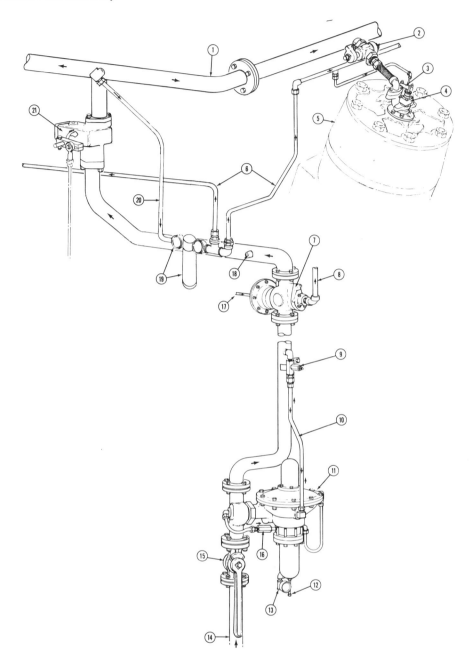

FIGURE 4–6   *Fuel piping system for two-stroke gas engine. 1—engine gas header; 2—load balance valve; 3—jet cell igniter; 4—gas injection valve; 5—cylinder head; 6—pilot gas header to igniters; 10—regulator feedback line; 11—gas pressure regulator; 12—starting pressure adjustment; 13—speed signal inlet pressure and gauge; 14—gas inlet, psi maximum; 15—manual gas shut-off valve; 16—regulator pilot filter; 17—fuel gas command signal; 18—gas pressure gauge tap; 19—igniter pilot gas filter and pressure gauges; 20—gas supply to pilot gas filter; 21—governor-operated gas regulating valve. (Source: Cooper-Bessemer Reciprocating, Grove City, PA.)*

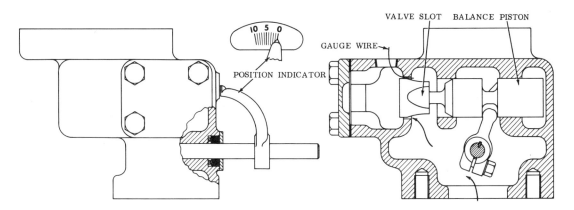

FIGURE 4–7   *Gas-regulating valve for modern gas engine. (Source: Cooper-Bessemer Reciprocating, Grove City, PA.)*

These design features give close regulation of gas flow and ensure steady operation and close regulation of engine speed at all loads.

Figure 4–8 illustrates the gas injection valve in cross section. Gas from the header is admitted to the injection valve through a cylinder-isolating plug valve. This valve is normally wide open and is used to restrict the flow to the injection valve to obtain load balance for equal distribution of load to the power cylinders.

The injection valve has a conical surface that seats on the valve seat insert. The valve is opened mechanically by a push rod and rocker arm operated by a cam attached to the crankshaft and is closed by the spring in the injection valve. Packing at the upper end of the gas valve stem prevents leakage of gas at this point.

The gas valve operating mechanism (Figure 4–9) consists of a rocker arm assembly, cam follower, and push rod with a hydraulic valve lifter. The rocker arm assembly is mounted on the cylinder head. An adjustable tappet is provided in the end of each rocker arm to adjust the hydraulic valve lifter. The cam follower is located in the engine base and is held in place by the push rod and crosshead bracket. Each follower consists of a crosshead and hardened steel roller that rides on a gas cam attached to the crankshaft. The push rod and hydraulic valve lifter assembly connects the rocker arm and cam follower. This assembly consists of a two-section push rod, the lower section being a tubular steel rod and adapter, with the hydraulic valve lifter installed in the lower end. A push rod guide supports the upper end of the lower section of the push rod at the point where it protrudes through the base. The upper section of the push rod is also tubular steel and is connected to the lower section by a ball pivot.

Cooling System

Cooling the engine is accomplished by two separate systems: the jacket water system, which circulates through the engine jackets and heat exchangers, and the aftercooler water system, which circulates through the heat exchanger, aftercoolers, and lube oil cooler.

One of the most important factors involved in the design of the cooling system is an adequate supply of clean water, free from sediment and scale forming ingredients, since even a very thin layer of scale or dirt on any heat transfer surface will act as an insulator, which may cause overheating and breakage. It is preferable to circulate a large volume of water accompanied by a small temperature rise than to circulate a small volume of water accompanied by a large temperature rise. A large volume results in

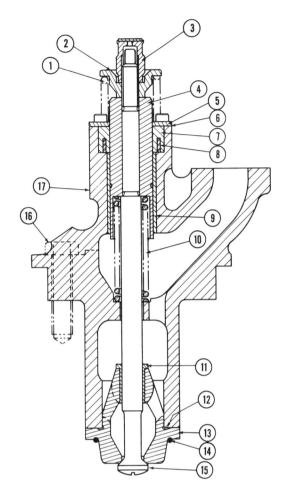

FIGURE 4–8   *Gas injection valve cross section.*
(Source: Cooper-Bessemer Reciprocating, Grove
City, PA.)

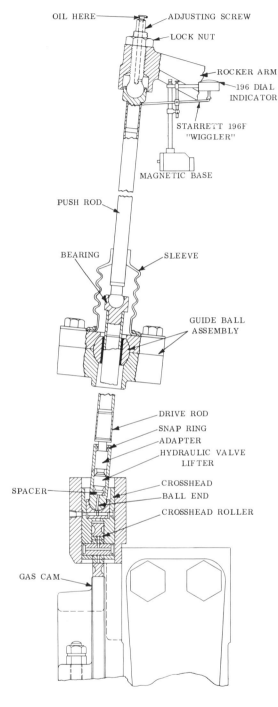

FIGURE 4–9   *Gas valve operating*
mechanism. (Source: Cooper-Bessemer
Reciprocating, Grove City, PA.)

higher velocity through the system and retards the formation of scale and deposits of
sediment in the jackets. Likewise, a low temperature rise means more uniform
temperature at all points and less possibility of casting strains from this source. For
these and other reasons, a closed cooling system is recommended. In such a system,
a minimum of makeup water is required. Therefore, treated water that removes scale-
forming ingredients is not expensive.

There are numerous piping arrangements that can be used, and these will vary according to the number of engines installed, cooling equipment used, and other individual requirements. Figure 4–10 is a typical diagram of the cooling system for an engine with a built-in jacket water pump and a motor-driven aftercooler water pump.

Tracing the flow of the jacket water system starting with the water pump, it is directed to the cooling equipment. This may be a cooling tower, radiator, or any other type of suitable equipment. A three-way thermostatically operated proportioning valve is located ahead of the cooling equipment to maintain the proper jacket water inlet temperature to the engine by bypassing a portion or all of the water around the cooling equipment. The water flows from the cooler, or bypass, to the engine inlet header.

Cooling water enters the engine through the inlet header located in the vee of the engine between the two cylinder banks. From the header, water enters the bottom of the cylinder jacket and passes upward around the ports and enters the cylinder head jacket through outside jumper connections. Outlets from the cylinder heads connect with the water discharge headers. A jumper connects the two outlet headers at the flywheel end to give a common outlet connection. Turbocharger cooling water is supplied from the inlet header at the flywheel end of the engine, circulates through the turbocharger, and discharges into the engine outlet header.

From the outlet header, the jacket water flows to the standpipe where makeup water is added when necessary. From the standpipe, water returns to the suction of the jacket water pump where it is again recirculated. The standpipe should be high enough to maintain a positive head at the suction of the pump. Its diameter should be large enough to limit the downward flow velocity to 0.5 feet per second, thus allowing any entrained air bubbles to rise to the surface.

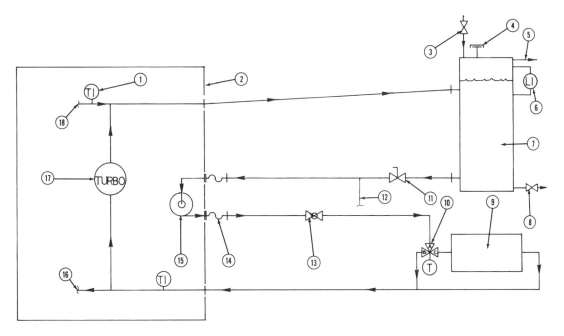

FIGURE 4–10 *Typical jacket water cooling system for a gas engine. 1—water outlet temperature; 2—outline of engine; 3—standpipe fill line; 4—vent; 5—standpipe overflow; 6—sightglass; 7—standpipe; 8—drain; 9—jacket water cooler; 10—thermostatic valve; 11—plug valve; 12—to intercooler water system (balance line); 13—gate valve; 14—pump flexible pipe connections; 15—jacket water pump; 16—to engine water inlet header; 17—turbocharger; 18—engine water outlet header. (Source: Cooper-Bessemer Reciprocating, Grove City, PA.)*

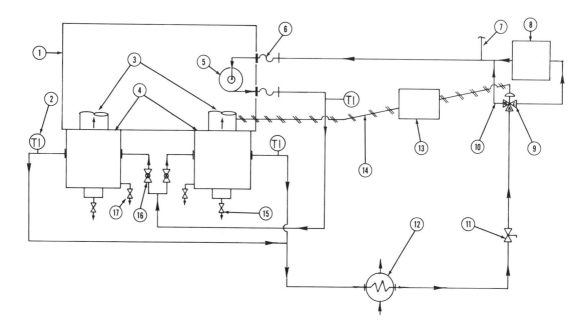

FIGURE 4–11   *Typical intercooler water system diagram for a gas engine. 1—outline of engine; 2—intercooler water outlet temperature; 3—engine air inlet manifold; 4—air intercoolers; 5—intercooler water pump; 6—pump flexible pipe connections; 7—balance line (supply) from jacket water system; 8—intercooler water cooler; 9—temperature control valve; 10—cooler bypass; 11—plug valve; 12—oil cooler; 13—temperature controller; 14—temperature control signal; 15—intercooler vent; 16—globe valve; 17—intercooler water drain. (Source: Cooper-Bessemer Reciprocating, Grove City, PA.)*

All engines are equipped with two fin-tube-type aftercoolers. In some localities, additional cooling of the air may be required to enable the engine to carry rated load. Precooling is then recommended whereby the air is cooled before entering the turbocharger by passing it through an aquatower.

A separate cooling water system is required for the water circulated through the aftercoolers. Engine jacket water is not suitable, as the water inlet temperature should not exceed 120 °F for proper cooling. Higher water inlet temperatures to the aftercoolers will not cool the air sufficiently, and the engine will fail to carry rated load. Each aftercooler must receive ample cooling water for efficient operation.

Tracing the flow of the aftercooler water system (Figure 4–11), starting with the pump, water is discharged from the pump through the heat exchanger, the aftercoolers, the lube oil cooler, and then back to the pump where it is again recirculated.

The Power Train

Engines equipped with motor starting are cranked by air- or gas-driven reduction gear motors attached to a Bendix starter drive. Four- and six-cylinder engines have one starting motor; eight-, ten-, and twelve-cylinder engines have two. The pinion on the Bendix drive then engages the ring gear on the flywheel to crank the engine. After the engine "fires," the starting valve is closed, pressure to the motors is shut off, and the Bendix drive pinion gear disengages the flywheel ring gear. Pressure to the starting motors is filtered and then lubricated by an oil-fog-type lubricator.

The layshaft is located in the vee between the power cylinders. It transmits power from the crankshaft to the auxiliary drive and is chain driven by the crankshaft at the

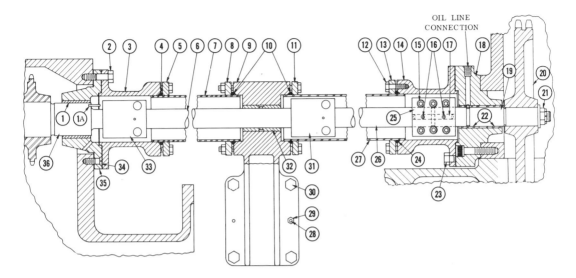

FIGURE 4–12 *Power train layshaft for a two-stroke gas engine. (Source: Cooper-Bessemer Reciprocating, Grove City, PA.)*

flywheel end. The shaft is constructed in two sections and supported at both ends and in the middle by bronze bearings (Figure 4–12, items 1, 22, 32). It is enclosed throughout its length by a tubular housing that is oil- and dust-tight. The layshaft chain tightener is located in the engine base at the flywheel end.

The main speed control governor, lubricators, and ignition timer are mounted on the auxiliary end drive cover and are driven by the layshaft. The governor is driven directly off the end of the layshaft and the timer is chain driven from the layshaft. The lubricators are chain driven from the timer drive. The positioning of these components is seen in Figure 4–1, foreground.

The Turbocharger

A turbocharger consists of a centrifugal blower and turbine mounted on a common shaft surrounded by five major castings. It is mounted on a diesel or gas engine and is driven by the engine exhaust gases. The exhaust gases, due to their elevated temperatures and high velocity, transmit enough energy to the turbine and blower to force 50 to 100 percent more air into the engine for scavenging and combustion. This additional air "supercharges" the engine, making it possible to burn more fuel to produce additional power. The scavenging air flow provides cooling for the cylinder head, cylinder walls, and piston. For this reason, a greater amount of fuel can be burned without harmful effects to the engine and turbocharger due to excessive heat.

The turbocharger output is proportional to engine load. It automatically slows down if the engine load is decreased and speeds up and delivers more air if engine load is increased or if barometric pressure drops. It maintains engine operation at or near optimum air-fuel ratio, resulting in high engine efficiency over a very wide operating range. The turbocharger operates in one direction only, regardless of engine rotation. Turbochargers used on "pure turbocharged" engines are equipped with air-assist nozzles on the blower diffuser. Air discharged through these nozzles, at the time of engine starting, assists in purging the engine and delivers sufficient energy to the turbocharger to maintain the required air-fuel ratio until exhaust energy becomes sufficient to drive the turbocharger.

Engine Control System

The control system of the engine has two purposes: first, to ensure safety, and second, to provide optimum operation. The pneumatically operated safety shutdown controls will stop the engine and indicate (on the control panel) the system in which the unsafe condition occurred. Any one of the safety shutdown devices (some are optional) will vent control air pressure from the fuel gas shutoff and vent valve to stop the engine should any of the following unusual malfunctions occur:

1. High jacket water temperature.
2. High air manifold temperature.
3. Force-feed lubricator, power cylinder, or compressor cylinder failure.
4. Low lubricating oil pressure.
5. High lubricating oil temperature.
6. Engine overspeeding.
7. High main bearing temperature.
8. High connecting rod bearing temperature.
9. Excessive turbocharger vibration.
10. Excessive engine vibration.
11. Low crankcase oil level.
12. Low aftercooler water pressure.
13. Low jacket water pressure.

Any number of these sensing devices as required by the installation can be used in the system to stop the engine via fuel gas shutoff and vent valves.

Engine speed is controlled by the main governor. The governor most commonly used is a Woodward hydraulic relay type. However, other types may be used according to service requirements and user preference. The governor is mounted on the auxiliary drive housing and driven through a set of bevel gears. The bevel gears are lubricated from the engine lube oil system. The governor is a self-contained unit with its own lubricating system. Governor signals are transmitted to the fuel gas accelerating valve to control the flow of gas to the headers in accordance with engine requirements. An increase in load causes a decrease in engine RPM, which in turn causes the governor to further open the fuel gas-regulating valve to admit more fuel gas to the engine. A decrease in load has the opposite effect. The result is that constant speed is maintained regardless of load and speed conditions.

Fuel gas header pressure and inlet manifold pressure change with load and speed. The manifold pressure regulator balances fuel gas pressure against air manifold pressure. Any change in sensed pressures will correspondingly move a pilot valve that directs oil under pressure to an actuator. This actuator repositions the exhaust bypass butterfly valve, thus maintaining the correct air manifold pressure in accordance with engine requirements.

Chapter 5

Mechanical Drive
Steam Turbines*

Steam turbines for mechanical drive applications were among the first real machines to usher in the Age of Industrialization. In 1629, an Italian inventor, Giovanni de Branca, envisioned a boiling pot whose nozzle opening was aimed against the paddle wheels of a roasting spit. Clearly, then, the use of expanding steam to turn a shaft is anything but a new idea.

By now, hundreds of thousands of steam turbines are installed and working in every conceivable type of process plant the world over. As shown in Table 5.1, inlet steam pressures range from a few pounds per square inch (psi) to over 2000 psi (140 bar) and power output covers the field from a single kilowatt (kw) in the emergency lube oil pump driver to almost 100,000 kw in large compressor drive applications at modern petrochemical plants. This should be no surprise, since the ability to efficiently convert large amounts of heat energy into mechanical work makes the steam turbine the logical choice for many industrial drive applications. Its reliability, smoothness of operation, and versatility also contribute to its popularity.

Before discussing turbine selection, let's review how a steam turbine converts the heat energy of steam into useful work.

The nozzles and diaphragms in a turbine are designed to direct the steam flow into well-formed, high-speed jets as the steam expands from inlet to exhaust pressure. These jets strike moving rows of blades mounted on the rotor. The blades convert the kinetic energy of the steam into rotation energy of the shaft.

In a reaction turbine, the steam expands in both the stationary *and* moving blades. The moving blades are designed to utilize the steam jet energy of the stationary blades and to act as nozzles themselves. Because they are moving nozzles, a reaction force—produced by the pressure drop across them—supplements the steam jet force of the stationary blades. These combined forces cause rotation.

To operate efficiently, the reaction turbine must be designed to minimize leakage around the moving blades. This is done by making most internal clearances quite small. The reaction turbine also usually requires a thrust balance piston (similar to large centrifugal compressors) due to the large thrust loads generated.

Because of these considerations, the reaction turbine is less often used for mechanical drive in the United States, despite its higher initial efficiency. However, reaction turbines are quite often used in other parts of the world. This text will explain them later. Moreover, since impulse and reaction turbines share many construction details and nomenclature, the reader is encouraged to become familiar with both.

---

*Sources: Elliott Company, a United Technologies Subsidiary, Jeanette, PA (impulse turbines), and Siemens Energy & Automation, Inc., Bradenton, FL (reaction turbines). Adapted by permission.

Table 5.1   Typical Steam and Power Conditions for Process Plant Turbines

Unit	Steam	Power
Small	150–400 psig; 500–750 °F (10–27 bar; 260–400 °C)	1–1,000 HP (0.75–750 kw)
Medium	400–600 psig; 750–825 °F (27–41 bar; 400–440 °C)	1,000–5,000 HP (750–3,750 kw)
Large	600–900 psig; 750–900 °F (41–62 bar; 400–482 °C)	5,000–60,000 HP (3,750–45,000 kw)
Very large	900–2,000 psig; 825–1000 °F (62–140 bar; 440–538 °C)	15,000–120,000 HP (11,200–90,000 kw)

Source: Elliott Company, a United Technologies Subsidiary, Jeannette, PA.

IMPULSE STEAM TURBINES

The impulse turbine has little or no pressure drop across its moving blades. Steam energy is transferred to the rotor entirely by the steam jets striking the moving blades (Figure 5–1).

Since there is theoretically no pressure drop across the moving blades (and thus no reaction), internal clearances are large and no balance piston is needed. These features make the impulse turbine a rugged and durable machine that can withstand the heavy-duty service of today's mechanical drive applications. Steam turbine materials are tabulated in Table 5.2.

Velocity-Compounded (Curtis) Staging

A Curtis stage (Figure 5–2) consists of two rows of moving blades. Stationary nozzles direct the steam against the first row; reversing blades (not nozzles) then redirect it to the second row.

The large pressure drop through the nozzle produces a high-speed steam jet. This high velocity is absorbed in a series of constant pressure steps (see below). The two rotating rows of blades make effective use of the high-speed jet, resulting in small wheel diameters and tip speeds, fewer stages, and a shorter, more rugged turbine for a given rating.

Pressure-Compounded (Rateau) Staging

Again referring to Figure 5–2, we observe how in Rateau staging the steam path is slightly different. Here the heat energy of the steam is converted into work by stationary nozzles (diaphragms) directing the steam against a *single* row of moving blades. As in a Curtis stage, pressure drops occur almost entirely across the stationary nozzles.

Turbine Configuration Overview

Single-flow condensing units (Figure 5–3) require less steam for a given horsepower than other types. They expand steam from inlet pressure to a pressure less than atmospheric. The exhaust pressure is maintained by a condenser, providing for recovery of the spent steam. A condensing unit thus minimizes the need for makeup water. Because of these advantages, the straight condensing turbine is much in demand, as evidenced by the literally thousands of units installed world-wide.

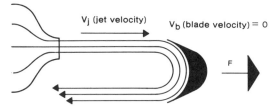

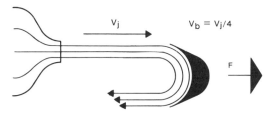

If the turbine rotor is locked, the steam jet exerts maxi-mum **force** on the blades, but no **work** is done since the blade doesn't move.

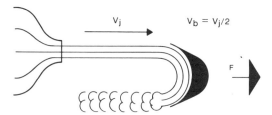

If the blade is moving at ¼ of the jet velocity, the force on the blade is reduced, but some **work** is done by moving the blades.

Maximum work is done when the blades are moving at ½ jet speed. Relative velocity of steam leaving blades is zero.

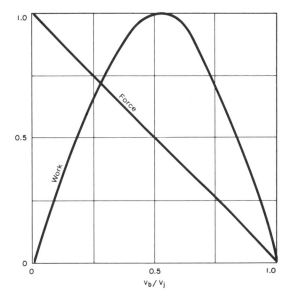

FIGURE 5–1   *The impulse principle. (Source: Elliott Company, A United Technologies Subsidiary, Jeannette, PA.)*

Table 5.2 Typical Standards Materials of Construction for Mechanical Drive Steam Turbines

	Material	*Commercial Specifications*
Steam chest and casing		
600 psi—750 °F/41 bar—399 °C	Cast carbon steel	ASTM A-216 Grade WCB
600 psi—825 °F/41 bar—440 °C	Carbon-molybdenum steel	ASTM A-217 Grade WC1
900 psi—900 °F/62 bar—482 °C	Chromium-molybdenum steel	ASTM A-217 Grade WC6
2,000 psi—950 °F/138 bar—510 °C	Chromium-molybdenum steel	ASTM A-217 Grade WC9
Exhaust casing		
Condensing and non-condensing		
(cast)	High-strength cast iron	ASTM A-278 Class 40
Non-condensing (cast)	Cast steel	ASTM A-216 Grade WCB
Fabricated	Steel	ASTM SA 285 Grade C
Nozzles	12% Chromium-stainless steel	AISI-405
Diaphragm centers		
Fabricated	Steel	ASTM SA 285 Grade C
Cast	High-strength cast iron	ASTM A-278 Class 40
Disks		
Forged	Chromium-nickel-molybdenum steel	AISI 4340
Cross-rolled plate	Constructional alloy steel	USS T-1
Integral with shaft	Chromium-nickel-molybdenum-vanadium steel	ASTM A-470 Class 4, 7, or 8
Blades	12% Chromium-stainless steel	AISI Type 403
Shroud bands	12% Chromium-stainless steel	AISI Type 410
Damping wire	15% Chromium steel	Inconel X750
Shaft		
Built-up	Chromium-molybdenum steel	AISI 4140
Integral	Chromium-nickel-molybdenum-vanadium steel	ASTM A-470 Class 4, 7, or 8
Bearing shells	Steel	ASTM SA 285 Grade C
Bearing liners	Bonded tin-base babbitt	ASTM B-23 Alloy #7
Labyrinth seals		
Shaft sleeves up to 750 °F/399 °C	Carbon steel	ASTM A-179
751–875 °F/400–468 °C	Nickel-chromium-molybdenum-steel	AISI 4340
Stationary baffles	Chromium-molybdenum-steel forgings	AISI 4140
Governor valves	12% Chromium-stainless steel	AISI Type 410
Governor valve stems and seals	12% Chromium-stainless steel, nitrided	AISI Type 416
Governor valve seats	12% Chromium-stainless steel	AISI Type 416
Bar lift rods and bushings	12% Chromium-stainless steel, nitrided	AISI Type 416
Steam strainer screen	Stainless steel	AISI Type 321
Bearing housings	Ductile iron	ASTM A-536 Grade 60-45-12

Source: Elliott Company, a United Technologies Subsidiary, Jeannette, PA.

Double-flow condensing units (Figure 5–4) are very similar to single-flow units except that the last-stage flow is divided between two rows of blades. This enables a double-flow turbine to operate at higher horsepowers and speeds than single-flow units of similar steam conditions.

Automatic extraction units are schematically represented in Figure 5–5. Extraction turbines are used when there is a need for process steam at a pressure between turbine inlet and exhaust pressures. They are designed to simultaneously maintain the desired

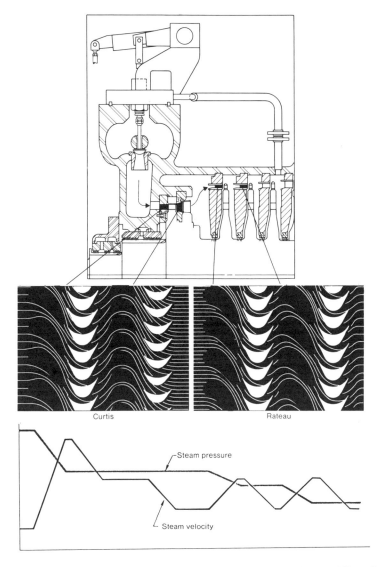

FIGURE 5–2 *Steam flow through turbine stages. S = stationary. (Source: Elliott Company, A United Technologies Subsidiary, Jeannette, PA.)*

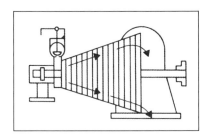

FIGURE 5–3 *Single-flow condensing turbine. (Source: Elliott Company, A United Technologies Subsidiary, Jeannette, PA.)*

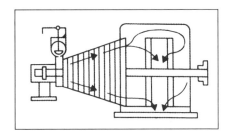

FIGURE 5–4 *Double-flow condensing turbine. (Source: Elliott Company, A United Technologies Subsidiary, Jeannette, PA.)*

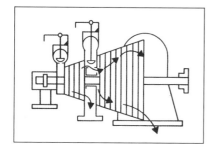

FIGURE 5–5   *Automatic extraction turbine.
(Source: Elliott Company, A United Technologies Subsidiary, Jeannette, P.A.)*

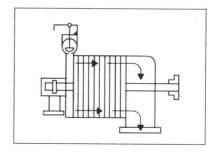

FIGURE 5–6   *Noncondensing steam turbine. (Source: Elliott Company, A United
Technologies Subsidiary, Jeannette, P.A.)*

extraction steam pressure and the speed of the driven machine. They can do this even
though the demand for extraction steam and the horsepower requirements of the driven
unit may vary over a wide range.

Noncondensing steam turbines (Figure 5–6) exhaust steam at greater than atmospheric pressure and while it still contains a great deal of energy. It can therefore be
used for other purposes in a given plant. Units of this type are much more compact
than condensing turbines of the same horsepower due to the smaller volume of steam
handled at the exhaust end.

Figure 5–7 depicts three typical rotor categories: single-flow condensing, double-
flow condensing, and noncondensing.

Turbine Components

Casing Overview.   A good casing should be just thick enough to contain the steam
pressure for which it is designed. Overly thick walls act as heat sinks and can restrain
expansion and contraction and lead to premature cracking.

Thin, contoured walls prevent large gradients between inside and outside "skin"
temperatures. A turbine casing designed this way conducts heat rapidly. This protects
against cracking of the nozzle partitions (critical area) and assures long life for casings
subjected to load changes and start-stop operation.

This casing philosophy is carried a step further on turbines designed for 2000 psi
and 950°F (140 bar and 510°C). The nozzle chambers and partitions are free of both
the front and side walls. The result, in effect, is a double-casing that relieves the internal stress in an area where there can be large inside-outside temperature differences.

Casing Construction.   A typical steam turbine casing consists of a steam chest, intermediate barrel section, and separate exhaust casing. To prevent leakage at the high-
pressure end, the steam chest and barrel section are often cast as one piece. This
permits the vertical casing joint to be made at the low-pressure end.

The high-pressure end of the turbine is supported by the steam end bearing housing. This housing is flexibly supported to permit axial expansion caused by temperature
changes. The exhaust casing is centerline-supported on pedestals that maintain alignment with the driven machine while allowing for lateral expansion. Figures 5–8 and
5–9 illustrate important casing details.

Steam Turbine Rotors.   Steam turbine rotors are either of the solid or built-up type.
Because each has its advantages, the user should not be tied to any one method of rotor
construction. Operating conditions often call for a solid rotor, but sometimes the user
may choose between the two or accept a recommendation from capable manufacturers.

FIGURE 5–7 *Principal rotor categories: Single-flow condensing (A, B), double-flow condensing (C, D), noncondensing (E, F). (Source: Elliott Company, A United Technologies Subsidiary, Jeannette, PA.)*

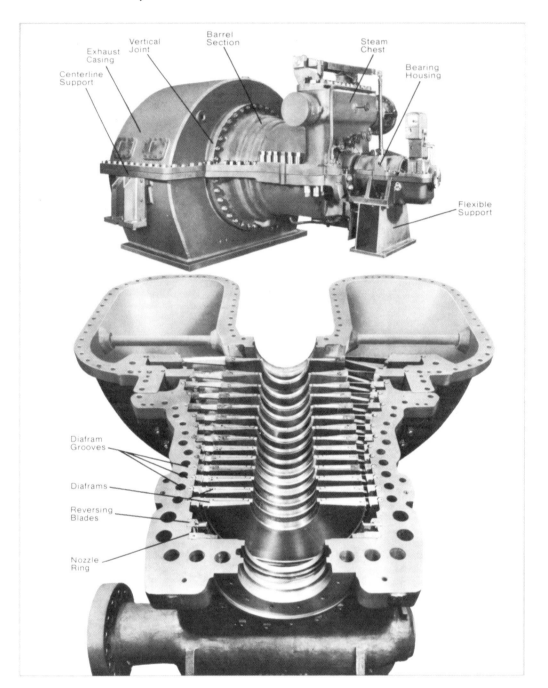

FIGURE 5–8   *Steam turbine casing components. (Source: Elliott Company, A United Technologies Subsidiary, Jeannette, PA.)*

With solid rotor construction, the shaft and disks are one piece. These rotors are generally used in units operating at high temperatures and/or high speeds.

Built-up rotors (Figure 5–10) cost less to make, less to buy, and are easier to repair if damaged. They are generally used in units operating at lower temperatures and speeds. Speeds are generally limited by the shrink fit required to overcome centrifugal growth of the disk.

FIGURE 5–9   *Steam end (2000 psi; 140 bar) showing steam space between nozzle chambers and front and side walls. This allows nozzle chambers to expand and contract freely in response to load and temperature changes. (Source: Elliott Company, A United Technologies Subsidiary, Jeannette, PA.)*

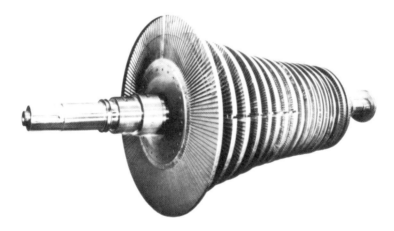

FIGURE 5–10   *Typical built-up single-flow rotor with 14-inch (360-mm) last-stage blades. Every disk is shrunk and keyed to the shaft. (Source: Elliott Company, A United Technologies Subsidiary, Jeannette, PA.)*

With either solid or built-up construction, shaft and disk materials must be selected to match the user's speed and temperature conditions. The shaft must be accurately proportioned to make sure that critical speeds are safely outside the operating speed range. Disk profiles should be designed to minimize centrifugal stresses, thermal gradients, and blade loading at the disk rims.

FIGURE 5–11   *Combination solid and built-up rotor for 28,000-HP (21-mW) steam turbine. (Source: Elliott Company, A United Technologies Subsidiary, Jeannette, PA.)*

Combination solid and built-up rotors are also available. Figure 5–11 shows a seventeen-stage, 28,000 HP (21 MW) rotor weighing 17,000 lbs. (7700 kg) and embodying the combination principle.

Blades.   The size and configuration of turbine blades depend on the operating conditions imposed on them. Machinery engineers have been designing blades to match horsepower, speed, and all types of steam conditions for many years. This continuing accumulation of blade knowledge has provided a wide selection of tested and proved blade designs. The blade performance record of reputable manufacturers usually reflects substantial design and manufacturing experience.

Figure 5–12 illustrates first-stage blades that have to withstand punishment of steam at its highest pressure and temperature. The blade at left is used for 2000 psi (140 bar) service; others are for lower pressures. A chromium stainless steel alloy, selected for its strength, erosion/corrosion resistance, and damping qualities is used on all blades. Last-stage blading is configured to efficiently expand large volumes of steam. Consequently, the blades are often considerably larger than those located in the higher pressure regions of a steam turbine.

Nozzle Rings and Diaphragms.   A stage in multistage turbines consists of both rotating and stationary blades. The stationary blades can be part of a nozzle ring or a diaphragm. In either case, their function is to direct steam onto the rotating blades, turning the rotor and producing mechanical work.

The first-stage nozzle ring is made by milling steam passages into stainless steel blocks, which are then welded together (Figure 5–13). Other manufacturing methods are, of course, available.

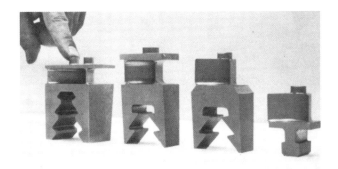

FIGURE 5-12   *First-stage blades for steam turbines. (Source: Elliott Company, A United Technologies Subsidiary, Jeannette, PA.)*

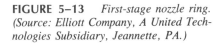

Milled Steam Passage

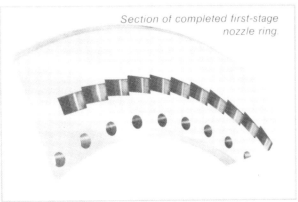

Section of completed first-stage nozzle ring.

FIGURE 5-13   *First-stage nozzle ring. (Source: Elliott Company, A United Technologies Subsidiary, Jeannette, PA.)*

The nozzles in the intermediate pressure stages are formed from stainless steel nozzle sections and inner and outer bands. These are then welded to a round center section and an outer ring, as shown in Figure 5-14.

Low-pressure diaphragms of condensing turbines are often made by casting the stainless nozzle sections directly into high-strength cast iron. This design (Figure 5-15) includes a moisture catcher to trap condensed droplets of water and keep them from re-entering the steam path. These diaphragms can also be completely fabricated, if desired, at extra cost.

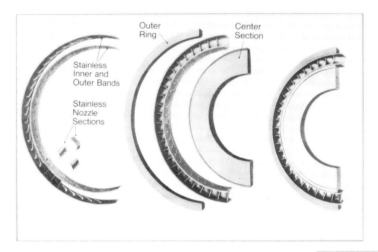

Outer Ring

Center Section

Stainless Inner and Outer Bands

Stainless Nozzle Sections

FIGURE 5–14 *Intermediate pressure diaphragm. (Source: Elliott Company, A United Technologies Subsidiary, Jeannette, PA.)*

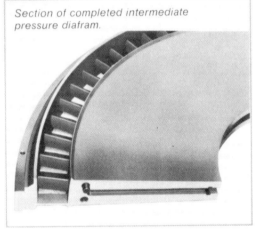

Section of completed intermediate pressure diafram.

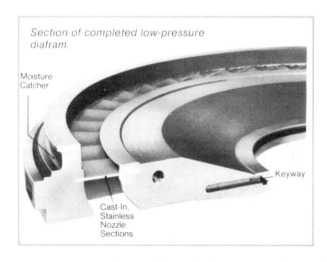

Section of completed low-pressure diafram.

Moisture Catcher

Keyway

Cast-In, Stainless Nozzle Sections

FIGURE 5–15 *Low-pressure diaphragm. (Source: Elliott Company, A United Technologies Subsidiary, Jeannette, PA.)*

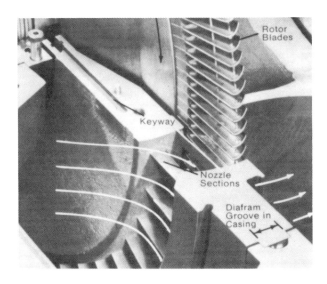

FIGURE 5–16   *Passage of steam through nozzles in diaphragm. (Source: Elliott Company, A United Technologies Subsidiary, Jeannette, PA.)*

Figure 5–16 illustrates steam turbine diaphragms as ultimately installed. The figure shows how the steam goes through the nozzles and strikes the rotor blades, causing rotation. Each diaphragm is located in its own groove in the casing. The keyway assures alignment of the diaphragm halves and helps minimize steam leakage across the split line.

Thrust Bearings.   Thrust bearings in modern steam turbines must withstand greater loads and higher speeds. The double-acting, self-leveling, fully equalizing Kingsbury type best meets these demands. It is long-wearing, requires little maintenance, and is capable of handling variable speeds.

As the wedge-shaped oil film is formed between the thrust collar and the surface of each shoe (Figure 5–17), the upper and lower leveling plates assure that the thrust load is distributed evenly among the shoes. The base ring then transmits this thrust load to the bearing housing. Stops prevent the base ring from rotating with the shaft.

Thrust bearings are generally sized to provide thrust capacity well in excess of that required for normal operating conditions.

Journal Bearings.   The turbine rotor runs in positively aligned, pressure-lubricated journal bearings that are horizontally split and lined with babbitt of the highest quality. Pressure-dam or tilting shoe bearings are used when required to assure bearing stability throughout the speed range.

Pressure-dam journal bearings (Figure 5–18) are generally used in high-speed units. This type of bearing is produced from a standard liner in which "dams" have been machined to create a "pressure pad" of oil. This pad forces the shaft to remain in position in the lower bearing liner, thus maintaining proper shaft attitude angle.

Tilting shoe journal bearings, shown in Figure 5-19 with five shoes (or pads), help ensure rotor stability.

Labyrinth Seals.   Experience has proved the labyrinth seal to be most reliable for steam turbines. It forms an effective seal without contact between its component parts, which means turbines will rarely if ever be shut down for seal maintenance.

FIGURE 5-17 *Self-leveling thrust bearing (Kingsbury-type) installed in modern steam turbine. (Source: Elliott Company, A United Technologies Subsidiary, Jeannette, PA.)*

The function of the seal, simply put, is to keep steam in and air out of the turbine where the rotor passes through the casing. It restricts steam leakage along the shaft by using the sealing strips on the shaft sleeve to form a ''labyrinth'' that hinders the flow of steam (Figure 5-20).

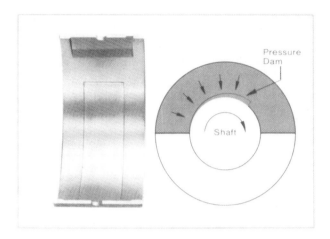

FIGURE 5–18 *Pressure-dam journal bearing. (Source: Elliott Company, A United Technologies Subsidiary, Jeannette, PA.)*

FIGURE 5–19 *Tilting-shoe journal bearing. (Source: Elliott Company, A United Technologies Subsidiary, Jeannette, PA.)*

In noncondensing or back-pressure turbines, the labyrinth seal must prevent steam leakage *out of* the casing at both inlet and exhaust because these pressures are greater than atmospheric. In condensing turbines, the pressure at the exhaust end is less than atmospheric. In this case, the labyrinth seal and sealing steam are used to prevent the flow of air *into* the casing.

Extraction Control

The use of extraction steam has gained wide acceptance in industry today. An extraction turbine can supply steam at some constant pressure between turbine inlet and exhaust pressures. This flexibility can be a great asset when designing a plant steam balance.

To automatically supply extraction steam, a second steam chest and a pressure control system are added to the speed control (governing) system of the turbine. This creates, in effect, two turbines on one shaft.

Extraction steam pressure and turbine speed are controlled by the extraction regulator as follows:

Constant Extraction Demand. Let's assume that extraction steam demand is constant, but the load on the turbine is reduced slightly and it begins to increase speed. The governor calls for reduced steam flow to the turbine.

The regulator is designed to reduce the steam flow proportionally through both the steam inlet valves and extraction valves. Extraction flow, therefore, remains unchanged, while both inlet and exhaust flows are reduced.

This diminished steam flow through the turbine decreases the power developed, and the turbine slows to its set speed. (Figure 5–21A).

FIGURE 5–20   *Labyrinth seal details. (A) Labyrinth seal showing how high and low sealing strips combine with the stationary baffle to hinder the flow of steam. The spring allows the stationary baffle to move away from the shaft if a rub occurs. The heat generated is absorbed by the stationary baffle. This protects the shaft and minimizes rotor damage. (B) Leakage flow and sealing steam arrangement for steam end of condensing turbine. Conditions that follow are typical values: (1) First-stage pressure equals 300 psia (21 bar) at full load. (2) Leakoff to turbine stage at approximately 125 psia (8.7 bar). (3) Leakoff to turbine stage at approximately 40 psia (2.8 bar). (4) Sealing steam at start-up is approximately 18 psia (1.25 bar). (5) Steam and air drawn to gland condenser at approximately 13.5 psia (0.9 bar). (6) Small amount of atmospheric air drawn in. (Source: Elliott Company, A United Technologies Subsidiary, Jeannette, PA.)*

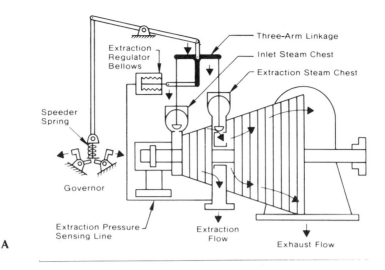

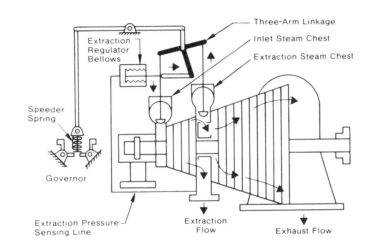

FIGURE 5–21   *Constant extraction demand (A) versus constant load output (B). (Source: Elliott Company, A United Technologies Subsidiary, Jeannette, PA.)*

Constant Load. Let's assume that the load on the turbine is constant, but there is a change in the process and the demand for extraction steam decreases. The extraction pressure increases, thus compressing the bellows in the regulator and rotating the three-arm linkage. This simultaneously closes the inlet valves and opens the extraction valves.

The inlet and extraction flows are thus decreased, while the exhaust flow is increased. Speed holds at setpoint because the three-arm linkage is proportioned to maintain a constant total horsepower developed by the rotor during these changes in extraction flow (Figure 5–21B).

In actual operation, both speed and extraction steam demands are constantly changing, and the sequences described are occurring simultaneously.

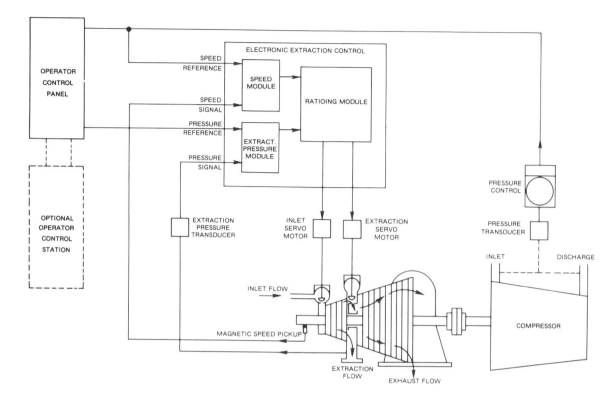

FIGURE 5–22   *Electronic extraction control for steam turbine. (Source: Elliott Company, A United Technologies Subsidiary, Jeannette, PA.)*

When excess steam is available, it can be ''induced'' into the turbine and expanded to exhaust pressure. Major manufacturers of steam turbines can supply automatic extraction units, automatic induction units, or a combination of the two.

It should also be stressed that fully electronic controls are sometimes an attractive feature that could be considered instead of the more traditional, time-tested mechanical controls. Many of the 1980-vintage steam turbines are fitted with highly reliable electronic controls.

An electronic extraction control system for a turbine driving a compressor is shown in Figure 5–22. Speed and pressure signals are fed to the control and electrically compared with the speed and pressure reference set points. Corrective signals are generated and sent to the inlet and extraction servo-motors. The optional operator control station is just one of a wide variety of optional features available with this type of control.

Trip Devices.   Steam turbines must be protected against excessive speed by automatic overspeed trip devices. These devices are either of the mechanical or electronic type. Mechanical overspeed devices are based on the centrifugal force principles as shown in Figures 5–23 and 5–24. Upon actuation of the trip lever, hydraulic oil is dumped from a pressurized cylinder that normally holds inlet steam or trip valves open. When depressurization occurs, these valves close instantly.

Electronic overspeed devices operate on a somewhat similar principle. Their actuation results in the opening of a solenoid dump valve that depressures the hydraulic cylinder that normally keeps the inlet and trip valves open.

FIGURE 5–23   *Eccentric overspeed trip device. (Source: Elliott Company, A United Technologies Subsidiary, Jeannette, PA.)*

FIGURE 5–24   *Disk-type overspeed device. (Source: Elliott Company, A United Technologies Subsidiary, Jeannette, PA.)*

REACTION STEAM TURBINES

Our introductory comments on steam turbines made the point that steam turbines extract heat energy from the steam and convert it into mechanical work. Heat and mechanical work are both forms of energy, and, therefore, can be converted from one to the other.

First, heat energy is converted into velocity (kinetic) energy. In this first step, steam expands in a nozzle discharging at high velocity. The total heat (enthalpy) of the steam is converted to kinetic energy (velocity).

Second, a steam turbine does mechanical work by virtue of the steam velocity that strikes moving blades. There are two types of stages to convert this velocity energy into usable work, namely, impulse and reaction. With an impulse design, the pressure drop is taken across the nozzles only, whereas with reaction design the drop is evenly distributed across the nozzle and blades. If the nozzle is fixed and the force of the jet is directed against a crescent shape (the blade), the jet's impulse force pushes the blade in the direction of the jet. Further, the leaving velocity is less than the initial jet velocity. Such an example would be a water hose jetting water against a wall. If the nozzle is free to move, the reaction pushes against the nozzle and it will travel in a direction opposite to the jet. The leaving velocity is greater than the jet velocity. An example would be a revolving lawn sprinkler.

The basic characteristics of impulse-type turbines and reaction-type turbines can be summarized as follows. In *impulse turbines* we find the following:

- Large bucket clearances permit quicker loading with minimal danger from thermal stress.
- Since most of the pressure drop is in stationary nozzles, efficiency does not depend on blade clearances.
- Low thrust generation.

In *reaction turbines,* we note the following:

- Expansion takes place in stationary nozzles and moving blades.
- More bucket sealing (sealing strips) is required at the end of blades.
- Efficiency is theoretically greater but depends on close clearances.
- The so-called greater efficiency of a reaction machine over an impulse machine in practice becomes a function of the staging (number of stages, efficiency of stages available, etc.). A reaction machine will have more stages because to keep losses to a minimum, the pressure drop per stage must be kept low.
- Reaction turbines require larger thrust bearings and/or a balancing piston.

Reaction staging is similar in concept to Rateau staging except that pressure drop occurs through both the stationary nozzle and matching blades. Figure 5–25 shows reaction staging and can readily be compared with Rateau staging.

A cutaway view of a typical reaction-type turbine is shown in Figure 5–26. It highlights the principal components as follows.

Casing and Guide Blade Carriers

To achieve satisfactory performance with the high steam pressures involved and to simplify erection and overhauls, a double-shell casing is used. The outer shell is split horizontally at the machine axis, the top half incorporating the integrally cast admission chest. This is in the form of a transverse tube with an opening at each end for assembly. The emergency stop valve body is welded to one end of the tube.

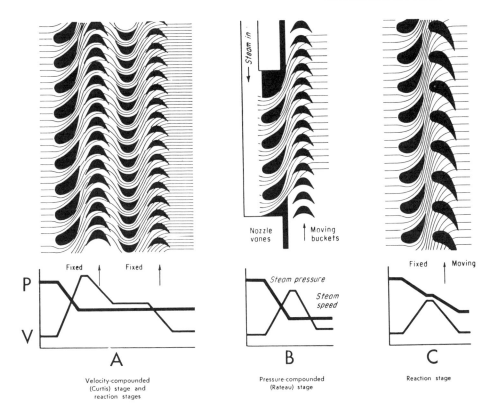

FIGURE 5–25 *Different types of stages employed in modern steam turbines. (Source: Elliott Company, A United Technologies Subsidiary, Jeannette, PA.)*

The two halves of the casing are flanged and bolted together. The casing flange bolts are heated to a predetermined temperature when being fitted. The cooling of the bolts prestresses them exactly to a predetermined value.

The guide blades are fitted in carriers. The front carrier is mounted in the one-piece inner shell, while the real carrier is mounted directly in the outer shell. The longer machines incorporate an additional carrier. The unsplit inner shell allows the useful feature of a split outer casing to be retained while at the same time offering the advantages of the "barrel" type of construction. The inner shell is shown in Figure 5–27. The front guide blade carrier is split axially and clamped together by a conical ring. It is located centrally in the inner casing by a bayonet coupling that is pinned to prevent it from turning.

Figure 5–28 shows the completely assembled rear blade carrier, which is kinematically supported in the outer casing. The guide blade carriers are aligned in the bottom half of the casing before final assembly.

The steam pressure in the annular space between the inner and outer shells is approximately half the initial steam pressure. This means that both the inner and outer shells are subjected to relatively low stresses. Since there is only a small difference in the temperature of the steam inside and outside the inner shell when the turbine is running normally, thermal stresses are also minimal. The largely symmetrical shape ensures that thermal expansion is uniform at all sections.

The fact that the front part of the outer casing is subjected to only half the initial steam pressure and the rear part of the casing to the exhaust pressure means that flanges and bolts can be of a size consistent with their reduced stress. The inner shell rests

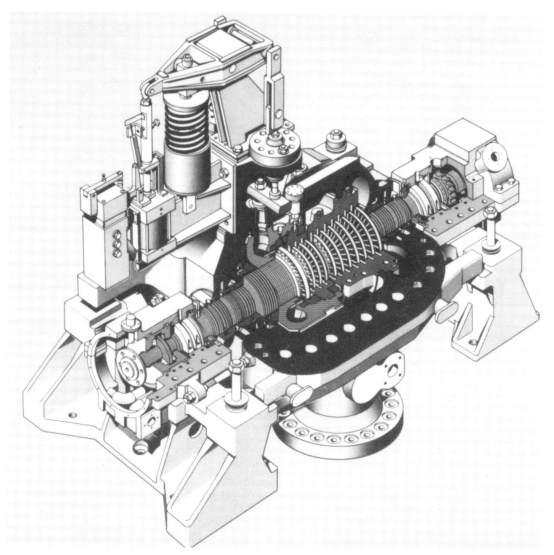

FIGURE 5–26   *Cutaway view of a typical reaction-type steam turbine. (Source: Siemens Energy & Automation, Inc., Medium Power Generation Division, Bradenton, FL.)*

FIGURE 5–27   *One-piece high-pressure inner casing for reaction turbine. (Source: Siemens Energy & Automation, Inc., Medium Power Generation Division, Bradenton, FL.)*

FIGURE 5–28   *Rear guide blade carrier for reaction turbine. (Source: Siemens Energy & Automation, Inc., Medium Power Generation Division, Bradenton, FL.)*

FIGURE 5–29   *Fitted eccentric pin for internal component alignment of reaction turbine. (Source: Siemens Energy & Automation, Inc., Medium Power Generation Division, Bradenton, FL.)*

on its integrally cast brackets in the bottom half of the outer casing. Eccentric pins (Figure 5–29) are incorporated to adjust the position of the inner shell in relation to the outer casing and to align the rear guide blade carrier in the outer casing.

Self-sealing, L-section rings free to expand in any direction are used for the seal between the admission chest and the inner shell.

The initial steam entering the various nozzle boxes of the admission chest gives rise to a downward thrust on the inner casing. This thrust is partially compensated by

FIGURE 5–30   *Vertical thrust balancing chambers with L-ring sealing to casing bottom half. (Source: Siemens Energy & Automation, Inc., Medium Power Generation Division, Bradenton, FL.)*

FIGURE 5–31   *Turbine rotor complete with inner casing and front guide blade carrier. (Source: Siemens Energy & Automation, Inc., Medium Power Generation Division, Bradenton, FL.)*

allowing steam from the two outer nozzle boxes to pass through holes in the bottom of the inner casing and build up an opposing force on the underside of the casing in two small chambers of appropriate cross-sectional area sealed from the outer casing by L-section rings (Figure 5–30).

The exhaust, gland, and drain connections are made to the bottom half of the outer casing with the result that only the initial steam line must first be removed to lift the top half of the casing. Since the bottom half remains in its aligned position on the pedestals, the amount of work involved in overhauls and inspections is considerably reduced. Once the top half of the casing has been lifted, the rotor, complete with the inner casing and front guide blade carrier, can be taken out (Figure 5–31). The combined weight of these parts is only approximately one eighth of the total weight of the turbine.

Rotor and Blading

The solid rotor shaft and control stage wheel are forged of one piece. The individual blades, which are machined from the solid, have integral shrouding and a T-root that fits into a groove in the rotor shaft. The drum blading is approximately 50 percent reaction.

 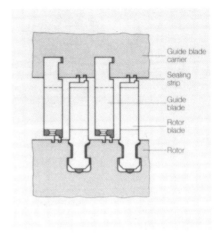

FIGURE 5-32 *Tip sealing of blade by integral shrouding. (Source: Siemens Energy & Automation, Inc., Medium Power Generation Division, Bradenton, FL.)*

Blade Tip Sealing

The rotor blades are milled from the solid together with the shrouding. After they have been fitted in the rotor groove, the shrouding forms a complete ring, which is then skimmed. A circumferential step is formed in the surface of the shrouding, which, together with the sealing strips caulked into the casing opposite the shrouding, produces an efficient sealing effect. Riveted shrouding is used for the guide blades as they are not subjected to centrifugal force. A labyrinth sealing effect is again produced by caulking sealing strips in the rotor opposite the shrouding.

Figure 5-32 shows some shrouded blading and also a diagrammatic section of the tip sealing employed. The radial clearance may be kept extremely small to reduce losses without adversely affecting operational reliability. Should maloperation cause contact to occur, the sealing strips will be rubbed away without producing a dangerous rise in temperature, and they can be replaced at the next overhaul without disturbing the blading in any way.

Shaft Glands

The points where the shaft passes through the casing are sealed by means of labyrinth glands, which are composed of sealing strips in the stationary part projecting into grooves in the turbine rotor shaft (Figure 5-33). Any steam leaking through the gland is drawn off at an intermediate stage to a region of lower pressure in the turbine or to a special condenser in order to limit the amount of steam discharged into the atmosphere.

Casing Supports, Journal and Thrust Bearings

The top half of the casing is supported on four brackets, two resting on the front bearing pedestal and two on the rear (Figure 5-34). Vertical alignment devices are placed between the feet and the pedestal surfaces to adjust the height accurately. The rear bearing pedestal is bolted rigidly to the foundation. When the casing expands as the temper-

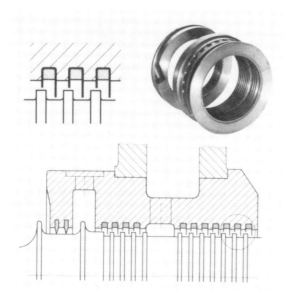

FIGURE 5–33   *Labyrinth shaft gland. (Source: Siemens Energy & Automation, Inc., Medium Power Generation Division, Bradenton, FL.)*

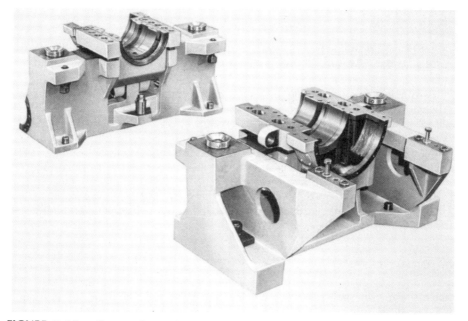

FIGURE 5–34   *Front and rear bearing pedestals for reaction turbine. (Source: Siemens Energy & Automation, Inc., Medium Power Generation Division, Bradenton, FL.)*

ature rises, it slides on the front bearing pedestal. In order to locate the casing to the bearing pedestals in the longitudinal and transverse directions, both pedestals have guide rails that engage with guide forks on the casing bottom half fitted with horizontal alignment devices.

The journal bearings are of the multiwedge type. They are white-metal lined. The oil-film wedges uniformly spaced around the circumference hold the rotor in a stable position.

The position of the bearing housings in the pedestals can be adjusted with alignment devices to align the rotor accurately. This alignment ensures that the correct radial

FIGURE 5–35   *Double-acting segmented thrust bearing. (Source: Siemens Energy & Automation, Inc., Medium Power Generation Division, Bradenton, FL.)*

clearance is obtained in the blading and labyrinth seals. The axial position of the rotor within the casing is determined by the thrust bearing in the front bearing housing. It is of the double-acting segmented type (Figure 5–35). The bearing segments are also lined with white metal.

As its temperature rises, the rotor expands in the direction of the exhaust hood. Since the casing is fixed at the rear bearing housing, the front bearing housing carrying the thrust bearing expands in the opposite direction. Hence, only the very much smaller differential expansion between rotor and casing is temporarily effective. The thrust bearing accepts any residual thrust that has not been eliminated by the balance piston.

Valves and Governing System

The initial steam is admitted to the admission chest through the emergency stop valve (Figure 5–36). During normal operation, this valve is held open against spring load by hydraulic pressure. In the event of a defect, the trip oil circuit is vented and the springs close the valve in a fraction of a second.

A steam strainer is incorporated in the valve body. It is comprised of corrugated steel strip wound spirally on edge on a former. In comparison with the perforated type of strainer, the wound type is considerably stronger, has a greater flow area and a smaller aperture size.

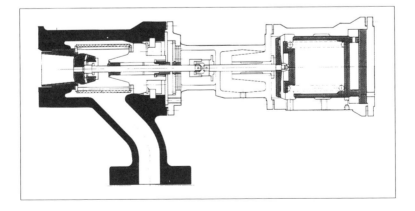

FIGURE 5–36   *Emergency stop valve incorporating a steam strainer. (Source: Siemens Energy & Automation, Inc., Medium Power Generation Division, Bradenton, FL.)*

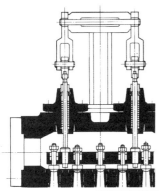

FIGURE 5-37   *Steam turbine control valves. (Source: Siemens Energy & Automation, Inc., Medium Power Generation Division, Bradenton, FL.)*

FIGURE 5-38   *Valve diffusors and L-section sealing rings in the top half casing. (Source: Siemens Energy & Automation, Inc., Medium Power Generation Division, Bradenton, FL.)*

The nozzle control system of standard turbines makes extensive use of lever mechanisms. The five nozzle control valves are suspended from a beam (Figure 5-37). The beam is raised and lowered on two spindles through a system of levers by a front-mounted hydraulic servomotor. This arrangement means that there are only two spindle glands in the admission chest. Both spindles move whenever there is any control action so that the possibility of seizure is virtually eliminated. The distance between the top of the adjusting bushing in the beam and the underside of the backing nut at the top of the valve spindle varies from valve to valve. Thus when the beam is lifted, the valves open in sequence, allowing steam to reach individual nozzle segments progressively. The valve diffusors in the top half of the casing are shown in Figure 5-38.

APPENDIX 5:

Steam Turbines

There are many factors affecting overall blade performance. They should all be carefully weighed before a design is released for production. The discussion here will be limited to just two of these factors: frequency and stress analysis. We shall see how they can affect the final design of a blade.

ENGINEERING

Stress Analysis

The stress levels in each blade must often be given strong consideration. As explained later, stress is best evaluated by a modified Goodman diagram, which provides a ratio of allowable cyclic stress to the maximum steam bending stress. This ratio should have certain minimum values for each stage under consideration.

There are many factors affecting the mechanical integrity of a blade. The damping of the stage, caused by the inherent damping qualities of the material, together with the method of attaching the blades to the disk, is one. Disk configuration and the type of shrouding are other factors.

A Campbell diagram, described below, is merely another one of the tools for designing blades. It cannot alone determine if a blade will operate reliably.

Blade Frequency

The simplest method for blade frequency analysis has been to use the natural frequencies of individual blades. This can be misleading because blades vibrate differently in assembled packets.

Individual blade frequencies are easy to obtain. However, static packet testing will yield more complete data. Dynamic packet testing is clearly the most accurate method, but it has been and always will be extremely expensive.

The most widely-used frequency consideration is the Campbell diagram. Mr. Campbell experimented with disk frequencies back in the early days of steam turbines. There were numerous disk failures then, caused primarily by disk resonance at wheel critical speeds.

The industry has since coined the name "Campbell diagram" to mean "interference diagram." These are useful in showing when natural blade frequencies coincide, or interfere, with various operating harmonics such as multiples of running speed or nozzle passing frequencies (NPF).

Many engineers feel that if *any* interference exists, the blade will fail—but is this true? Experience indicates it is almost impossible to design a multistage turbine with no interference. The turbine must therefore be designed to operate reliably in certain modes of resonance.

Example

To prove the point that frequency is not the only factor to consider in the design of a turbine blade, let's review a blade in terms of frequency and stress. The example is an 0.847-inch (21.5-mm) blade on an 18-inch (460-mm) diameter.

The first Campbell diagram, Figure 5A-1, shows no interference of the fundamental tangential mode with NPF. Keep in mind that these data are based on individual blades. Realize, also, that the various "modes" simply describe vibration behavior in different directions or at multiples of a fundamental vibratory frequency.

The second interference diagram, Figure 5A-2, is on the basis of static packet data and the situation worsens; the fundamental axial mode intersects the NPF. *(Packets describe several blades joined by a shroud band or similar means.)*

Centrifugal force will tend to increase the natural frequencies, but this particular blade will still be operating in resonance. The conclusion might be that this blade will

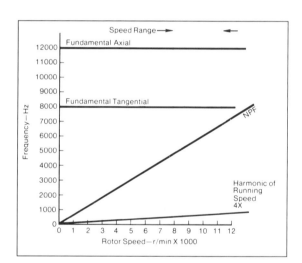

FIGURE 5A-1   *Campbell diagram for an individual steam turbine blade. (Source: Elliott Company, A United Technologies Subsidiary, Jeannette, PA.)*

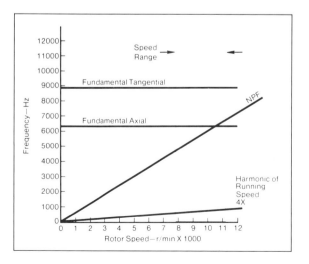

FIGURE 5A-2   *Cambell diagram for an eight-blade "packet." (Source: Elliott Company, A United Technologies Subsidiary, Jeannette, PA.)*

fail. This is not necessarily so. The stress in the blade has not been considered, and it is a fact that failure occurs only when the operating stress exceeds the endurance limit, regardless of what causes the stress.

Modified Goodman Diagrams

Many experienced manufacturers rely on a modified Goodman diagram for stress evaluation. For the ultimate stress, the strength of the blade material at the *design operating temperature* is used. The endurance limit used is based on the corrected ultimate stress and other considerations such as the fatigue notch factor. A straight line is then constructed between the ultimate stress and the corrected endurance limit.

Theory states that failure will occur when the total stress in the material is to the right of the failure line. The steady state stress is the sum of the centrifugal stress and steam bending stress. This total steady state stress is plotted on the abscissa. A straight horizontal line is then drawn over to the failure line, a vertical line is constructed, and this becomes the allowable alternating stress. This is shown in Figure 5A–3.

The allowable alternating stress, divided by the steam bending stress, is the Goodman factor. In theory, the steady state stress can equal the ultimate stress, without failure, when the alternating stress is zero. Further, the alternating stress can equal the corrected endurance limit, without failure, when the steady state stress is zero. Blade reliability is thus seen to be a combination of steady state stresses and alternating stresses.

Past operating history has established safe, minimum Goodman factors. The blade in question has a 25.8 Goodman factor, as indicated in Figure 5A–4. This is well above

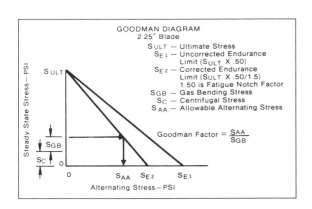

FIGURE 5A–3 *Modified Goodman diagram used for blade stress evaluation. (Source: Elliott Company, A United Technologies Subsidiary, Jeannette, PA.)*

FIGURE 5A–4 *Data used in modified Goodman diagram. (Source: Elliott Company, A United Technologies Subsidiary, Jeannette, PA.)*

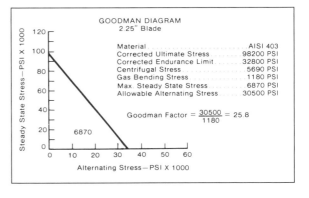

the minimum requirements for this stage. Many manufacturers would feel safe in recommending it for this application even though a Campbell diagram shows frequency interference. They can support their recommendation by showing that this stage has been in actual operation, with both the fundamental tangential and axial modes at nozzle passing frequency, for many years. The example adds emphasis to what was said at the beginning: many factors affect reliability, and it behooves the designer to look at all of them.

Steam Consumption (Approximate Steam Rates)

Figure 5A–5 and Table 5A.1 can be used to determine an approximate turbine efficiency when horsepower, speed, and steam conditions are known. In many instances, an *approximate* efficiency may well serve your purpose, since the parameters mentioned may change.

As can be seen in the example, a 25,000 HP, 5000 r/min turbine using steam at 600 psi/750°F/4-inch Hg abs. has an approximate efficiency of 77 percent.

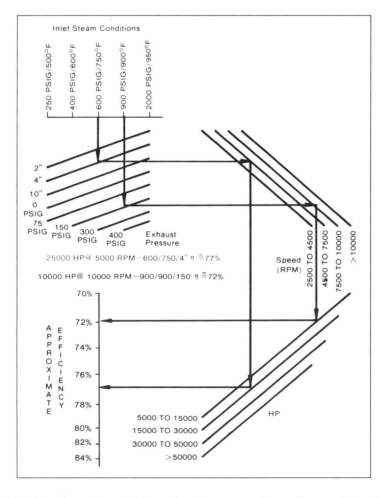

FIGURE 5A–5   *Approximate efficiency chart for steam turbines. (Source: Elliott Company, A United Technologies Subsidiary, Jeannette, PA.)*

Table 5A.1 Theoretical Steam Rates (lbs/kwhr)*

	250 psig 550°F	400 psig 750°F	600 psig 750°F	600 psig 850°F	900 psig 900°F	1,500 psig 900°F	2,000 psig 950°F
2″ Hg abs.	8.78	7.36	7.08	6.66	6.26	6.08	5.84
4″ Hg abs.	9.67	7.98	7.64	7.17	6.69	6.48	6.20
0 psig	14.57	11.19	10.40	9.64	8.74	8.26	7.78
50 psig	26.75	17.56	15.36	13.98	12.06	10.94	10.07
100 psig	42.40	23.86	19.43	17.64	14.50	12.75	11.55
200 psig	—	43.51	29.00	26.33	19.45	15.84	13.96
300 psig	—	—	43.72	39.70	25.37	18.94	16.19
400 psig	—	—	—	—	33.22	22.32	18.49
600 psig	—	—	—	—	63.40	30.75	23.63

Note: Interpolate, where necessary, for approximate values.

*From Theoretical Steam Rate Tables, copyright 1969 by A.S.M.E.

The TSR for steam conditions not tabulated can be found by using the equation:

$$TSR = 3413 \text{ Btu/kwhr}/\Delta H_i \text{ Btu/lb}$$

where ΔH_i = isentropic enthalpy drop from inlet conditions to exhaust pressure.

For example, find TSR for steam conditions of 1250 psig/900°F/175 psig using the Mollier diagram.

At 1250 psig (1265 psia) and 900°F
  H inlet = 1439 Btu/lb
  S   = 1583 Btu/lb °F

At 175 psig (190 psia) and an S of 1583 the isentropic exhaust enthalpy, H exhaust, would be 1227 Btu/lb.

ΔH_i would therefore be:

ΔH_i = H inlet − H exhaust
        = 1439 − 1227
        = 212 Btu/lb

The TSR would then be:

TSR = 3413 Btu/kwhr/212 Btu/lb
     = 16.1 lb/kwhr

Source: Elliott Company, a United Technologies Subsidiary, Jeannette, PA.

Applying this approximate efficiency to the theoretical steam rate (TSR) results in a steam rate and steam flow as follows:

$$TSR \text{ 600 psi/750°F/4-in Hg abs.} = 7.64 \text{ lb/kwhr}$$

$$\text{Approximate efficiency } = \eta a = 77\%$$

$$\text{Approximate steam rate (ASR)} = TSR/\eta a = (7.64 \text{ lb/kwhr}) \ (.746 \text{ kw/HP})/.77$$

$$= 7.40 \text{ lb/HP-hr}$$

$$\text{Approximate steam flow} = ASR \times HP = (7.40 \text{ lb/HP-hr})(25,000 \text{ HP})$$

$$= 185,000 \text{ lb/hr}$$

A Mollier diagram can be used to determine the TSR for steam conditions not tabulated.

Figure 5A–6 can be used to find approximate steam rates for turbines operating at part-load and speed. For example, find the approximate steam rate when the 25,000 HP, 5000 r/min turbine we've discussed is operated at 20,000 HP and 4500 r/min.

$$\% \text{ HP} = 20,000 \text{ HP}/25,000 \text{ HP} = 80\%$$

$$\% \text{ r/min} = 4500 \text{ r/min}/5000 \text{ r/min} = 90\%$$

From the curve, the HP correction is 1.04 and the r/min correction 1.05.

$$\text{Total correction is } 1.04 \times 1.05 = 1.09.$$

The part-load steam rate is therefore $7.40 \times 1.09 = 8.06$ lb/HP-hr.

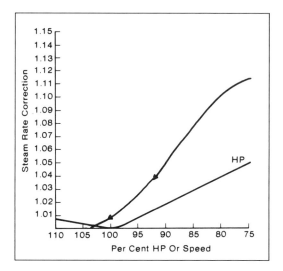

FIGURE 5A–6   *Part-load/speed correction curves. (Source: Elliott Company, A United Technologies Subsidiary, Jeannette, PA.)*

Turbine Selection

Staging, Pressures, and Temperatures

The following examples illustrate the performance of various combinations of impulse staging. It should be understood, of course, that stage selection and overall turbine efficiency are affected by many important considerations other than stage efficiency. Speed limitations, mechanical stresses, leakage and throttling losses, windage, bearing friction, and reheat—all these must be factored into the ultimate turbine design. That's the job of the factory specialist.

The approximate efficiencies of a Curtis stage and a Rateau stage are shown in Figure 5A–7.

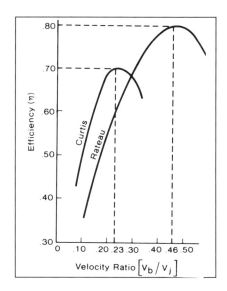

FIGURE 5A–7   *Velocity ratio versus efficiency. (Source: Elliott Company, A United Technologies Subsidiary, Jeannette, PA.)*

Table 5A.2   English-Metric Conversion Table

To Obtain	Multiply	By
kg/cm²	psig	0.0703
Atmospheres	psig	0.06804
kg/cm²	Inches of mercury (″Hg abs.)	0.03453
mm of mercury	Inches of mercury (″Hg abs.)	25.4
kw	HP	0.746
°C	°F −32	0.556
cm	Inches	2.54
kg	Pounds	0.454

Source: Elliott Company, a United Technologies Subsidiary, Jeannette, PA.

Basic Formulae

(Refer to Table 5A.2 for English to metric conversions.)

$$V_b = \frac{\pi DN}{720} \qquad V_j = 224 \sqrt{h_1 - h_2} = 224 \sqrt{\Delta H}$$

where V_b = Pitch line (blade) velocity, ft/s

   D = Pitch diameter of wheel, inches (base diameter plus height of blade)

   N = Rotative speed, r/min

   V_j = Steam jet velocity, ft/s

   h_i = Inlet steam enthalpy, Btu/lb

   h_2 = Isentropic exhaust steam enthalpy, Btu/lb

   h_{2e} = Stage exit steam enthalpy, Btu/lb

   ΔH = Isentropic heat drop, Btu/lb ($h_1 - h_2$)

   V_b/V_j = Velocity ratio, dimensionless

Example 1: Curtis Stage Performance

Conditions: 1500 psig (1515 psia); 950 °F
     5000 r/min, 25-in wheel diameter
     Assume 1-in blade height
Find isentropic heat drop and end point (see Figure 5A–9).

$$V_b = \frac{\pi DN}{720} = \frac{(3.14)(25 + 1)(5000)}{720} = 568 \text{ ft/s}$$

From Figure 5A–8, velocity ratio for optimum Curtis stage efficiency = .23

$$V_b/V_j = .23; \qquad V_j = \frac{V_b}{.23} = \frac{568}{.23} = 2470 \text{ ft/s}$$

$$V_j = 224 \sqrt{\Delta H} = 2470 \text{ ft/s}$$

$$\Delta H = 121.5 \text{ Btu/lb}$$

h_1 (from Mollier chart or steam table) = 1459.9 Btu/lb
h_2 = 1459.9 − 121.5 = 1338.4 Btu/lb
Exhaust pressure (from Mollier chart) = 590 psia

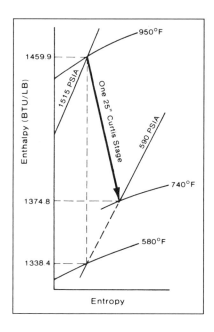

FIGURE 5A–8   *Entropy versus enthalpy chart for Example 1. (Source: Elliott Company, A United Technologies Subsidiary, Jeannette, PA.)*

Assuming a stage efficiency of 70 percent, the stage exit conditions are:

exhaust pressure ··· 590 psia

$h_{2e} = 1459.9 - (.7)(121.5)$ ····················· 1374.8 Btu/lb

Example 2: Rateau Stage Performance

Conditions: 400 psig (415 psia); 600 °F
    5000 r/min, 35-in wheel diameter
    Assume 1-in blade height
Find isentropic heat drop and end point (see Figure 5A–9).

$$V_b = \frac{\pi DN}{720} = \frac{(3.14)(35 + 1)(5000)}{720} = 785 \text{ ft/s}$$

From Figure 5A–7, velocity ratio for optimum Rateau stage efficiency = 46%

$$V_b/V_j = .46; \quad V_j = \frac{V_b}{.46} = \frac{785}{.46} = 1705 \text{ ft/s}$$

$$V_j = 224 \sqrt{\Delta H} = 1705 \text{ ft/s}$$

$$\Delta H = (V_j/224)^2 = (1705/224)^2 = 58.0 \text{ Btu/lb}$$

h_1 (from Mollier chart or steam tables) = 1305.7 Btu/lb
$h_2 = 1305.7 - 58.0 = 1247.7$ Btu/lb
Exhaust pressure (from Mollier chart) = 230 psia
Assuming a stage efficiency of 80 percent, the stage exit conditions are:

exhaust pressure ··· 230 psia

$h_{2e} = 1305.7 - (.8)(58.0)$ ····················· 1259.3 Btu/lb

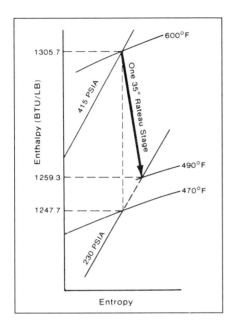

FIGURE 5A–9 *Entropy versus enthalpy chart for Example 2. (Source: Elliott Company, A United Technologies Subsidiary, Jeannette, PA.)*

FIGURE 5A–10 *Entropy versus enthalpy chart for Example 3. (Source: Elliott Company, A United Technologies Subsidiary, Jeannette, PA.)*

Example 3: Straight Rateau Staging

Conditions: 400 psig (415 psia); 600 °F
Exhaust pressure 100 psig (115 psia)
Find the number of 35-in diameter Rateau stages required, assuming optimum stage efficiency (see Figure 5A–10).
From Mollier chart, the isentropic heat available is:

$$1305.7 - 1184.2 = 121.5 \text{ Btu/lb}$$

Alternatively, using a TSR table or Polar Mollier chart:
TSR = 28.08 lb/kwhr
Using the basic definition of TSR:

$$\Delta H = h_1 - h_2 = \frac{3413}{\text{TSR}} = \frac{3413}{28.08} = 121.5 \text{ Btu/lb}$$

From Example 2, the optimum isentropic heat drop per Rateau stage = 58.0 Btu/lb
Approximate stages required = 121.5/58.0 = 2.1 (thermodynamically)
The turbine will require 2 Rateau stages.
Assuming 80% stage efficiency:
121.5 × .80 = 97.2 Btu/lb
1305.7 − 97.2 = 1208.5 Btu/lb

Example 4: Curtis and Rateau Staging

Conditions: 1500 psig (1515 psia); 950 °F
Exhaust pressure 150 psig (165 psia)

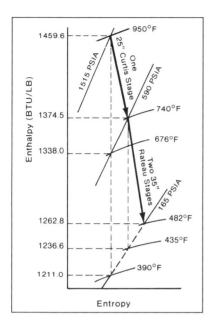

FIGURE 5A–11   *Entropy versus enthalpy chart for Example 4. (Source: Elliott Company, A United Technologies Subsidiary, Jeannette, PA.)*

Find the number of 35-in-diameter Rateau stages required when using one 25-in-diameter Curtis stage, assuming optimum stage efficiencies (see Figure 5A–11). From Example 1, we found that a Curtis stage removes 121.5 Btu/lb

$$1459.9 - 121.5 = 1338.4 \text{ Btu/lb pressure} = 590 \text{ psia}$$

Assuming 70% stage efficiency:
   121.5 × .70 = 85.05 Btu/lb
   1459.9 − 85.05 = 1374.8 Btu/lb
From this end point to 165 psia, the isentropic heat drop for the Rateau stages is
   1374.8 − 1239.0 = 135.8 Btu/lb

From Example 2, the optimum isentropic heat drop per Rateau stage = 58.0 Btu/lb
Approximate stages required = 135.8/58.0 = 2.34 (thermodynamically)
The turbine will therefore require one 25-in Curtis stage and two 35-in Rateau stages.
Assuming an overall Rateau efficiency of 81%, the end point will be
   1374.8 − .81(137.3) = 1264.8 Btu/lb

EXTRACTION TURBINE PERFORMANCE

Today's fuel costs demand that the maximum amount of energy be squeezed from each pound of steam generated. To help in this effort, when both process steam and shaft power are required, many plant designers are turning to extraction turbines.

The extraction turbine can substantially reduce the energy charged to the driven machine if process steam would otherwise be supplied through pressure-reducing valves. Even though back-pressure turbines can be used to supply process steam, they are rather inflexible, since the shaft power and process steam requirements must be closely matched. An extraction turbine, however, can cope with changes in these variables and satisfy the requirements of each over a broad range.

The diagram in Figure 5A–12 shows the performance map for a typical extraction turbine. Determining the shape of this diagram is a problem that often arises. Here is an example that demonstrates the procedure to follow in drawing an approximate extraction diagram.

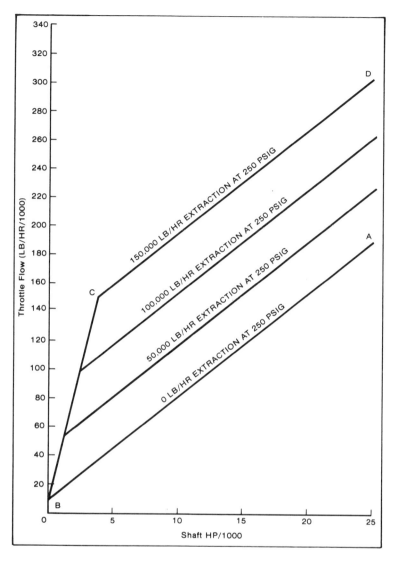

FIGURE 5A–12   *Performance map for typical extraction turbine. (Source: Elliott Company, A United Technologies Subsidiary, Jeannette, PA.)*

Assume: Shaft HP and speed: 25,000 HP at 4500 r/min
          Steam conditions: 600 psig/750 °F/4 in Hg abs.
          Extraction requirements: 150,000 lb/hr @ 250 psig
First tabulate the TSRs.
TSR (inlet to extraction)

$$600 \text{ psig}/750\,°\text{F}/250 \text{ psig} = 35.4 \text{ lb/kwhr}$$

TSR (inlet to exhaust)

$$600 \text{ psig}/750\,°\text{F}/4 \text{ in Hg abs.} = 7.64 \text{ lb/kwhr}$$

Now assume that the efficiency of the entire turbine (inlet to exhaust) is 75 percent and the efficiency of the inlet-to-extraction section is 70 percent.

Therefore:

approximate steam rate (SR) (inlet to extraction) will be

$$(35.4 \text{ lb/kwhr} \times .746 \text{ kw/HP}) \div .70 = 37.8 \text{ lb/HP-hr}$$

approximate SR (inlet to exhaust) will be

$$(7.64 \text{ lb/kwhr} \times .746 \text{ kw/HP}) \div .75 = 7.60 \text{ lb/HP-hr}$$

Point A is the first point to be located on the diagram by multiplying

$$\text{Total HP} \times \text{approximate SR (inlet to exhaust)} = 25,000 \text{ HP} \times 7.60 \text{ lb/HP-hr}$$
$$= 190,000 \text{ lb/hr}$$

Locate Point A at 25,000 HP and 190,000 lb/hr throttle flow.
Point B is located at zero HP and a throttle flow of 5 percent of A or

$$.05 \times 190,000 \text{ lb/hr} = 9500 \text{ lb/hr}$$

The zero extraction line results from connecting Points A and B.
Point C is located by dividing extraction flow requirement by approximate steam rate (inlet to extraction):

$$(150,000 \text{ lb/hr}) \div (37.8 \text{ lb/HP-hr}) = 3970 \text{ HP}$$

Locate Point C at 3970 HP and 150,000 lb/hr throttle flow. Now draw line C-D parallel to A-B and another line C to B. Label A-B "zero lb/hr extraction at 250 psig" and C-D "150,000 lb/hr extraction at 250 psig."

Notice that the general shape of the diagram and the slopes of the lines are determined mainly by the steam conditions used. The turbine investigated here could be built, but one must remember that we have dealt only with the thermodynamic aspects of this application. The mechanical aspects, such as blade stresses, nozzle flow limits, cooling steam, and other factors must also be checked.

This example can also be carried further to determine the number of stages in each section elsewhere. To do so requires finding the energy available in each portion of the turbine.

The energy available (ΔH_i) to the inlet-to-extraction section is as follows:

$$\Delta H_i = 3413 \text{ Btu/kwhr} \div \text{TSR (inlet to extraction)}$$
$$= 3413 \text{ Btu/kwhr} \div 35.4 \text{ lb/kwhr} = 96.4 \text{ Btu/lb}$$

$$V_b = \pi(25 + 1)(4500)/720 = 511 \text{ ft/s}$$

V_j for an ideal Curtis stage would therefore be:

$$V_j = \frac{V_b}{.23} = \frac{511}{.23} = 2220 \text{ ft/s}$$

V_j for an ideal Rateau stage would be:

$$V_j = \frac{V_b}{.46} = \frac{511}{.46} = 1110 \text{ ft/s}$$

With a ΔH_i of 96.5 Btu/lb, V_j for the inlet-to-extraction section with one stage would be:

$$V_j = 224 \sqrt{\Delta H} = 224 \sqrt{96.5} = 2200 \text{ ft/s}$$

This is seen to be very close to the 2220 ft/s for an ideal Curtis stage. We will therefore assume that the inlet-to-extraction section will contain one 25-in Curtis stage.

Now for the extraction-to-exhaust section. To find the energy available to this section we need the temperature of the steam entering this portion of the turbine (extraction steam temperature). The enthalpy of this steam will be as follows:

Inlet steam enthalpy $- \Delta H_i$ (inlet to extraction) $\times \eta$ (inlet to extraction) $=$

1378 Btu/lb $-$ 96.4 Btu/lb $\times$ .70 = 1378 Btu/lb $-$ 67.5 Btu/lb = 1310.5 Btu/lb

From a good Mollier diagram, at 250 psig and 1310.5 Btu/lb the extraction steam temperature is found to be close to 590 °F (say 600 °F).

The extraction-to-exhaust portion of this turbine therefore operates on steam conditions of 250 psig/600 °F/4 in Hg abs. TSR (extraction to exhaust) is 9.35 lb/kwhr. The energy available to the extraction-to-exhaust section is therefore:

$$\Delta H_i = (3413 \text{ Btu/kwhr}) \div (9.35 \text{ lb/kwhr}) = 365 \text{ Btu/lb}$$

The blade velocity for 35-in nominal diameter staging with a 1-in blade height will be:

$$V_b = \pi(35 + 1)(4500)/720 = 706 \text{ ft/s}$$

If all staging is of the Rateau type in this portion of the turbine:

$$V_j = 706/.46 = 1535 \text{ ft/s}$$

ΔH_i per stage is therefore:

$$\Delta H_i = \left(\frac{V_j}{224}\right)^2 = \left(\frac{1535}{224}\right)^2 = (6.85)^2 = 47.0 \text{ Btu/lb}$$

The number of Rateau stages in this section would therefore be:

$$\frac{\text{Total energy available}}{\text{Energy removed per stage}} = \frac{365}{47.0} = 7.77 \text{ (say 8)}$$

This turbine will, therefore, contain one 25-in diameter Curtis stage followed by eight 35-in diameter Rateau stages with the extraction opening after the Curtis stage.

STEAM BALANCE CONSIDERATIONS

The steam balance of a process plant can be quite complicated due to the multiple steam pressure levels often required.

Selecting a turbine to complement a particular steam balance is made easier, however, by the wide variety of turbines available. Condensing, back-pressure, or extraction/induction turbines can be used, as required, in designing both new plants and additions to existing plants.

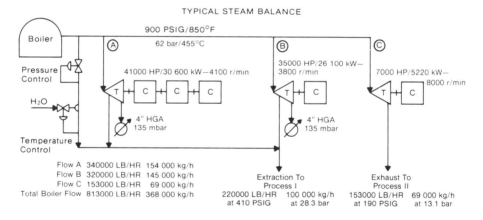

TYPICAL STEAM BALANCE

Flow A   266000 LB/HR   121 000 kg/h
Flow B   256000 LB/HR   116 000 kg/h
Flow C   50000 LB/HR   23 000 kg/h
Total Boiler Flow 522000 LB/HR   237 000 kg/h

FIGURE 5A–13   *Steam balance, Example 1. Steam is extracted from turbine A at 400 psig, and steam is extracted from turbine B at 255 psig. (Source: Elliott Company, A United Technologies Subsidiary, Jeannette, PA.)*

TYPICAL STEAM BALANCE

Flow A   340000 LB/HR   154 000 kg/h
Flow B   320000 LB/HR   145 000 kg/h
Flow C   153000 LB/HR   69 000 kg/h
Total Boiler Flow 813000 LB/HR   368 000 kg/h

FIGURE 5A–14   *Steam balance, Example 2. Three turbines all use steam from the 900 psig/850°F (62 bar/455°C) boiler. Steam is extracted from both larger units at 410 psig (28.3 bar) for process I, and the remainder is condensed at 4 in Hg abs. (135 mbar). Smaller back-pressure turbine exhausts steam at 190 psig (13.1 bar) for use in process II. (Source: Elliott Company, A United Technologies Subsidiary, Jeannette, PA.)*

Steam for process use, for example, can be supplied from the exhaust of a back-pressure turbine or from an extraction turbine. The choice would depend on the number of pressure levels involved, the design of the remainder of the plant, the number of turbines required, etc. This versatility simplifies the job of optimizing a steam balance.

The steam balance diagrams shown in Figures 5A–13 through 5A–15 illustrate how various types of turbines have been used to supply both shaft horsepower and steam for other uses.

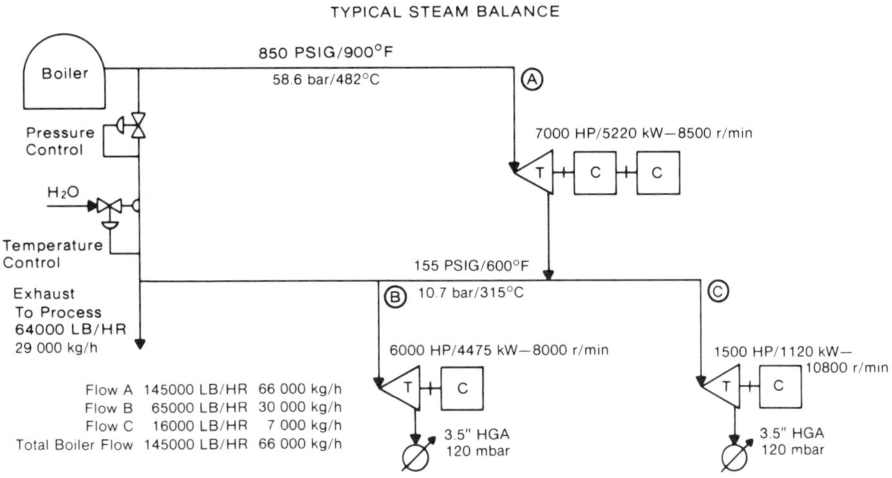

TYPICAL STEAM BALANCE

FIGURE 5A–15   *Steam balance, Example 3. Topping turbine concept was used in this plant, where exhaust from a high-pressure turbine as well as the low-pressure turbines is used to supply process demands. Exhaust steam from these low-pressure units is then condensed at 3.5 in Hg abs. (120 mbar). (Source: Elliott Company, A United Technologies Subsidiary, Jeannette, PA.)*

We note from Figure 5A–13 that two 1500 psig/950 °F turbines drive the large compressors in this application. Two different extraction pressures were used (400 psig and 255 psig), with lower pressure steam being supplied to a process and to the smaller turbine. Exhaust steam from all three turbines is then condensed at 4 in Hg abs.

Chapter 6

Turboexpanders*

A radial expansion turbine or ''expander'' is generally used to recover power from steam or other gases or to provide refrigeration for petrochemical plants, hydrocarbon separation plants, and similar processes. The shaft power and process cooling is provided by the nearly isentropic expansion of the gas from the inlet pressure to the outlet pressure. With the increasing cost of power and fuel and scarcity of petrochemical feedstocks, it is increasingly important that expanders be designed for maximum efficiency and reliability.

Most process-type expanders built in recent years have been of the radial type, due largely to the mechanical simplicity of the variable guide vane mechanism, the improved ability to expand high-energy streams efficiently in a single stage, and the general compactness of the design. The following discussion will deal with radial expanders, although many of the comments apply equally to axial designs.

Radial expanders are usually applied when one or more of the following is a consideration: refrigeration, power recovery, and power generation.

REFRIGERATION

Expanders can provide refrigeration by direct expansion of process gas, thus eliminating the need for closed-cycle refrigeration systems. Hundreds of such expanders are in operation in the cryogenic processing of hydrocarbon gases and air separation plants. Field experience has shown that, if required, substantial liquid condensation can occur in the expander without damage of any kind.

POWER RECOVERY

Radial expanders can provide power recovery from pressure reduction in liquid or gas streams, such as purge gas, waste gas, fuel gas, or natural gas pressure letdown. The expander can usually be controlled in a way that does not restrict overall plant operation.

POWER GENERATION

Radial expanders can form the heart of a closed- or open-cycle power generation system. Power cycles, such as the Brayton or Rankine cycle, using a working fluid matched to the requirements of the energy source and the expander can provide excel-

---

*Source: Mafi-Trench Corporation, Santa Maria, CA. Adapted by permission.

lent full- and part-load efficiency. Several Rankine cycle geothermal power plants are presently in operation.

EXPANDER DESIGN AND CONSTRUCTION

In the typical process-type expanders, the shaft power is absorbed and used by a single-stage centrifugal compressor. The compressor is mounted on the same shaft as the expander, providing a simple and compact design. An expander of this type is shown in Figure 6–1. The expander flow enters through the flange at the upper left, flows through the inlet guide vanes (nozzles) and expander wheel, and exits through the flange at the lower left. The compressor flow enters the flange at the right, is compressed, and exits at the top center. Bearings and seals are located between the two wheels on a short rigid shaft. This arrangement is typical of expanders used for natural gas processing, petrochemical processes, and expansion of air and nitrogen in air separation and nitrogen liquification plants.

A complete dual expander system including two expanders and all required auxiliary equipment and local control panel for a petrochemical process application is shown in Figure 6–2. Radial flow expanders for process applications must be capable of reliable and efficient operation over a relatively wide range of flow rates. To accomplish this, these machines incorporate variable inlet guide vanes or "nozzles," as shown in Figure 6–3. This design uses an externally mounted pneumatic actuator to control the opening of the guide vanes and therefore the expander flow and resultant power output. An internally mounted fulcrum mechanism translates the linear motion of the pneu-

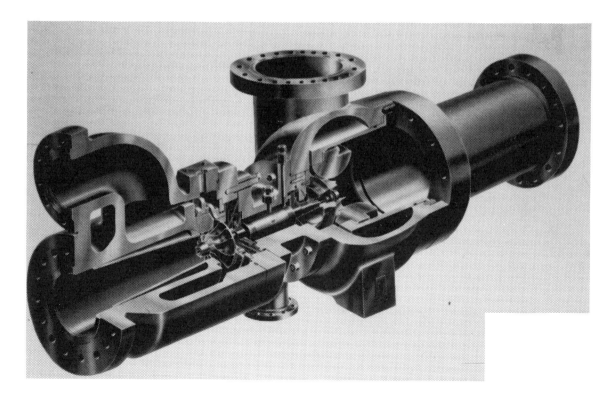

FIGURE 6–1   *Typical expander/compressor assembly. (Source: Mafi-Trench Corp., Santa Maria, CA.)*

FIGURE 6–2   *Expander system for a petrochemical process. (Source: Mafi-Trench Corp., Santa Maria, CA.)*

FIGURE 6–3   *Expander inlet guide vanes. (Source: Mafi-Trench Corp., Santa Maria, CA.)*

matic actuator into rotation of a ring that pivots each guide vane on a hardened pin. The reliable operation of this mechanism is critical to the control of the expander and the process, and therefore the design details and materials of construction must be carefully selected to avoid galling and excessive wear during operation.

Expander and compressor wheels are usually constructed of high-strength aluminum alloy. The low-density and relatively high-strength aluminum alloy is ideally suited to these wheels, as they operate at moderate temperature on relatively clean gas and the low-density alloy permits minimizing the weight of the wheels, which is desirable to avoid critical speed problems. A typical expander wheel is shown in Figure 6–4. This wheel was produced by machining from a solid aluminum alloy forging. This construction has been shown to be superior to wheels produced by welding, brazing, or casting.

The rotor assembly (expander wheel, shaft, and compressor wheel) for a typical cryogenic process expander is shown in Figure 6–5. The expander wheel is on the right and the compressor wheel on the left. The compact and rigid design of the rotor is apparent in this figure.

Labyrinth-type seals are used between the expander and compressor wheels and the oil-lubricated bearings. These seals prevent mixing of the process gas and lube oil by injecting filtered buffer gas (seal gas) in the middle of the labyrinth and allowing the gas to flow both toward the process and the bearing. The stationary and rotating elements of a stepped-type labyrinth seal for a process expander are shown in Figure 6–6.

Process expanders generally use a combination journal and thrust bearing similar to that shown in Figure 6–7. An essentially identical bearing assembly is located near the expander wheel and compressor wheel, providing thrust capacity in both directions. Since both the thrust and journal bearings are hydrodynamic, there is no bearing wear during normal operation.

FIGURE 6–4   *Typical process expander wheel. (Source: Mafi-Trench Corp., Santa Maria, CA.)*

FIGURE 6–5   *Rotary assembly for process expander/compressor. (Source: Mafi-Trench Corp., Santa Maria, CA.)*

FIGURE 6–6   *Expander labyrinth seal assembly.*
(Source: Mafi-Trench Corp., Santa Maria, CA.)

FIGURE 6–7   *Combination journal and thrust bearing. (Source: Mafi-Trench Corp., Santa Maria, CA.)*

TYPICAL APPLICATION

A simplified schematic diagram of a cryogenic turboexpander plant is shown in Figure 6–8. Hundreds of such plants are in operation throughout the world, extracting the heavier hydrocarbons from natural gas. The process is based on providing the required refrigeration by direct expansion of the process gas in a single-stage radial expander similar to that shown in Figure 6–1.

In this process, the gas is first dehydrated to prevent the formation of ice or hydrates within the cryogenic portion of the plant. Next, the gas is cooled by heat exchange with cold residue gas. This cooling usually results in some condensation. The condensed liquids are removed by a high-pressure separator before reaching the

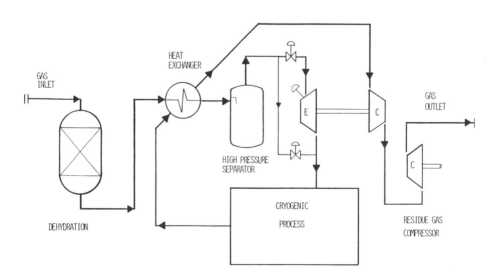

FIGURE 6–8   *Cryogenic turboexpander plant. (Source: Mafi-Trench Corp., Santa Maria, CA.)*

expander. The gas is then expanded through the expander, producing a relatively large temperature drop and substantial liquid formation. Residue gas from the cryogenic process is used to cool the expander inlet gas and is then compressed by the expander driven "boost" compressor. Additional residue gas compression is then required to increase the pressure to a level near the inlet to the plant.

OPERATION

Since essentially all of the turboexpander plant flow passes through both the expander and the boost compressor, it is possible to efficiently vary the plant flow rate using the expander inlet guide vanes. The expander shaft speed is allowed to vary freely in response to changing plant flow rate. This method of control has been shown to produce excellent off-design performance. During operation at reduced flow, the boost compressor will tend to operate near surge. It is therefore essential that the compressor be provided with an adequate surge control system. If the compressor is allowed to operate in surge for any significant length of time, damage will occur to the bearings and seals due to the resulting shaft vibrations. (See the chapter on centrifugal compressors for a more detailed description of surge.)

If, during operation, solid particles are carried into the expander, the centrifugal forces created by the expander wheel will tend to cause them to be centrifuged out and strike the underside of the inlet guide vanes. This can cause erosion damage to both the inlet guide vanes and the expander wheel blade tips. To minimize this type of damage, it is necessary to provide a 60- to 80-mesh screen upstream of the expander. The screen should be capable of withstanding a 100- to 200-psi differential pressure, and the differential pressure should be monitored continuously.

Noncontacting shaft vibration monitoring equipment has been shown to be very useful in monitoring the mechanical condition of expanders. It is important to keep accurate logs of vibration. These data can be particularly useful in evaluating the expander condition after a major process upset or expander trip.

As with all high-speed rotating equipment, the cleanliness and quality of the lube oil and seal gas should be carefully maintained. Routine sampling and spectrographic analysis of the lube oil to detect the buildup of water, particulate, or trace contaminants is recommended.

Any condition that causes an expander trip shuts down the expander by rapidly (in less than one-half second) closing the expander inlet trip valve. If the trip is due to an interruption in lube oil supply, the oil flow during coastdown must be provided by an accumulator. The accumulator must be adequately sized, and the precharge pressure must be properly maintained.

It is possible that an expander will slowly rotate or "windmill" due to leaky process valves on the expander or compressor during shutdown. If the lube system is off during this time, damage to the bearings can occur. This type of damage is more common than might be expected because the rotating speed during windmilling may be so slow that it does not indicate on the electronic tachometers on process expanders.

MAINTENANCE

Expander operating reliability has been shown to be strongly dependent on the quality of routine maintenance. As a minimum, the following items should be included in any maintenance program.

- Maintain proper expander data logs to be used for trend analysis.
- Provide adequate surge protection for the compressor.

- Regularly sample and analyze the lube oil for contamination.
- Provide a 60- to 80-mesh expander inlet screen and differential pressure monitor.
- Install a compressor discharge check valve to prevent back flow on shutdown.
- Maintain process valves for tight sealing to prevent expander "windmilling" during shutdown.
- Maintain proper filtration of seal gas.
- Assure that the expander inlet trip valve closes in less than one-half second on trip.
- Verify all expander alarm and shutdown functions at least yearly.

Chapter 7

Centrifugal Pumps*

Centrifugal pumps are the very heart and pulse of most process plants. Literally hundreds of thousands are in use in the United States alone, and modern plants are unthinkable without them.

Actually, the first centrifugal pump was built around 1680, but in the United States it was not until 1818 that the first practical pump of this type was constructed. Reynolds developed the turbine-type centrifugal pump in England in 1887. Since centrifugal pumps are essentially high speed machines, it was not until the advent of high speed electric motors and steam turbines in the early twentieth century that centrifugal pumps came into popular use.

There are two distinct types of centrifugal pumps: the turbine-type pump, which uses diffusers or guide vanes in the casing for the conversion of the velocity to pressure energy, and the volute-type centrifugal pump.

Since the volute-type centrifugal pump is most commonly used, our text will primarily deal with volute-type characteristics. Turbine pumps are generally built in vertical column configuration and will be given further consideration later in this chapter.

Centrifugal pumps are designed and manufactured for many different pumping services. These widely varying requirements are best accommodated by the specific construction features of a given pump type. Important pump types, services, and typical ratings are listed in Table 7.1.

CONVENTIONAL PROCESS PUMPS

Figure 7–1 depicts a typical American National Standards Institute (ANSI) process pump, which is typical of conventional process pumps. ANSI standards for pumps are dimensional standards that facilitate pump and component interchangeability. As illustrated in Figure 7–2, this standard dimension process pump is furnished with a fully open impeller, generally preferred for solids handling and for stringy or abrasive-containing pumpage. Two different sealing arrangements are shown: soft packing above the pump centerline, and a mechanical face seal below the pump centerline. Figure 7–3 illustrates an enhanced ANSI pump with an elastomer bellows seal shown above the pump centerline and a conventional multispring seal shown below the centerline. There are literally hundreds of mechanical-seal types available to serve the numerous different pumps and pumped fluids. Some of these will be described later in this text.

---

*Source: Goulds Pumps, Inc., 240 Fall St., Seneca Falls, NY. Adapted by permission.

Table 7.1

Pump Type	Typical Services	Typical Ratings
ANSI Process	Corrosive/abrasive liquids, slurries, and solids, high temperature, general purpose pumping process and transfer.	Q to 4500 GPM (1022 m³/h) H to 730 ft (222 m) T to 700 °F (371 °C) H to 375 PSIG (2586 kPa)
Nonmetallic Chemical Process	Severe corrosives.	Q to 800 GPM (182 m³/h) H to 490 ft (149 m) T to 300 °F (150 °C) P to 225 PSIG (1550 kPa)
Self-priming Process	Corrosive/abrasive liquids, slurries, and suspensions, high temperature, industrial sump, mine dewatering, tank car unloading, bilge water removal, filter systems, chemical transfer.	Q to 1500 GPM (340 m³/h) H to 375 ft (114 m) T to 500 °F (260 °C) Suction Lifts to 25 ft (7.6 m)
In-line Process	Process, transfer and general service. Corrosive and volatile liquids. High temperature services.	Q to 1500 GPM (340 m³/h) H to 700 ft (207 m) T to 500 °F (260 °C) P to 375 PSIG (2586 kPa)
Canned motor	Zero leakage services: toxic liquids, refrigerants, liquified gas, high temperature heat transfer, explosive liquids, liquids sensitive to atmosphere, carcinogenic and other hazardous services.	Q to 2500 GPM (568 m³/h) H to 1400 ft (427 m) T to 700 °F (371 °C) P to 450 PSIG (3103 kPa)

Table 7.1 (continued)

Pump Type	Typical Services	Typical Ratings
API Process (Horizontal)	High temperature and high pressure services, offsite, transfer, heat transfer liquids.	Q to 7500 GPM (1700 m^3/h) H to 1100 ft (335 m) T to 800 °F (427 °C) P to 870 PSIG (6000 kPa)
API Process (In-line)	Petrochemical, chemical, refining, offsite, gasoline plants, natural gas processing, general services.	Q to 7500 GPM (1700 m^3/h) H to 750 ft (229 m) T to 650 °F (343 °C) P to 595 PSIG (4100 kPa)
Paper Stock/High Capacity Process	Paper stock, solids and fibrous/stringy materials, slurries, corrosive/abrasive process liquids.	Q to 28,000 GPM (6360 m^3/h) H to 350 ft (107 m) T to 450 °F (232 °C) P to 285 PSIG (1965 kPa)
	Medium consistency (8 to 14%) paper stock.	Q to 1800 TPD (1650 MTPD) H to 400 ft (125 m) T to 250 °F (120 °C)
Horizontal (Abrasive Slurry)	Corrosive/abrasive services. Coal, fly ash, mill scale, bottom ash, slag, sand/gravel, mine slurries. Large solids.	Q to 10000 GPM (2273 m^3/h) H to 350 ft (107 m/Stage) T to 400 °F (204 °C) P to 300 PSIG (2068 kPa) Spherical solids to 4″ (102 mm)
Axial Flow	Continuous circulation of corrosive/abrasive solutions, slurries and process wastes. Evaporator and crystallizer, reactor circulation, sewage sludge recirculation.	Q to 200,000 GPM (35,000 m^3/h) H to 30 ft (9 m) T to 350 °F (180 °C) P to 150 PSIG (1034 kPa) Solids to 9″ (228 mm)

Table 7.1 (continued)

Pump Type	Typical Services	Typical Ratings
Large Solids Handling (Horizontal) 	Pumps for extra demanding municipal and industrial services; large pulpy and fibrous solids, sewage, abrasives.	Q to 100,000 GPM (22,700 m³/h) H to 240 ft (73 m) T to 202 °F (43 °C) P to 300 PSIG (2065 kPa) Solids to 10″ (254 mm)
(Vertical Dry Pit) 		
Double suction 	Cooling tower, raw water supply, booster service, primary and secondary cleaner, fan pump, cooling water, high lift, low lift, bilge and ballast, fire pumps, river water, brine, sea water, pipelines, crude.	Q to 72,000 GPM (16,300 m³/h) H to 570 ft (174 m) T to 350 °F (177 °C) P to 275 PSIG (1896 kPa)
Multi-stage 	Refinery, pipeline, boiler feed, descaling, crude oil charging, mine pumping, water works . . . other high pressure services. Water, cogeneration, reverse osmosis, booster service, boiler feed, shower service. Boiler feed, mine dewatering and other services requiring moderately high heads.	Q to 3740 GPM (850 m³/h) H to 6000 ft (1824 m) T to 375 °F (190 °C) P to 2400 PSIG (16,546 kPa)

Table 7.1 (continued)

Pump Type	Typical Services	Typical Ratings
Low Flow/High Head Multi-Stage Moderate speed  	Reverse osmosis descaling, high pressure cleaning, process water transfer, hydraulic systems, spraying systems, pressure boosters for hi-rise buildings, all low flow applications where efficiency is critical.	Q to 280 GPM (64 m³/h) H to 2600 ft (792 m) T to 400 °F (204 °C) P to 1100 PSIG (7584 kPa)
Submersible • Wastewater • Solids Handling • Slurry	Flood and pollution control, liquid transfer, sewage and waste removal, mine dewatering, sump draining. Large stringy or pulpy solids. Abrasive slurries.	Q to 4000 GPM (910 m³/h) H to 210 ft (65 m) T to 140 °F (60 °C) Solids to 2″ (50 mm)
Vertical Submerged (Submerged Bearing and Cantilever) • Process • Solids Handling • Slurry 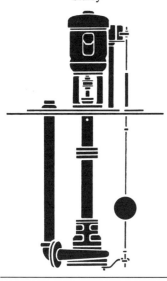	Industrial process, sump drainage, corrosives, pollution control, molten salts, sewage lift, wastewater treatment, extremely corrosive abrasive slurries, large or fibrous solids.	Q to 7500 GPM (1703 m³/h) H to 310 ft (95 m) T to 450 °F (232 °C) Solids to 10″ (254 mm)

Table 7.1 (continued)

Pump Type	Typical Services	Typical Ratings
Vertical Turbine 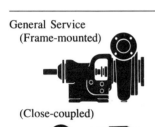	Irrigation, fire pumps, service water, deep well, municipal water supply, mine dewatering, cooling water, seawater and river water intake, process, utility circulating, condenser circulating, ash sluice, booster, petroleum/refiner, boiler feed, condensate, cryogenics, bilge, fuel oil transfer, tanker and barge unloading.	Q to 150,000 GPM (34,065 m³/h) H to 3500 ft (1070 m) T to 700 °F (371 °C)
General Service (Frame-mounted) (Close-coupled)	Close Coupled and frame-mounted pumps for water circulation, booster, OEM packages, irrigation, chemical process, transfer, and general purpose pumping.	Q to 2100 GPM (477 m³/h) H to 400 ft (122m) T to 300 °F (149 °C)

Figure 7–4 depicts a *between-bearing pump* with a double-flow impeller. Chosen for high-flow capability and balanced axial thrust, double-flow impellers are widely used in large or heavy-duty process pumps. Where maximum accessibility to pump parts and flanges is needed, the user may opt for vertical mounting, as illustrated in Figure 7–5.

Multistage horizontal split case pumps (Figure 7–6) are used for a wide range of moderate- to high-pressure services in process plants. Their typical internal construction features are seen in Figure 7–7. The pump suction nozzle is on the left. Pumpage leaving the third impeller is routed to the suction of the fourth impeller, located at the opposite end of the pump. This internal flow arrangement results in axial rotor thrust balance by hydraulic means. The external piping ensures pressure equalization between the space to the right of the balancing drum near the entry to stage 4, and to the left of the suction eye of stage 1.

A *multistage centrifugal pump with barrel-type outer casing* is depicted in Figure 7–8. These pumps are primarily used for high-pressure and extreme-pressure light hydrocarbon liquids, although certain boiler feedwater services often use this casing style also.

Conventional *low flow-high head centrifugal pumps* are typically configured as shown in Figure 7–9. The multistaging is achieved by adding modular elements that are designed for maximum interchangeability and minimum spare parts requirements.

FIGURE 7-1   *Standard dimension process pump (ANSI). (Source: Goulds Pumps, Inc., Seneca Falls, NY.)*

The particular model illustrated here achieves sealing of the casing by the use of O-rings and long external tie bolts. An alternative execution, which uses a containment casing instead of the tie bolts, is shown in Figure 7-10.

High-speed pumps for low-flow high-head services are substantially different from conventional low-flow high-head centrifugals and merit special coverage. These pumps are described later in this chapter.

The construction features of *submersible wastewater pumps* are shown in Figure 7-11. These pumps obviously have to be capable of occasional solids ingestion, which makes it necessary to design and build the impeller with suitable features.

Close-coupled and *frame-mounted pumps* are primarily designed for general purpose pumping. Quite similar to the close-coupled vertical pump, close-coupled and frame-mounted pumps have the impeller placed on the electric motor shaft.

Large *solids-handling pumps* are manufactured in a variety of configurations. Figure 7-12 shows a horizontally arranged model. Accessibility for service and general ease of maintenance are important for this pump category.

In *self-priming process pumps,* priming and air separation are accomplished within the pump casing. The pump is designed with two volutes; these are separate stationary channels into which the rotating impeller pushes the pumpage exiting from the impeller tip. During the priming cycle, the lower volute functions as the intake while the upper volute discharges liquid and entrained air into a separation chamber. Air is separated and expelled through the pump discharge while liquid circulates into the lower volute. Once air is completely exhausted from the suction region and liquid fills the impeller eye, the pump is primed and functions as a conventional pump, with both volutes acting

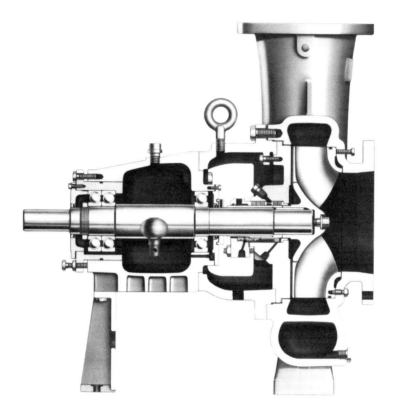

FIGURE 7–2   *Pump cross section showing typical seal areas. (Source: Goulds Pumps, Inc., Seneca Falls, NY.)*

as discharges. As shown in Figure 7–13, the casing is designed so that an adequate volume of liquid for repriming is always retained in the pump, even if liquid is allowed to drain back to the source of supply from both discharge and suction.

The function of a dual volute design is shown in Figure 7–14. The dual volute casing design is ideal where pumps must periodically operate at capacities above or below design capacity or at uninterrupted high head. Essentially, this design equalizes radial forces and lessens radial reaction on shaft and bearings. This equalization or balancing of radial forces is accomplished by dividing the liquid discharged by the impeller into two half-capacity volutes with two cutwaters, set 180° apart. Radial forces on the shaft and bearings are equally opposed.

In-line process pumps are vertically oriented pumps with the casing designed to bolt directly into the piping system. They require a minimum of support from a relatively small foundation or similar structure and have proven to be as reliable and easy to maintain as conventional, horizontally oriented centrifugal pumps.

Figure 7–15 shows an in-line pump with flexibly coupled electric motor shaft to pump shaft connection. The pump has its own bearing support whereas the so-called close-coupled in-line pump shown in Figure 7–16 uses a rigid coupling sleeve and has its rotor supported by the electric motor bearings only. Although the flexibly coupled and close-coupled styles are generally equally reliable, the flexibly coupled version should be preferred from an ease-of-maintenance point of view. A third variation of the in-line pump construction has the pump impeller placed on the motor shaft end. This style is found less often in process plants.

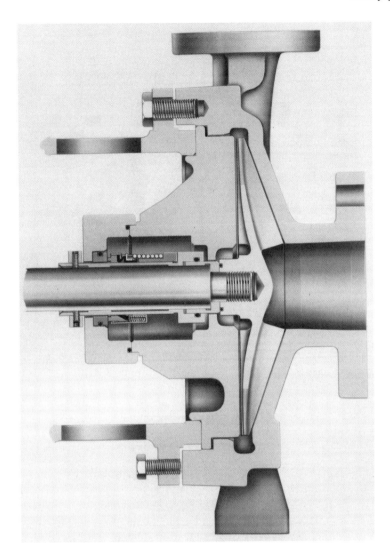

FIGURE 7–3 *Enhanced ANSI pump showing oversized seal housing and two different mechanical seals. (Source: Goulds Pumps, Inc., Seneca Falls, NY.)*

American Petroleum Institute (API) process pumps get their name from an API standard (API-610) that specifies the requirements for this heavy-duty pump. While statistics show that properly applied ANSI pumps have a useful life and reliability matching that of API pumps, the latter nevertheless has some construction features that make it the proper choice in certain high-risk applications.

API pumps differ from ANSI in the following respects:

- API pumps have greater corrosion allowance.
- They have higher permissible nozzle loads.
- API pumps have more available stuffing box space.
- Wear rings are furnished in API pumps. They are not always supplied with ANSI pumps.

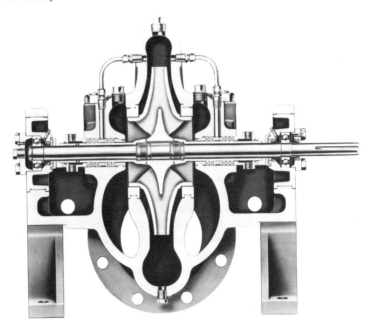

FIGURE 7–4   *Between-bearing pump with double-suction impeller (single stage). (Source: Goulds Pumps, Inc., Seneca Falls, NY.)*

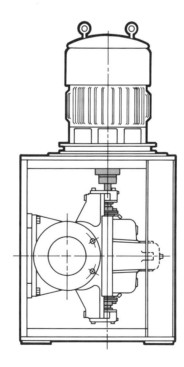

FIGURE 7–5   *Vertically mounted double-suction pump. (Source: Goulds Pumps, Inc., Seneca Falls, NY.)*

FIGURE 7–6   *Multistage horizontally split case process pump. (Source: Goulds Pumps, Inc., Seneca Falls, NY.)*

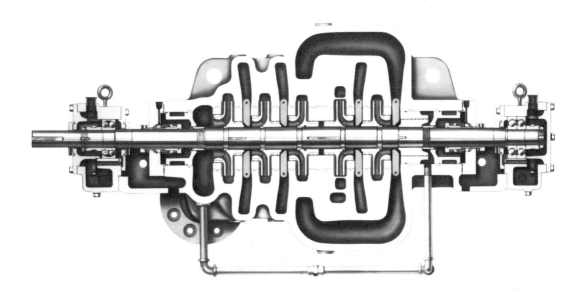

FIGURE 7–7   *Internal component arrangement of a five-stage horizontally split case pump. (Source: Goulds Pumps, Inc., Seneca Falls, NY.)*

- API pumps are centerline-mounted; ANSI pumps are often foot-mounted.
- Bearing housings in API pumps are generally fitted with higher load capacity bearings and higher life expectancy end seals.

Figure 7–17 illustrates a typical API process pump.

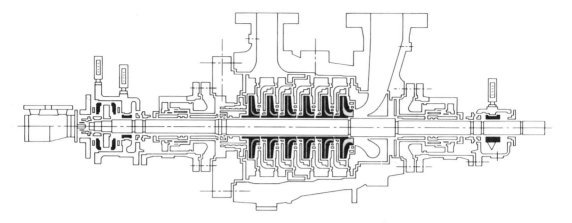

FIGURE 7–8   *Multistage barrel casing-type centrifugal pump. (Source: Sulzer Brothers, Winterthur, Switzerland.)*

FIGURE 7–9   *Conventional low-flow, high-head centrifugal multistage pump made up of modular elements. (Source: Goulds Pumps, Inc., Seneca Falls, NY.)*

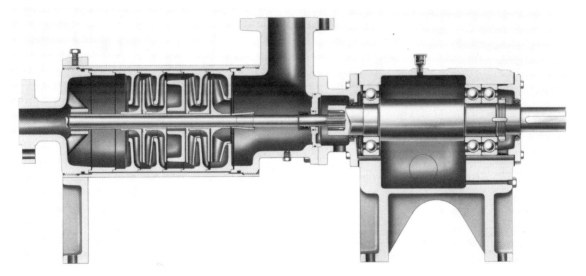

FIGURE 7–10   *Low-flow, high-head three-stage centrifugal pump with containment casing. (Source: Goulds Pumps, Inc., Seneca Falls, NY.)*

FIGURE 7–11   *Submersible wastewater pump. (Source: Goulds Pumps, Inc., Seneca Falls, NY.)*

Paper stock and *high-capacity process pumps* are typically configured as shown in Figures 7–18 and 7–19. Both pumps incorporate wear plates opposite the open side of the impeller. Maintainability and simplicity of construction are key requirements in these services. Figure 7–18 incorporates a repeller arrangement to oppose the outflow of liquid along the shaft. This arrangement consists of two stationary plates and a rotating part. Figure 7–19 depicts a model that incorporates special inducers to accommodate difficult pumpage.

A typical *slurry pump* is shown in Figure 7–20. Resistance to corrosion and wear is of great importance in slurry pumps, and simple construction aids in making the pumps maintainable. In some pumps, replaceable rubber lining is used on the wetted parts.

Much of the confusion in deciding when to specify a slurry pump arises from the lack of agreement on the meaning of the word *slurry*. This is due in large part to the nearly infinite number of solid-liquid mixes. In place of the many academic slurry definitions, this broader, more functional definition is offered: A slurry is any mixture of liquid and solids capable of causing significant pump abrasion, clogging, or mechanical failure due to high loads or impact shocks.

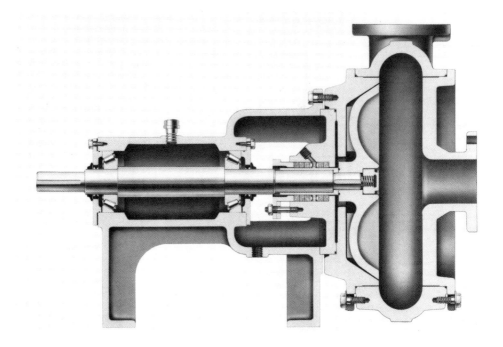

FIGURE 7–12   *Large solids handling pump with vortex impeller. (Source: Goulds Pumps, Inc., Seneca Falls, NY.*

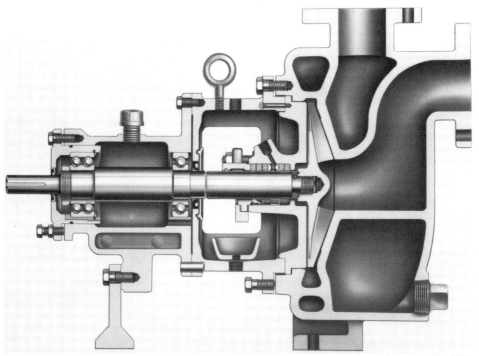

FIGURE 7–13   *Self-priming process pump. (Source: Goulds Pumps, Inc., Seneca Falls, NY.)*

FIGURE 7–14   *Principle of dual volute design. The dual volute casing design is ideal where pumps must periodically operate at capacities above or below design capacity or at uninterrupted high head. Essentially, this design equalizes radial forces and lessens radial reaction on shaft and bearings. This equalization or balancing of radial forces is accomplished by dividing the liquid discharged by the impeller into two half-capacity volutes with two cut-waters set 180° apart. Radial forces on the shaft and bearings are equally opposed. (Source: Goulds Pumps, Inc., Seneca Falls, NY.)*

FIGURE 7–15   *Vertical in-line pump with flexibly coupled electric motor. (Source: Goulds Pumps, Inc., Seneca Falls, NY.)*

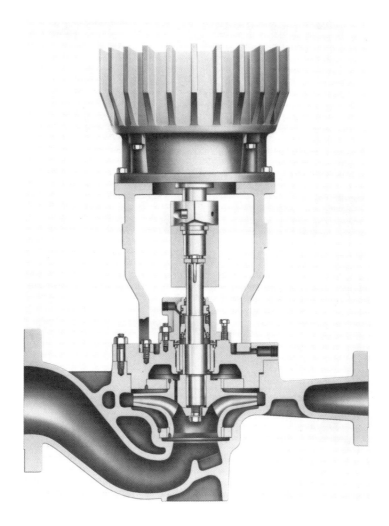

FIGURE 7–16   *Rigid spacer-coupled vertical in-line pump. (Source: Goulds Pumps, Inc., Seneca Falls, NY.)*

Under some circumstances, it may seem superfluous to consider the "when" of slurry pump selection. Obviously, a pump employed to move "deliberate" slurries such as mine tailings or chemical concentrates must be designed and constructed with exceptional strength, abrasion resistance, and solids-passage ability.

But what about a pump employed to supply large quantities of water from a sandy river for cooling purposes? In such "accidental" slurries, the transport of liquids is the prime purpose—the presence of solid materials is not intended (or, sometimes, even recognized). Nevertheless, failure to use a slurry pump for this type of application can frequently result in excessive maintenance, parts usage, and downtime costs.

The "when" of slurry pump selection might best be answered by a rule of thumb that says that whenever the fluid to be pumped contains more solids than are found in potable water, at least consider the use of a slurry pump.

There are many features that set a slurry pump apart from a standard, general service centrifugal pump. Outwardly, there are few differences, although the slurry pump is usually larger in size. Internally, however, there are many characteristics that make a slurry pump a very specialized breed.

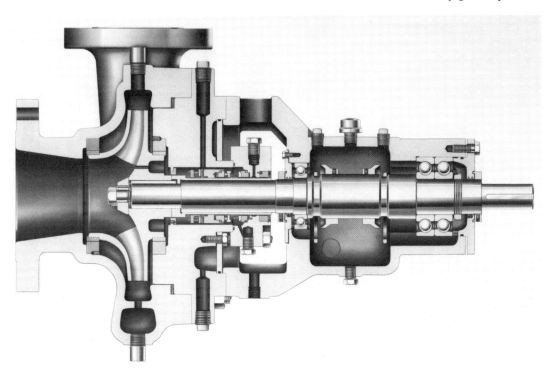

FIGURE 7–17 *Single-stage back pull-out-type centrifugal process pump complying with API SPEC 610. (Source: Goulds Pumps, Inc., Seneca Falls, NY.)*

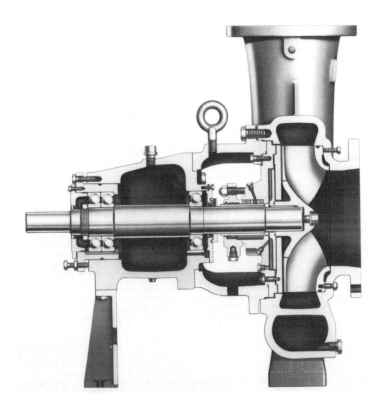

FIGURE 7–18 *Heavy-duty process pump for paper stock incorporating repeller arrangement to reduce load on stuffing box area. (Source: Goulds Pumps, Inc., Seneca Falls, NY.)*

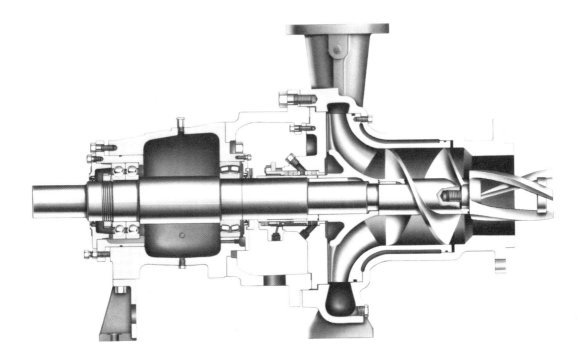

FIGURE 7–19 *Paper stock pump incorporating special inducers. (Source: Goulds Pumps, Inc., Seneca Falls, NY.)*

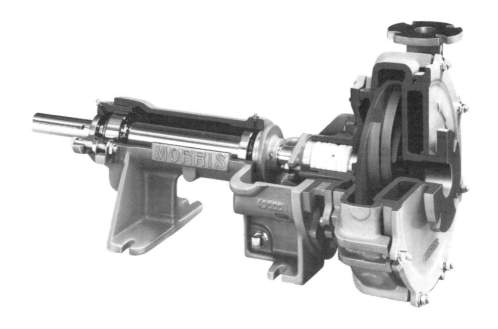

FIGURE 7–20 *Typical slurry pump. (Source: Goulds Pumps, Inc., Seneca Falls, NY.)*

Wall thicknesses of wetted-end parts (casing, impeller, etc.) are greater than those used in conventional centrifugal pumps. The cutwater, or volute tongue (the point on the casing at which the discharge nozzle diverges from the casing), in the casing is generally less pronounced in order to minimize the effects of abrasion. Flow passages through both the casing and impeller are large enough to permit solids to pass without clogging the pump. Slurry pumps are available in a variety of materials of construction to best handle the abrasive, corrosive, and impact requirements of nearly any solids-handling application.

Because the gap between the impeller face and suction liner will increase as wear occurs, the rotating assembly of the slurry pump must be capable of axial adjustment to maintain the manufacturer's recommended clearance. This is critical if design heads, capacities, and efficiencies are to be maintained. Other specialized features include extra-large stuffing boxes, replaceable shaft sleeves, and impeller back vanes that act to keep solids away from the pump stuffing box.

Both radial and axial-thrust bearings on the slurry pump are generally heavier than for standard centrifugals, owing to the demands imposed by slurries of high specific gravity. Although impeller back vanes (used to lower stuffing box pressures) do actually reduce axial thrust, these vanes can wear considerably in abrasive services. Consequently, the bearings must be of ample capacity to handle thrust loads by themselves. Balancing holes through the impeller should not be used to reduce axial thrust, since they can either clog or initiate excessive localized impeller wear.

Nearly all slurry pumps have larger diameter impellers than units for pumping clear liquids, enabling heads and capacities to be met at reduced rotational speed.

Low-speed operation is one of the most important wear-reducing features of a slurry pump. In fact, experience shows that abrasive wear on any given pump rises at least with the third power of RPM increase.

An analysis of the static profile of the slurry pump will help determine the solids-passage ability, abrasion resistance, and mechanical strength required of the pump. The most important elements in the static profile can be assigned to four categories:

1. Size of the solids: What are the largest particles the pump must handle? Are these solids similar or random in size?
2. Nature of the solids: Are they pulpy or hard, light or dense, round or jagged? Are they abrasive or corrosive?
3. Nature of the liquid: How corrosive is the liquid? Will it lubricate the solids and reduce abrasion?
4. Concentration of the solids: It is the ratio of solids to liquids that determines how the characteristics of the solids will influence the slurry as a whole.

These four static characteristics create unique demands, requiring specific pump design and construction features. For example, Figure 7–21 shows a pump model designed to handle wastes, light slurries, and random large solids. Unlike the slurry pump discussed earlier, this unit does not use wear liners. The emphasis here is on very large flow passages through the casing and impeller. Because such units are generally used for pumping sewage, light slurries, and relatively nonabrasive industrial wastes, certain wear-reducing design features can be compromised to increase hydraulic efficiency.

When chemical sludges or wastes containing large solids must be pumped, a vortex pump is often the best answer. Because its impeller is fully recessed into the rear of the casing, a relatively small pump can be used to handle liquids containing very large solids. A vortex pump was shown earlier in Figure 7–12.

Still other slurries may exchange the problems of large solids for the equally difficult pumping idiosyncrasies associated with high concentrations of small solids. More often than not, such slurries present extreme abrasion problems. Typical are those

FIGURE 7-21   *Pump designed to handle industrial and municipal wastewater. (Source: Goulds Pumps, Inc., Seneca Falls, NY.)*

associated with lime slurry pumping, the handling of ore concentrates, kaolin clay, or cement slurries. Figure 7-22 shows an extremely heavy-duty slurry pump ideal for such applications.

Figure 7-23 represents only one of numerous styles of *vertical sump and process pumps.* This particular model has the discharge piping attached to the bowl assembly. Also, this model is shown with externally connected tubing for the lubrication of lineshaft bearings. Depending on the nature of the service, a process plant may be best served by vertical industrial turbine pumps similar to those shown in Figure 7-24.

The vertical pump shown in Figure 7-24 has either a fabricated or cast discharge head and either a threaded or flanged column. It is designed for clean, noncorrosive liquids, at low to medium pressures. This model is often selected when lowest initial cost is of prime consideration. Its principal applications are irrigation, fire water services, service water and deep well pumping, drainage, and municipal water supply.

The use of a flanged column on pumps of the type shown in Figure 7-24 facilitates maintenance, and a good selection of additional lineshaft bearing materials is often available for these pumps. They are primarily used in low- to medium-pressure effluent, oily wastewater, and mine dewatering applications.

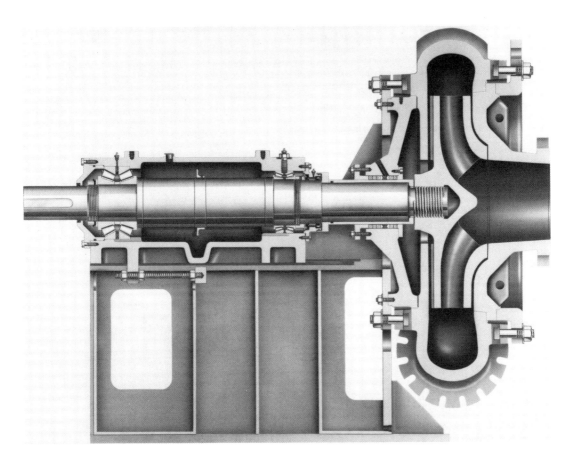

FIGURE 7-22 *Very heavy-duty pump for slurries with large concentrations of highly abrasive particles. (Source: Goulds Pumps, Inc., Seneca Falls, NY.)*

Vertical industrial turbine pumps that incorporate both a fabricated discharge head and flanged column are designed for high-pressure applications, when ease of maintenance is a prime consideration, or when alloy materials of construction are required for corrosive and/or erosive services. These pumps would also be suitable for a wide range of pumping temperatures, if used in industrial processes. However, they are primarily used in cooling water, sea and river water intake, utility circulating water, condensate, and ash sluice water services.

A vertical can-type pump is depicted in Figures 7-25 and 7-26. Using a fabricated discharge head and barrel and a flanged column, the pump is designed for low net positive suction head (NPSH) available and subatmospheric suction pressure services. Typical applications are pipeline boosters, product unloading, refinery blending, injection/secondary recovery, ammonia transfer, condensate, cryogenic, and liquid natural gas (LNG) transfer duties.

Figure 7-27 shows a vertical marine pump that uses a fabricated discharge head and flanged column. Pumps of this type are often designed to be self-priming and to efficiently unload or strip product tankers and tank barges. They have also been applied as ship firewater pumps, ballast pumps, bilge pumps, and fuel oil transfer pumps.

In our next illustration, Figure 7-28, we see a vertical industrial submersible pump. This version is used for deep settings or where the use of a lineshaft pump is impractical, e.g., in irrigation, service water, and deep well supply situations.

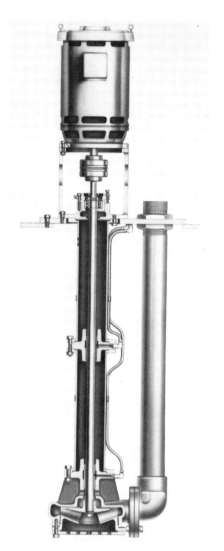

FIGURE 7–23 *Vertical sump and process pump. (Source: Goulds Pumps, Inc., Seneca Falls, NY.)*

FIGURE 7–24 *Typical vertical turbine pump with principal application in water services. (Source: Goulds Pumps, Inc., Seneca Falls, NY.)*

FIGURE 7–25 *Vertical can-type turbine pump, shown with fabricated head principally used in hydrocarbon processing services. (Source: Goulds Pumps, Inc., Seneca Falls, NY.)*

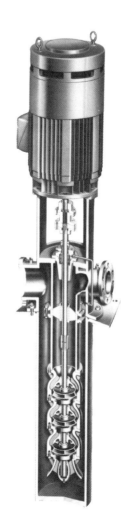

FIGURE 7–26 *Cross section of vertical can-type pump. (Source: Goulds Pumps, Inc., Seneca Falls, NY.)*

FIGURE 7–27 *Vertical marine pump with fabricated discharge head, flanged column, and right-angle gear drive. (Source: Goulds Pumps, Inc., Seneca Falls, NY.)*

FIGURE 7–28   *Vertical industrial submersible pump for deep-well applications. (Source: Goulds Pumps, Inc., Seneca Falls, NY.)*

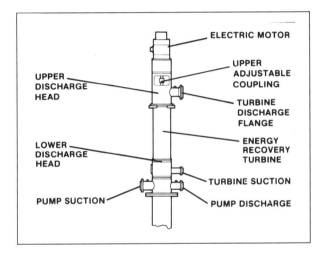

FIGURE 7–29   *Vertical energy recovery turbine—a vertical pump in reverse rotation. (Source: Goulds Pumps, Inc., Seneca Falls, NY.)*

Finally, there is also a close cousin to the vertical pump—the vertical energy recovery turbine (Figure 7–29). The recovery turbine takes high-pressure liquid and converts it into rotating energy that can be used to power other pumps, or other rotating equipment.

Virtually all of the vertical industrial pumps discussed for process plant applications include a variety of features and/or options that are of interest. Take, for instance, the bowl assembly, which is the heart of the vertical turbine pump. The impeller- and diffuser-type casings are designed to deliver the head and capacity that a system requires for optimum efficiency. The fact that the vertical turbine pump can be multi-staged allows maximum flexibility both in the initial pump selection and in the event that future system modifications require a change in the pump rating. Submerged impellers allow a pump to be started without priming.

A variety of material options allows the selection of a pump best suited for even the most severe services. The many bowl assembly options available assure that the vertical turbine pump satisfies the users' needs for safe, efficient, reliable, and maintenance-free operation. Figure 7–30 depicts the more important ones.

There exists also a large number of column options (Figure 7–31), discharge heads (Figure 7–32), coupling arrangements (Figure 7–33), and sealing flexibility options (Figure 7–34).

CHOICE OF SEMI-OPEN OR ENCLOSED IMPELLERS

Available in alloy construction for a wide range of corrosive/abrasive services.

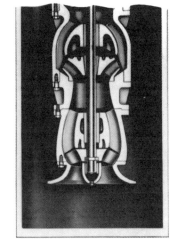

LOW NPSH FIRST STAGE IMPELLERS

For low $NPSH_A$ applications. Both large eye and mixed flow first stages available; minimizes pump length.

STRAINERS

Basket or cone strainers are available to provide protection from large solids.

KEYED IMPELLERS

Keyed impellers are standard on 18″ and larger sizes; furnished all pumps for temperatures above 180° F (82° C) and on cryogenic services. Regardless of size, keyed impellers provide ease of maintenance and positive locking under fluctuating load and temperature conditions.

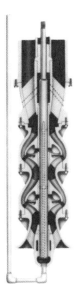

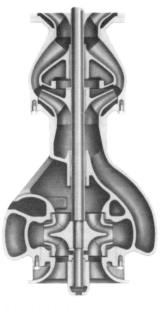

DUAL WEAR RINGS

Available for enclosed impellers and bowls; permits re-establishing initial running clearances and efficiency at lower cost. Hard facing of wear surfaces available for longer life. Wear rings can be flushed when solids are present in pumpage.

RIFLE DRILLING/ DISCHARGE BOWL

Rifle drilling of bowl shafts available for bearing protection on abrasive services.

Discharge bowl included with enclosed lineshaft construction.

DOUBLE SUCTION FIRST STAGE

The dual volute casing and double suction impeller can be installed alone or as a first stage with turbine stages added. In either event, double suction results in reduced $NPSH_R$. Another benefit is reduced axial downthrust. Goulds Vertical Double Suction (Model VDS) is ideally suited for steel mills, mine dewatering and river water, or, as a first stage in hotwell condensate and pipeline booster applications where NPSH is critical.

FIGURE 7–30   *Assembly options for vertical pump bowls. (Source: Goulds Pumps, Inc., Seneca Falls, NY.)*

Threaded Column

Threaded column is used whenever initial cost is primary consideration. Pipe ends are machine-faced for butt fit between sections to maintain alignment. Threaded column is used where pump length requires numerous column sections such as a deep well application.

OPEN LINESHAFT ASSEMBLY

Product lubricated open lineshaft construction using rubber bearings. Shaft sections joined by sleeve-type couplings.

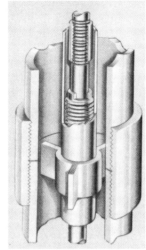

ENCLOSED LINESHAFT ASSEMBLY

Oil lubricated lineshaft for bearing lubrication of long set pumps or water flushing of lineshaft bearings for short or long set pumps in abrasive service.

Flanged Column

Column sections are provided with flanged ends incorporating registered fits for ease of alignment during assembly. Facilitates disassembly where corrosion is a problem. On 12″ and smaller sizes, the bearing retainer registers the fit between the column section flanges. On 14″ and larger sizes the bearing retainer is welded into the column section.

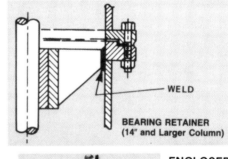

WELD

BEARING RETAINER
(14″ and Larger Column)

OPEN LINESHAFT

Flanged column/product lubricated lineshaft is recommended for ease of maintenance or whenever a special bearing material is required.

Keyed lineshaft coupling available in all sizes for ease of maintenance. Various bearing materials available. Renewable shaft sleeve or hard facing of shaft available for longer life.

BEARING RETAINER
(4″ to 12″ Column)

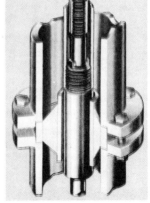

ENCLOSED LINESHAFT

The lineshaft is protected by water flushing the enclosing tube bearings on corrosive/abrasive services. Oil lubricated lineshaft available on long settings.

Alignment is attained by register fit between the flange faces.

FIGURE 7–31 *Vertical pump column assembly options. (Source: Goulds Pumps, Inc., Seneca Falls, NY.)*

The discharge head functions to change the direction of flow from vertical to horizontal and to couple the pump to the system piping in addition to supporting and aligning the driver. Discharge head accommodates all modes of drivers including hollow shaft and solid shaft motors, right angle gears, vertical steam turbines, etc. Optional sub-base can be supplied. Goulds offers three basic types for maximum flexibility.

FABRICATED DISCHARGE HEAD

For pressures exceeding cast head limitations or services that require alloy construction such as high or low temperature or corrosive services. Segmented elbow available for efficiency improvement. Large hand holes for easy access. Base flange can be machined to match ANSI tank flange.

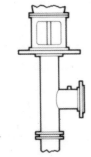

BELOW GROUND DISCHARGE HEAD

Used whenever VIT pump is required to adapt to an underground discharge system.

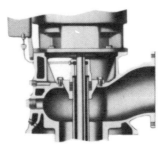

CAST IRON DISCHARGE HEAD

Used for low pressures (not exceeding 175 psi) and/or when low initial cost is primary consideration.

FIGURE 7–32 *Discharge head arrangements for vertical column pumps. (Source: Goulds Pumps, Inc., Seneca Falls, NY.)*

RIGID FLANGED COUPLING (Type AR)

To couple pump to vertical hollow shaft driver. Impeller adjustment is performed on adjusting nut located on top of motor.

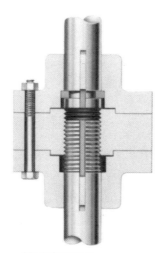

ADJUSTABLE COUPLING (Type A)

For vertical solid shaft driver. Impeller adjustment made by using adjustable plate in the coupling.

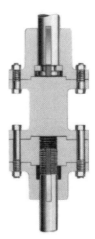

ADJUSTABLE SPACER COUPLING (Type AS)

Same function as type A coupling with addition of spacer. Spacer may be removed for mechanical seal maintenance without disturbing driver.

FIGURE 7–33 *Coupling arrangements for vertical column pumps. (Source: Goulds Pumps, Inc., Seneca Falls, NY.)*

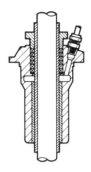

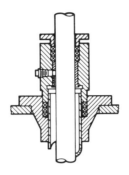

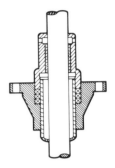

PACKED BOX

Whenever packing lubrication leakage
can be tolerated and the discharge
pressure does not exceed 300 psi, a
packed box may be used. Optional
headshaft sleeve available to protect
shaft.

WATER FLUSH

Water flush tube connection is supplied when pressurized water is introduced into the enclosing tube for
bearing protection on abrasive
services.

OIL LUBRICATED

Oil lubricated option is recommended
when water elevation would cause
the upper lineshaft bearings to run
without lubrication during start-up.
Oil is fed thru tapped opening and
allowed to gravitate down enclosing
tube lubricating bearings.

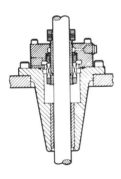

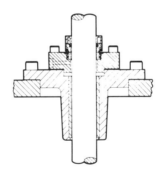

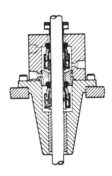

INSIDE MOUNTED SEALS

Most popular method — used for
medium to high pressures. Cartridge
style for ease of installation and
maintenance.

OUTSIDE MOUNTED SEALS

Provides a method of no-leak sealing
for low pressure applications.

TANDEM SEALS

Two seals mounted in-line. Chamber
between seals can be filled with a
buffer liquid and may be fitted with
a pressure sensitive annunciating
device for safety.

FIGURE 7-34   *Sealing flexibility options for vertical column pumps. (Source: Goulds Pumps, Inc.,
Seneca Falls, NY.)*

Circulator, or *axial flow pumps* are shown in Figures 7–35 and 7–36. Although
not strictly centrifugal flow pumps, these high-flow, low-head machines are worthy of
mention.

FIGURE 7–35 *Axial flow or elbow-type (circulator) pump used for high-volume, low-head applications in process plants. (Source: Goulds Pumps, Inc., Seneca Falls, NY.)*

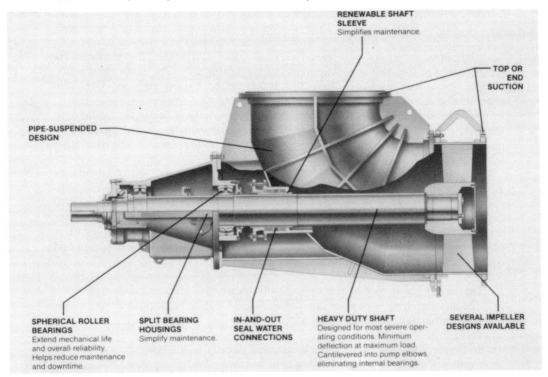

FIGURE 7–36 *Principal components of axial flow pumps. (Source: Goulds Pumps, Inc., Seneca Falls, NY.*

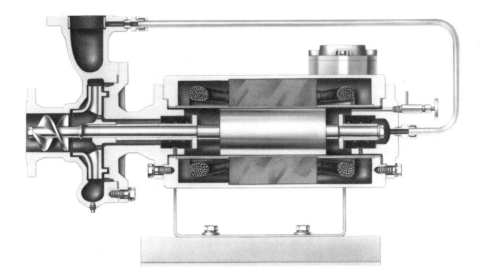

FIGURE 7–37   *Canned motor pump cross section. Note inducer for effective lowering of NPSH requirement. (Source: Goulds Pumps, Inc., Seneca Falls, NY.)*

CANNED MOTOR AND SEALLESS MAGNET-DRIVE CENTRIFUGAL PUMPS*

Canned motor and sealless drive pumps were developed to contain hazardous, valuable, toxic, or carcinogenic pumpage. These designs avoid the use of mechanical shaft seals and confine the pumpage within a hermetically sealed space.

The canned motor design (Figure 7–37) comprises a single shaft that combines the functions of a motor rotor and pump rotor in a single assembly. The motor rotor is surrounded by a stainless steel sleeve that is permeated by the magnetic flux lines generated by the surrounding stator windings.

A large number of variations of the standard canned motor design are available to the user. Figures 7–38 and 7–39 give an overview that includes canned motor pump models suitable for hot fluids, abrasive liquids, and pumpage close to the vaporization temperature.

A typical sealless magnet drive pump is illustrated in Figure 7–40. While unique when compared with conventional-design centrifugal pumps, magnet-drive pumps represent a combination of standard components and proven concepts. Figure 7–40 depicts a typical sealless magnet-drive pump with a separately mounted electric motor drive. In this installation, the base, the electric motor, and the motor coupling are identical to parts used in conventional pumps. The differences between these pumps and conventional pumps are concentrated in two areas:

1. Driving torque is transmitted magnetically rather than mechanically.
2. The impeller drive shaft rides in bushings housed within the pump enclosure rather than on bearings mounted externally.

In Figure 7–40, the drive motor is coupled directly to the outer magnet ring by a conventional motor-to-pump coupling. The overhung load of the outer magnet ring

*Sources: The Kontro Company, Inc., Orange, MA (magnet-drive), and Goulds Pumps, Inc., Seneca Falls, NY (canned motor pumps). Adapted by permission.

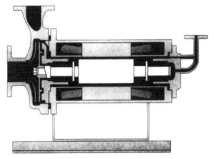

Suitable for handling volatile liquids; ammonia, freon and other liquified gases.

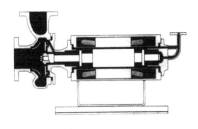

Same as R-Type but uses an adapter between pump and motor. Allows for greater pump/motor flexibility.

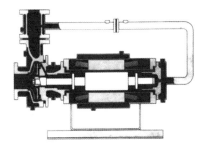

Suitable for liquids with high melting point.

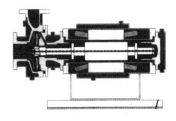

Similar to K-S Type but better suited for fluids with relatively low melting point.

FIGURE 7–38 *Canned motor pump variations for special fluid conditions. (Source: Goulds Pumps, Inc., Seneca Falls, NY.)*

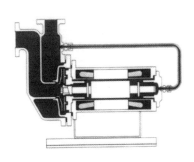

Used for sump and unloading services.

Handles fluids with large amount of fine solids. S-Type with external flushing also available.

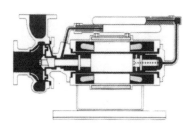

Suitable for handling high temperature fluids; heat transfer.

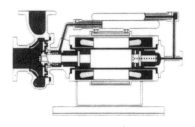

Suitable for handling liquids with small amount of fine solids.

FIGURE 7–39 *Canned motor pumps for unusual fluid conditions. (Source: Goulds Pumps, Inc., Seneca Falls, NY.)*

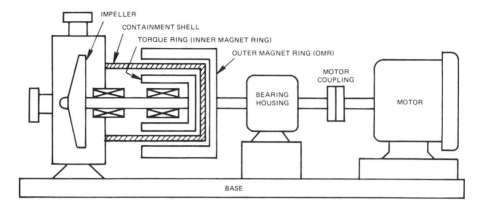

FIGURE 7–40 *Sealless magnet drive pump with a separately mounted electric motor. In this installation, the base, the electric motor, and the motor coupling are identical to parts used in conventional pumps. The difference between sealless pumps and conventional pumps occurs in two areas: (1) driving torque is transmitted magnetically rather than mechanically; and (2) the impeller drive shaft rides in bushings housed within the pump enclosure rather than by bearings mounted externally. (Source: Kontro Pump, Orange, MA.)*

is carried by the bearings in the bearing housing. Figure 7–40 also shows that the impeller is mounted on the same shaft as the inner magnet rotor, sometimes called the torque ring. The impeller drive shaft is carried by one or two bushings that are within the pumping enclosure. Note that the pump enclosure is formed by the pump casing and the containment shell. The driving torque of the electric motor is transmitted to the pump impeller by the magnetic coupling of the outer magnet ring and the torque ring without breaching the pumping enclosure. It is this magnetic coupling that replaces the mechanical seals of conventional centrifugal pumps. The efficiency obtainable with canned motor and magnet drive pumps is below that of well-designed, conventional centrifugal pumps. Also, canned motor and magnet-drive pumps may not be available in size ranges much over 100 kw. Nevertheless, they represent viable options that must be evaluated on a case-by-case basis.

HIGH-SPEED CENTRIFUGAL PUMPS*

Development of the High-Speed Concept

The term *high speed* is generally used to classify equipment that operates above two pole motor speeds. Centrifugal pump designs falling into this category have gained considerable acceptance since 1960. The increasing popularity of high-speed pumps has coincided with the expanding need for higher pressures in the process and general industries since World War II. At the same time, improved technology, manufacturing techniques, and materials have facilitated the transition from design theory to production hardware.

The developed head in centrifugal pumps is a function of the tip speed of the impeller and/or the number of impellers employed. There are three principal methods that can be used to achieve higher pressures, and in some cases a combination of these methods may be utilized:

---

*Source: Sundstrand Fluid Handling, Arvada, CO. Adapted by permission.

- Increasing the size of the impeller to increase its peripheral speed. This is a simple and effective method, but only to a point. The practical design limit for impeller diameters is 13 to 16 inches at two pole speeds.
- Using a number of staged impellers. Although continued development of multistage pumps has resulted in hydraulic efficiencies approaching single-stage efficiencies, the complexity of the design and close impeller clearances results in high first cost, loss of performance as the clearances wear, and often high maintenance requirements if the pump is subjected to difficult service conditions.
- Increasing rotational speed. The practical speed limitations for pumps directly driven by induction motors is 3600 RPM with 60 cycle power and 3000 RPM with 50 cycle power. The rotational speed of electric-driven pumps can, however, be increased by using either higher frequency power or a speed-increasing gearbox. Since with few exceptions, higher frequency power has not yet become an economically viable approach, the use of gear-driven single-stage, high-speed centrifugal pumps has become widespread.

This segment of our text deals primarily with a unique pump design that has been adapted for operation at higher speeds to produce high heads at low to moderate flow rates. Although high-speed pumps are now widely used in industry, comparatively few design discussions have appeared in print.

An early commercial application of this design concept emerged in 1959, when a high-speed single-stage centrifugal pump was used in aircraft service to augment the thrust in jet engines during takeoff. This pump rotated at 11,000 RPM, delivering 80 gallons per minute of water to the combusters at 400 psi to increase the mass flow rate through the engine, thereby increasing the thrust by 15 percent during takeoff. The unit weighed only 8½ pounds, including step-up gearing from the 6500 RPM power takeoff pad to pump speed. Some 250 units were produced for this service.

By 1962, the first integrally geared high-speed process pumps were finding their way into the petrochemical and refining industries. In subsequent decades, their use has been greatly expanded, and they are now available from 1 to 2500 HP, utilizing speeds to 32,000 RPM, and producing heads to 12,000 feet. Most commonly, these products consist of a single stage but may employ two or three stages to satisfy the need for extreme heads or the combination of high head and low net positive suction head available (NPSHA). By the early 1970s, the high-speed pump concept had been extended to include a wide variety of services across a broad spectrum of industries.

Unique Design Advantages

The increasing popularity of integrally geared high-speed centrifugal pumps is due to a number of factors:

- Shaft dynamics. As shaft speeds are increased, the size of the components required for any given condition of service grows smaller. The smaller, more compact design results in shorter shaft spans, lower shaft deflection, and improved shaft dynamics.
- Reduced size and weight. A high-speed pump with a single six-inch impeller can exhibit the same performance as a multistage pump that uses as many as 40 stages with the same size impellers. The size and weight reduction can be as much as 5 to 1, which of course translates to smaller and less expensive mounting foundations.
- Fewer parts required. High-speed pumps generally have one stage and occasionally two or three stages. This can mean a significant reduction in the number of "wetted" components. In processes requiring exotic metallurgies, this can provide a substantial capital cost advantage as well as less supporting inventory and lower repair parts costs.

FIGURE 7–41   *Removable high-speed shaft assembly for low-flow high-head pump. (Source: Sundstrand Fluid Handling, Arvada, CO.)*

- High-speed gear-driven pumps can be designed to incorporate many common parts. Seals, bearings, and housings are typically common across a given product line. Only the pump case, impeller, and/or gear ratio need to be changed to provide a wide range of performance. In a plant requiring many of the same types of pumps, the number of spare parts can be reduced.
- Reduced maintenance. Another benefit realized by the user is simplified maintenance resulting from the reduced size and quantity of parts. Additionally, some high-speed pump designers have allowed for the complete removal and installation of all fluid end components, seals, and bearings in a single modular high-speed shaft assembly (see Figure 7–41).
- Performance consistency. The head, flow, and efficiency of many high-speed pumps is resistant to efficiency and head degradation, since open radial bladed impellers do not require close clearances. In applications where consistency of performance over the life of the pump is desired, a high-speed pump with large running clearances can be very desirable.

Hydraulic Capabilities

High-speed pumps are manufactured in both single and multistage configurations. Radial vaned open impellers are optimum for low specific speed applications (see Figure 7–42) from $N_S = 150$ to $N_S = 850$. This hardware is capable of achieving 6000 feet in a single stage. Higher flow units typically use Francis vaned impellers with wear rings when the specific speed ranges from 850 to 1860. As impeller speed is increased to meet a given set of operating conditions, the specific speed and pump efficiency

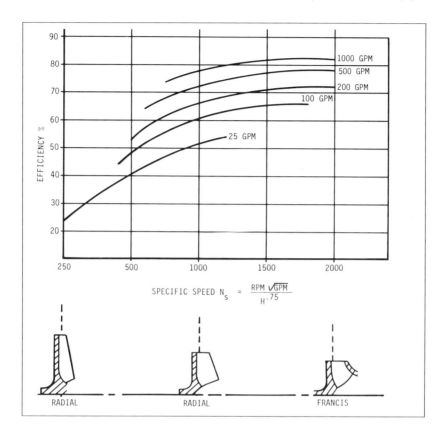

FIGURE 7–42   *Effect of specific speed on high-speed pump efficiency. (Source: Sundstrand Fluid Handling, Arvada, CO.)*

increase, while torque decreases for the same horsepower requirement. Accordingly, shaft stress, gear, and radial bearing loads improve with higher speeds.

The user should not be unduly concerned with increased wear or stress due to higher rotational speeds. The maximum stress level in an impeller is a function of the impeller tip speed. As shown in Figure 7–43, the stresses at 200 feet per second (FPS) are the same regardless of whether the speed is produced by a 3600 RPM twelve-inch impeller or by a 23,000 RPM two-inch impeller, and shaft deflection and bearing loading are minimized using unusually low overhang ratios and small impeller weights.

As impeller speed increases, the NPSH required for stable operation also increases. Often an axial flow inducer with good suction performance is used in series with the impeller to lower NPSH requirements. It is attached directly to the shaft in place of the impeller nut. On two- or three-stage machines, the first stage can often be geared at a lower speed for lower net positive suction head required (NPSHR), while the subsequent stages do most of the work.

Mounting Arrangements

The unique design of the high-speed pump lends itself to a variety of mounting configurations. Since the pump first appeared in the industrial market as a single-stage vertical in-line type, the full range of possibilities has been explored. Today these pumps exist in the following forms to serve a wide range of general industrial and process

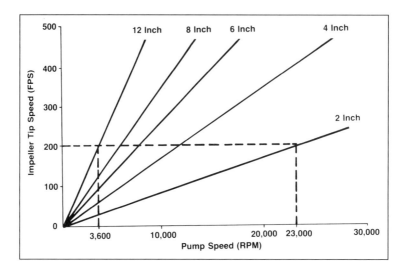

FIGURE 7–43 *Relationship of pump rotative speed and impeller diameter to tip speed. (Source: Sundstrand Fluid Handling, Arvada, CO.)*

markets: vertical in-line; horizontal single-stage, two-stage, and three-stage, with both single-step and two-step speed-increasing gearboxes. High-speed pumps are often available in either close-coupled or frame-mounted configurations. The close-coupled design eliminates the need for coupling alignment and occupies the least amount of floor space, while the frame-mounted units are used whenever conventional driver packages are selected.

Applications

Process applications for high-speed centrifugal pumps exist wherever there is a need for medium or higher pressures. Their widespread use is based on adaptability to many diverse requirements. High-speed pumps are an essential part of processes utilized in the production of such end items as plastics, pharmaceuticals, petrochemicals, synthetic rubber, and paper. The technology incorporated in these pumps makes them especially suited to lower flow, high-head applications, displacing reciprocating two-stage and multistage pumps as the preferred product.

Users apply high-speed pumps to process applications for numerous reasons, but the primary deciding factor is economics. Economic evaluations typically include first cost, installed cost, operating cost, maintenance cost, and overall evaluated cost. Each determining factor must be based on the user's specific situation. The primary reasons that high-speed pumps are often selected over reciprocating, single-stage, two-stage, or multistage centrifugals are lower first cost, lower installed cost, and occasionally lower maintenance costs. Operating costs will generally approach those of other centrifugal pumps but are almost always higher than positive displacement pumps with their inherently better mechanical efficiencies. The performance area where the high-speed pump has the greatest advantage is in the low-flow range.

Pump mechanical requirements vary, depending on the critical nature and location of the particular service. In severe or hazardous applications, API-610 requirements may be necessary, and high-speed pumps are available in this category. For less critical services, however, general service pumps should also be considered. Although many of the same design features are available in both types, the significantly lower cost associated with non-API designs encourages their use in less critical services.

High-speed pumps can be used as the primary pump, as an installed spare for an existing pump, as a boost pump piped in series with another pump, or as a support pump for a seal flush or lube oil. They are utilized in both continuous and standby operation.

It is not unusual to find high-speed pumps feeding a variety of systems where flow demands are constantly changing. These pumps are especially suited to high-pressure washdown and shower services where multiple sprayers are turned on and off as the system demands change. The controls required to operate these systems are simple and reliable, allowing operation of the pump over most of its performance envelope down to flows as low as 15 percent of the best efficiency point flow.

Process plant applications are as diverse as the industry served. Beginning with power systems, high-speed centrifugal pumps are applied in boiler feedwater, condensate return, desuperheater or attemporator, gas turbine NOX supression, and reverse osmosis applications.

Process systems use high-speed pumps in a variety of services including but not limited to transfer, seal flush, waste injection or disposal, blending, sampling, recycling, descaling, metering, waste disposal, reactor feed, booster, pipeline, charge, reflux, circulation, bottoms, flare drum knockout, and high-pressure washdown. Some typical fluids pumped include water, caustics, ammonia, carbamate, fuel oil, naphtha, acids, a majority of hydrocarbons, and chemicals too numerous to mention. As evidenced by the wide variety of applications, high-speed pumps are a proven product with years of reliable operating experience.

System Controls

The control of high-speed centrifugal pumps is similar to most conventional centrifugal pumps. When specifying the control system, it is important to consider the allowable operating range of the pump and its hydraulic characteristics, as well as the hydraulic requirements of the process.

There are generally two objectives that need to be kept in mind when designing a control system. One is to protect the pump from damage that can be incurred from operating outside its design operating range. A second is to provide the controls that will enable the pump to meet the needs of the process.

Centrifugal pumps tend to operate over a wide flow range with relatively slight variation in pressure in comparison with positive displacement pumps. The maximum and minimum operating limits for centrifugal pumps with flat performance curves are normally based on flow rather than pressure. Thus the protective controls should be designed to measure and control flow rate rather than discharge pressure (Figure 7–44).

Maximum Flow Limit

Volute-type centrifugal pumps have the lowest bearing radial loads at the design flow rate or best efficiency point. As the flow through the pump is increased or decreased from the best efficiency point, the radial hydraulic loads increase. Also, as flow velocities increase, the potential for impeller cavitation increases. The power also increases with flow. Operation at excessive flows can lead to bearing failures, high shaft stresses and possible failures, and cavitation damage to impellers and casings.

Attention to the maximum flow limit of the pump and knowledge of the process hydraulic characteristics when the pump is specified can result in a process that is self limiting and without need of special controls to prevent excessive flow through the pump. Figure 7–45 shows a pump curve that has been matched to the process needs at maximum capacity. The initial startup should be carefully planned to allow pipes to empty and vessels to be filled gradually, preventing water hammer or overloading the pump.

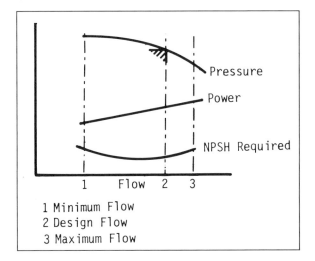

FIGURE 7–44   *Typical centrifugal pump performance. (Source: Sundstrand Fluid Handling, Arvada, CO.)*

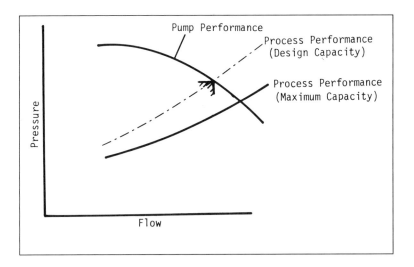

FIGURE 7–45   *Pump curve matched to process requirements at maximum capacity. (Source: Sundstrand Fluid Handling, Arvada, CO.)*

Minimum Flow Limit

As flow increases from the design point, bearing radial loads generally increase and efficiency decreases. If the flow decreases enough, recirculation can occur and the pump becomes hydraulically unstable. Extensive damage can be done if a pump is allowed to operate for long periods in an unstable condition.

As efficiency decreases at low flow, the rate of temperature rise of the fluid increases. This can be a concern in applications with low available NPSH.

The specific minimum flow limit depends partially on the pump and partially on the process (assuming adequate bearing capacity). Advertised performance curves generally show minimum flow that is expected with ideal fluid properties and proper

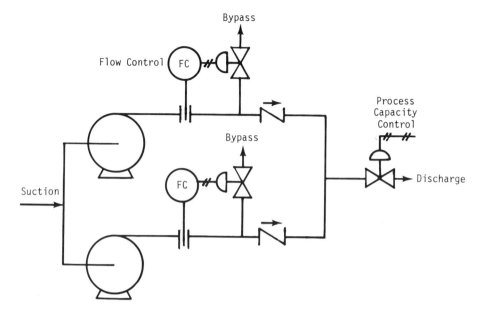

FIGURE 7–46 *Parallel pumps with individual flow-modulated bypasses for minimum flow protection. (Source: Sundstrand Fluid Handling, Arvada, CO.)*

inlet and discharge piping. Minimum flow controls should always be checked by observing the pump in operation with the minimum flow control functioning.

Near the minimum flow point, most centrifugal pumps have nearly constant pressure with respect to flow. To prevent operation below the minimum flow limit, the first choice for the measured variable is flow. A control system that prevents operation above a particular maximum discharge pressure does not necessarily ensure minimum flow protection.

When centrifugal pumps are operated in parallel, individual minimum flow control is necessary. A check valve should be installed in the discharge line of each pump. This is to prevent one pump from driving the other pump off its performance curve if both pumps are operating and the process is modulated to a low-capacity condition (see Figure 7–46).

Modulating a bypass line is the normal method for preventing minimum flow. If the bypass line discharges immediately into the pump suction, the fluid temperature will rise because of the power being dissipated in the pump bypass loop. If prolonged operation with this arrangement is expected, then a maximum temperature trip should be considered (Figure 7–47).

Suction Pressure Limit

Occasionally a process scheme may allow suction pressure to vary. If variations can cause the NPSH available to fall below that required by the pump, or the inlet pressure to rise above the maximum rated pressure for either the pump or the seal system, appropriate limiting controls will be required.

Maintenance Considerations

Routine maintenance on high-speed centrifugal pumps consists of two activities: periodic inspection and periodic service.

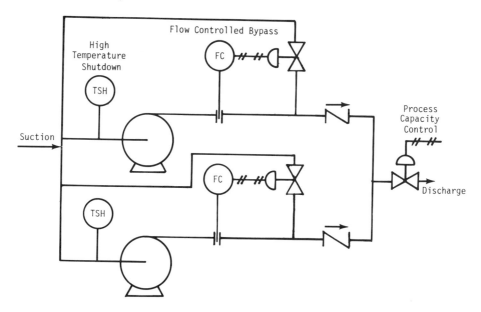

FIGURE 7–47   *Parallel pumps with individually controlled recycle lines. (Source: Sundstrand Fluid Handling, Arvada, CO.)*

Periodic Inspection

The items checked and the frequency of checking will vary with the specific design of the pump and its duty. The common ingredient in high-speed pumps is the integral speed-increasing gear box. Lube oil level, lube pressure, and temperature are normal inspection items on all of these pumps, requiring monitoring at least once per week. The need for periodic inspections will vary with the type of auxiliary equipment installed on the pump.

Where there are additional auxiliary systems supporting the pump, there are normally automatic alarm and shutdown devices that help to simplify inspection tasks. Proper functioning of protective alarm and shutdown devices should be verified periodically. If such tests are to be made while the equipment is running, it is best to specify this provision when pump and auxiliary equipment are initially purchased.

It is best if the periodic inspections include record keeping. Records can show trends that can help in the planning of service work, keeping maintenance costs low.

Periodic Service

Lubrication requirements are normally identified in the manufacturer's instruction manual. The driver, coupling, and gearbox each have their own lubrication needs. Each machine is designed to be run using lubricants with certain specific properties, so the manufacturer's recommendation should be considered when lubricants are chosen. Dibasic ester and poly-alpha olefin synthetic lubes are often advantageous, and knowledgeable gear manufacturers will specify these modern fluids.

Many high-speed gear pumps incorporate a modular high-speed rotor assembly that is easily removed for inspection and maintenance.

Contaminants in the pumped fluid or in the bearing lubricant can penetrate through the film that separates the moving and stationary parts, and also may cause wear.

Seal life is primarily determined by the seal environment. Most pumps have features that allow the user to control the seal environment to provide maximum seal life. If seal life of less than six months occurs, then system modifications can often extend the life. If life greater than one or two years is observed, then system modifications to further improve the life will not likely be cost effective. Overall, seals and bearings are considered wearing items needing periodic maintenance. The frequency of this maintenance can vary significantly with the type of duty the pump serves.

When bearings are replaced on high-speed pumps, the manufacturer's recommendations should be followed. In addition to the load capacity of high-speed ball bearings, internal clearances, contact angle, tolerance class, and retainer design are all important factors. Modern pump manufacturers specify bearings that operate well within the manufacturer's ratings, but careless substitutions can have disastrous results.

Machinery Condition Monitoring. High-speed pumps in critical service are often monitored for continuous determination of machine condition. Parameters most often monitored are vibration, lube pressure and temperature, and bearing temperature. Such monitoring is not normally considered mandatory by pump manufacturers for general pump service. However, contemporary manufacturers generally can provide optional provisions for monitoring these items when needed for pumps in critical service.

Vibration Monitoring. The three types of vibration monitoring most commonly used are noncontacting proximity probes, seismic casing vibration sensors, and acceleration-spike energy transducers. Noncontacting proximity probes measure shaft displacement (peak-to-peak). These probes are normally installed inside the speed-increasing gearbox to measure displacement of the output shaft relative to the gearbox housing. They are normally installed by the manufacturer and are ordered with the pump at purchase. Casing vibration sensors normally measure housing velocity amplitude. Such instruments can be either permanently installed on the gearbox or can be obtained as portable units that are periodically taken from one machine to another. Acceleration-spike energy monitoring is often done with portable data terminals and is one of the most effective ways to obtain early warning of incipient defects in rolling element bearings. With either permanently mounted or portable types of instruments, it is best to take readings periodically and to monitor trends. Permanently installed instruments are normally connected to automatic alarm and shutdown controls.

Lube Pressure and Temperature. When pressurized lubrication systems are used on speed-increasing gearboxes, lube oil pressure and temperature are often monitored. If pressures fall or temperatures rise, the equipment can be shut down automatically to prevent or minimize damage.

Bearing Temperature. On pumps with journal bearings, temperature sensors can be imbedded in the bearings. Temperature sensors are usually either thermocouples or resistance temperature detectors (RTDs). Bearing temperature monitoring can provide early warning of loss of lubricating properties, reduction in lubricant flow, bearing failure, or loss of lubricant cooling.

APPENDIX 7A

Centrifugal Pump Fundamentals

Head

The pressure at any point in a liquid can be thought of as being caused by a vertical column of the liquid which, due to its weight, exerts a pressure equal to the pressure at the point in question. The height of this column is called the "static head" and is expressed in terms of feet of liquid.

The static head corresponding to any specific pressure is dependent upon the weight of the liquid according to the following formula:

$$\text{Head in Feet} = \frac{\text{Pressure in psi} \times 2.31}{\text{Specific Gravity}}$$

A Centrifugal pump imparts velocity to a liquid. This velocity energy is then transformed largely into pressure energy as the liquid leaves the pump. Therefore, the head developed is approximately equal to the velocity energy at the periphery of the impeller. This relationship is expressed by the following well known formula:

$$H = \frac{v^2}{2g}$$

Where H = Total head developed in feet.
  v = Velocity at periphery of impeller in feet per sec.
  g = 32.2 Feet/Sec.^{2}

We can predict the approximate head of any centrifugal pump by calculating the peripheral velocity of the impeller and substituting into the above formula. A handy formula for peripheral velocity is:

$$v = \frac{RPM \times D}{229} \qquad \text{Where D = Impeller diameter in inches.}$$

The above demonstrates why we must always think in terms of feet of liquid rather than pressure when working with centrifugal pumps. A given pump with a given impeller diameter and speed will raise a liquid to a certain height regardless of the weight of the liquid, as shown in Fig. 1.

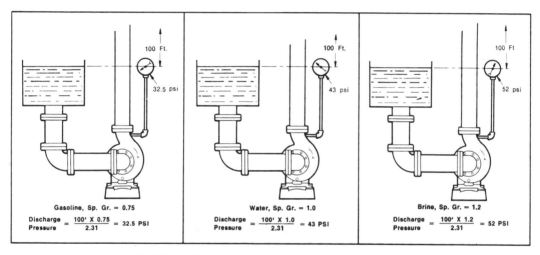

Fig. 1 Identical Pumps Handling Liquids of Different Specific Gravities.

All of the forms of energy involved in a liquid flow system can be expressed in terms of feet of liquid. The total of these various heads determines the total system head or the work which a pump must perform in the system. The various forms of head are defined as follows.

SUCTION LIFT exists when the source of supply is below the center line of the pump. Thus the STATIC SUCTION LIFT is the vertical distance in feet from the center line of the pump to the free level of the liquid to be pumped.

SUCTION HEAD exists when the source of supply is above the centerline of the pump. Thus the STATIC SUCTION HEAD is the vertical distance in feet from the centerline of the pump to the free level of the liquid to be pumped.

STATIC DISCHARGE HEAD is the vertical distance in feet between the pump centerline and the point of free discharge or the surface of the liquid in the discharge tank.

TOTAL STATIC HEAD is the vertical distance in feet between the free level of the source of supply and the point of free discharge or the free surface of the discharge liquid.

The above forms of static head are shown graphically in Fig. 2a & b

FRICTION HEAD (h$_f$) is the head required to overcome the resistance to flow in the pipe and fittings. It is dependent upon the size and type of pipe, flow rate, and nature of the liquid.

VELOCITY HEAD (h_v) is the energy of a liquid as a result of its motion at some velocity V. It is the equivalent head in feet through which the water would have to fall to acquire the same velocity, or in other words, the head necessary to accelerate the water. Velocity head can be calculated from the following formula:

$$h_v = \frac{V^2}{2g}$$ where $g = 32.2$ ft/sec.^{2}
$V =$ liquid velocity in feet per second.

The velocity head is usually insignificant and can be ignored in most high head systems. However, it can be a large factor and must be considered in low head systems.

PRESSURE HEAD must be considered when a pumping system either begins or terminates in a tank which is under some pressure other than atmospheric. The pressure in such a tank must first be converted to feet of liquid. A vacuum in the suction tank or a positive pressure in the discharge tank must be added to the system head, whereas a positive pressure in the suction tank or vacuum in the discharge tank would be subtracted. The following is a handy formula for converting inches of mercury vacuum into feet of liquid.

$$\text{Vacuum, ft. of liquid} = \frac{\text{Vacuum, in. of Hg} \times 1.13}{\text{Sp. Gr.}}$$

The above forms of head, namely static, friction, velocity, and pressure, are combined to make up the total system head at any particular flow rate. Following are definitions of these combined or "Dynamic" head terms as they apply to the pump.

TOTAL DYNAMIC SUCTION LIFT (h_s) is the static suction lift minus the velocity head at the pump suction flange plus the total friction head in the suction line. The total dynamic suction lift, as determined on pump test, is the reading of a gage on the suction flange, converted to feet of liquid and corrected to the pump centerline*, minus the velocity head at the point of gage attachment.

TOTAL DYNAMIC SUCTION HEAD (h_s) is the static suction head plus the velocity head at the pump suction flange minus the total friction head in the suction line. The total dynamic suction head, as determined on pump test, is the reading of the gage on the suction flange, converted to feet of liquid and corrected to the pump centerline*, plus the velocity head at the point of gage attachment.

TOTAL DYNAMIC DISCHARGE HEAD (h_d) is the static discharge head plus the velocity head at the pump discharge flange plus the total friction head in the discharge line. The total dynamic discharge head, as determined on pump test, is the reading of a gage at the discharge flange, converted to feet of liquid and corrected to the pump centerline*, plus the velocity head at the point of gage attachment.

TOTAL HEAD (H) or TOTAL DYNAMIC HEAD (TDH) is the total dynamic discharge head minus the total dynamic suction head or plus the total dynamic suction lift.

TDH = h_d + h_s (with a suction lift)
TDH = h_d − h_s (with a suction head)

*On vertical pumps the correction should be made to the eye of the suction or lowest impeller.

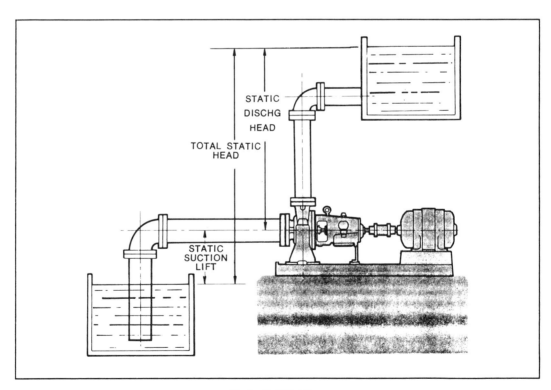

Fig. 2-a Suction Lift —
Showing Static Heads in a Pumping System Where the Pump
is Located Above the Suction Tank. (Static Suction Head)

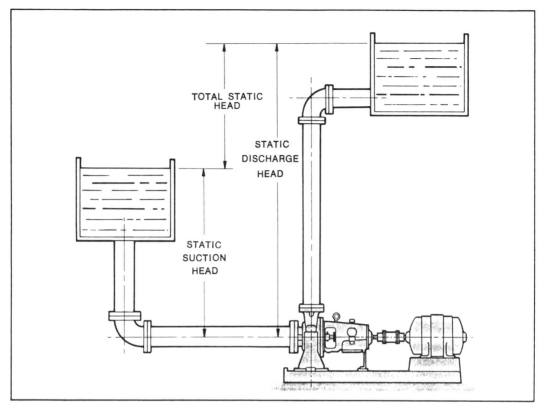

Fig. 2-b Suction Head —
Showing Static Heads in a Pumping System Where the Pump
is Located Below the Suction Tank. (Static Suction Head)

Capacity

Capacity (Q) is normally·expressed in gallons per minute (gpm). Since liquids are essentially incompressible, there is a direct relationship between the capacity in a pipe and the velocity of flow. This relationship is as follows:

$$Q = A \times V \text{ or } V = \frac{Q}{A}$$

Where A = Area of pipe or conduit in square feet.
V = Velocity of flow in feet per second.

Power and Efficiency

The work performed by a pump is a function of the total head and the weight of the liquid pumped in a given time period. The pump capacity in gpm and the liquid specific gravity are normally used in the formulas rather than the actual weight of the liquid pumped.

Pump input or brake horsepower (bhp) is the actual horsepower delivered to the pump shaft. Pump output or hydraulic horsepower (whp) is the liquid horsepower delivered by the pump. These two terms are defined by the following formulas.

$$whp = \frac{Q \times TDH \times Sp. Gr.}{3960}$$

$$bhp = \frac{Q \times TDH \times Sp. Gr.}{3960 \times Pump\ Efficiency}$$

The constant 3960 is obtained by dividing the number or foot pounds for one horsepower (33,000) by the weight of one gallon of water (8.33 pounds.)

The brake horsepower or input to a pump is greater than the hydraulic horsepower or output due to the mechanical and hydraulic losses incurred in the pump. Therefore the pump efficiency is the ratio of these two values.

$$Pump\ Eff = \frac{whp}{bhp} = \frac{Q \times TDH \times Sp. Gr.}{3960 \times bhp}$$

Specific Speed and Pump Type

Specific speed (N_s) is a non-dimensional design index used to classify pump impellers as to their type and proportions. It is defined as the speed in revolutions per minute at which a geometrically similar impeller would operate if it were of such a size as to deliver one gallon per minute against one foot head.

The understanding of this definition is of design engineering significance only, however, and specific speed should be thought of only as an index used to predict certain pump characteristics. The following formula is used to determine specific speed:

$$N_s = \frac{N\sqrt{Q}}{H^{3/4}}$$

Where N = Pump speed in RPM
Q = Capacity in gpm at the best efficiency point
H = Total head per stage at the best efficiency point

The specific speed determines the general shape or class of the impeller as depicted in Fig. 3. As the specific speed increases, the ratio of the impeller outlet diameter, D_2, to the inlet or eye diameter, D_1, decreases. This ratio becomes 1.0 for a true axial flow impeller.

Radial flow impellers develop head principally through centrifugal force. Pumps of higher specific speeds develop head partly by centrifugal force and partly by axial force. A higher specific speed indicates a pump design with head generation more by axial forces and less by centrifugal forces. An axial flow or propeller pump with a specific speed of 10,000 or greater generates its head exclusively through axial forces.

Radial impellers are generally low flow high head designs whereas axial flow impellers are high flow low head designs.

Values of Specific Speed, N_s

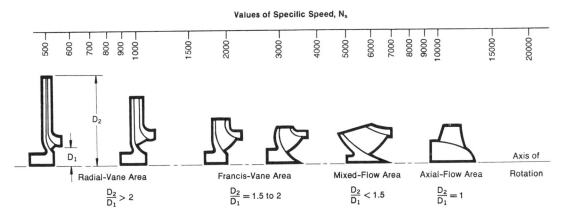

Fig. 3   Impeller Design vs Specific Speed

Net Positive Suction Head (NPSH) and Cavitation

The Hydraulic Institute defines NPSH as the total suction head in feet absolute, determined at the suction nozzle and corrected to datum, less the vapor pressure of the liquid in feet absolute. Simply stated, it is an analysis of energy conditions on the suction side of a pump to determine if the liquid will vaporize at the lowest pressure point in the pump.

The pressure which a liquid exerts on its surroundings is dependent upon its temperature. This pressure, called vapor pressure, is a unique characteristic of every fluid and increases with increasing temperature. When the vapor pressure within the fluid reaches the pressure of the surrounding medium, the fluid begins to vaporize or boil. The temperature at which this vaporization occurs will decrease as the pressure of the surrounding medium decreases.

A liquid increases greatly in volume when it vaporizes. One cubic foot of water at room temperature becomes 1700 cu. ft. of vapor at the same temperature.

It is obvious from the above that if we are to pump a fluid effectively, we must keep it in liquid form. NPSH is simply a measure of the amount of suction head present to prevent this vaporization at the lowest pressure point in the pump.

NPSH Required is a function of the pump design. As the liquid passes from the pump suction to the eye of the impeller, the velocity increases and the pressure decreases. There are also pressure losses due to shock and turbulence as the liquid strikes the impeller. The centrifugal force of the impeller vanes further increases the velocity and decreases the pressure of the liquid. The NPSH Required is the positive head in feet absolute required at the pump suction to overcome these pressure drops in the pump and maintain the liquid above its vapor pressure. The NPSH Required varies with speed and capacity within any particular pump. Pump manufacturer's curves normally provide this information.

NPSH Available is a function of the system in which the pump operates. It is the excess pressure of the liquid in feet absolute over its vapor pressure as it arrives at the pump suction. Fig. 4 shows four typical suction systems with the NPSH Available formulas applicable to each. It is important to correct for the specific gravity of the liquid and to convert all terms to units of "feet absolute" in using the formulas.

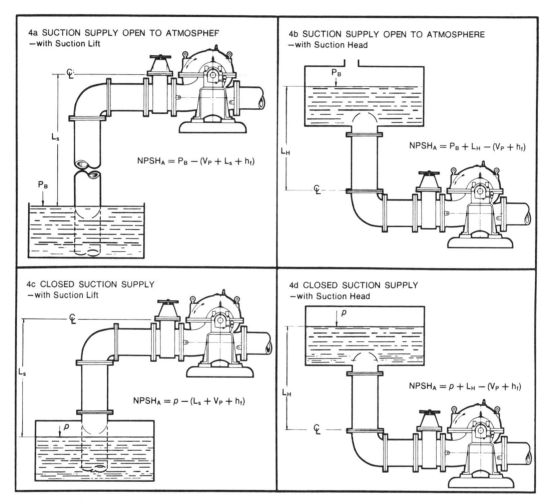

4a SUCTION SUPPLY OPEN TO ATMOSPHEF
—with Suction Lift

$$NPSH_A = P_B - (V_P + L_s + h_f)$$

4b SUCTION SUPPLY OPEN TO ATMOSPHERE
—with Suction Head

$$NPSH_A = P_B + L_H - (V_P + h_f)$$

4c CLOSED SUCTION SUPPLY
—with Suction Lift

$$NPSH_A = p - (L_s + V_P + h_f)$$

4d CLOSED SUCTION SUPPLY
—with Suction Head

$$NPSH_A = p + L_H - (V_P + h_f)$$

P_B = Barometric pressure, in feet absolute.
V_P = Vapor pressure of the liquid at maximum pumping temperature, in feet absolute.
p = Pressure on surface of liquid in closed suction tank, in feet absolute.

L_s = Maximum static suction lift in feet.
L_H = Minimum static suction head in feet.
h_f = Friction loss in feet in suction pipe at required capacity

Fig. 4 Calculation of system Net Positive Suction Head Available for typical suction conditions.

In an existing system, the NPSH Available can be determined by a gage reading on the pump suction. The following formula applies:

$$NPSH_A = P_B - V_P \pm Gr + hv$$

Where Gr = Gage reading at the pump suction expressed in feet (plus if above atmospheric, minus if below atmospheric) corrected to the pump centerline.

h_v = Velocity head in the suction pipe at the gage connection, expressed in feet.

Cavitation is a term used to describe the phenomenon which occurs in a pump when there is insufficient NPSH Available. The pressure of the liquid is reduced to a value equal to or below its vapor pressure and small vapor bubbles or pockets begin to form. As these vapor bubbles move along the impeller vanes to a higher pressure area, they rapidly collapse.

The collapse, or "implosion" is so rapid that it may be heard as a rumbling noise, as if you were pumping gravel. The forces during the collapse are generally high enough to cause minute pockets of fatigue failure on the impeller vane surfaces. This action may be progressive, and under severe conditions can cause serious pitting damage to the impeller.

The accompanying noise is the easiest way to recognize cavitation. Besides impeller damage, cavitation normally results in reduced capacity due to the vapor present in the pump. Also, the head may be reduced and unstable and the power consumption may be erratic. Vibration and mechanical damage such as bearing failure can also occur as a result of operating in cavitation.

The only way to prevent the undesirable effects of cavitation is to insure that the NPSH Available in the system is greater than the NPSH Required by the pump.

Pump Characteristic Curves

The performance of a centrifugal pump can be shown graphically on a characteristic curve. A typical characteristic curve shows the total dynamic head, brake horsepower, efficiency, and net positive suction head all plotted over the capacity range of the pump.

Figures 5, 6, & 7 are non-dimensional curves which indicate the general shape of the characteristic curves for the various types of pumps. They show the head, brake horsepower, and efficiency plotted as a per cent of their values at the design or best efficiency point of the pump.

Fig. 5 shows that the head curve for a radial flow pump is relatively flat and that the head decreases gradually as the flow increases. Note that the brake horsepower increases gradually over the flow range with the maximum normally at the point of maximum flow.

Mixed flow centrifugal pumps and axial flow or propeller pumps have considerably different characteristics as shown in Figs.

6 and 7. The head curve for a mixed flow pump is steeper than for a radial flow pump. The shut-off head is usually 150% to 200% of the design head. The brake horsepower remains fairly constant over the flow range. For a typical axial flow pump, the head and brake horsepower both increase drastically near shut-off as shown in Fig. 7.

The distinction between the above three classes is not absolute, and there are many pumps with characteristics falling somewhere between the three. For instance, the Francis vane impeller would have a characteristic between the radial and mixed flow classes. Most turbine pumps are also in this same range depending upon their specific speeds.

Fig. 8 shows a typical pump curve as furnished by a manufacturer. It is a composite curve which tells at a glance what the pump will do at a given speed with various impeller diameters from maximum to minimum. Constant horsepower, efficiency, and $NPSH_R$ lines are superimposed over the various head curves. It is made up from individual test curves at various diameters.

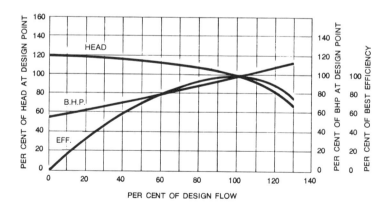

Fig. 5   Radial Flow Pump

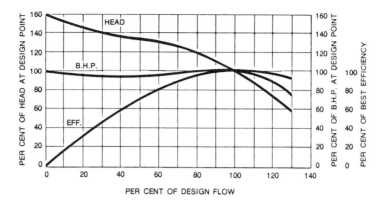

Fig. 6 Mixed Flow Pump

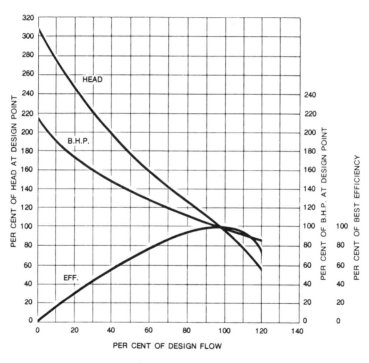

Fig. 7 Axial Flow Pump

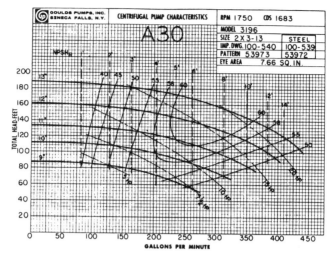

Fig. 8   Composite Performance Curve

Affinity Laws

The affinity laws express the mathematical relationship between the several variables involved in pump performance. They apply to all types of centrifugal and axial flow pumps. They are as follows:

1. With impeller diameter, D, held constant:

   A. $\dfrac{Q_1}{Q_2} = \dfrac{N_1}{N_2}$

   B. $\dfrac{H_1}{H_2} = \left(\dfrac{N_1}{N_2}\right)^2$

   C. $\dfrac{BHP_1}{BHP_2} = \left(\dfrac{N_1}{N_2}\right)^3$

   Where   Q = Capacity, GPM
   H = Total Head, Feet
   BHP = Brake Horsepower
   N = Pump Speed, RPM

2. With speed, N, held constant:

   A. $\dfrac{Q_1}{Q_2} = \dfrac{D_1}{D_2}$

   B. $\dfrac{H_1}{H_2} = \left(\dfrac{D_1}{D_2}\right)^2$

   C. $\dfrac{BHP_1}{BHP_2} = \left(\dfrac{D_1}{D_2}\right)^3$

When the performance (Q_1, H_1, & BHP_1) is known at some particular speed (N_1) or diameter (D_1), the formulas can be used to estimate the performance (Q_2, H_2, & BHP_2) at some other speed (N_2) or diameter (D_2). The efficiency remains nearly constant for speed changes and for small changes in impeller diameter.

EXAMPLE

To illustrate the use of these laws, refer to Fig. 8. It shows the performance of a particular pump at 1750 rpm with various impeller diameters. This performance data has been determined by actual tests by the manufacturer. Now assume that you have a 13″ maximum diameter impeller, but you want to belt drive the pump at 2000 rpm.

The affinity laws listed under 1 above will be used to determine the new performance, with N_1 = 1750 rpm and N_2 = 2000 rpm. The first step is to read the capacity, head, and horsepower at several points on the 13″ dia. curve in Fig. 10. For example, one point may be near the best efficiency point where the capacity is 300 gpm, the head is 160 ft, and the bhp is approx. 20 hp.

$$\frac{300}{Q_2} = \frac{1750}{2000} \qquad\qquad Q_2 = 343 \text{ gpm}$$

$$\frac{160}{H_2} = \left(\frac{1750}{2000}\right)^2 \qquad\qquad H_2 = 209 \text{ ft.}$$

$$\frac{20}{BHP_2} = \left(\frac{1750}{2000}\right)^3 \qquad\qquad BHP_2 = 30 \text{ hp}$$

This will then be the best efficiency point on the new 2000 rpm curve. By performing the same calculations for several other points on the 1750 rpm curve, a new curve can be drawn which will approximate the pumps performance at 2000 rpm, Fig. 9.

Trial and error would be required to solve this problem in reverse.

In other words, assume you want to determine the speed required to make a rating of 343 gpm at a head of 209 ft. You would begin by selecting a trial speed and applying the affinity laws to convert the desired rating to the corresponding rating at 1750 rpm. When you arrive at the correct speed, 2000 rpm in this case, the corresponding 1750 rpm rating will fall on the 13″ diameter curve.

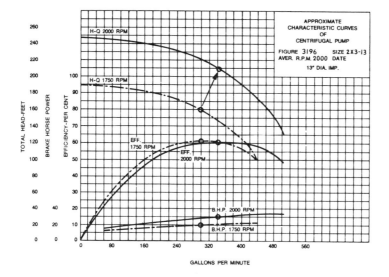

Fig. 9

System Curves

For a specified impeller diameter and speed, a centrifugal pump has a fixed and predictable performance curve. The point where the pump operates on its curve is dependent upon the characteristics of the system in which the pump is operating, commonly called the "System Head Curve". By plotting the system head curve and pump curve together, we can tell:

1. Where the pump will operate on its curve.
2. What changes will occur if the system head curve or the pump performance curve changes.

STATIC SYSTEM HEAD

Consider the system shown in Fig. 10. Since the lines are oversized and relatively short, the friction head is small compared to the static head. For this example, the system head will be considered as entirely static, with the friction neglected.

Assume the fluid being handled has 1.0 Sp. Gr. $NPSH_A$ is 13′. The flow requirement is 100 gpm. Since the system head is made up entirely of elevation and pressure differences, it does not vary with flow.

The normal system head is 250′ TDH (19′ elevation difference plus 231′ pressure difference). Since the discharge vessel pressure may vary ±3 psi, the system head will vary between 243′ and 257′.

Consider the application of a pump sized for 100 gpm at 250′ TDH, with a relatively flat performance curve as shown in Fig. 11. Note that the pump will shut off at 254′ TDH. At the maximum discharge tank pressure, the pump will stop delivering fluid, as the system head is greater than the pump TDH.

A second consideration associated with static system head is motor overload on pump runout. Again, consider Fig. 11 at the minimum system head of 243′. The pump under discussion will deliver 130 gpm against 243′ head. Horsepower requirements will increase from 8.9 BHP at 100 gpm to 12.0 BHP at 130 gpm. A 10 HP motor could be overloaded on this service.

NPSH problems may also arise when large increases in flow occur. At the rating of 100 gpm at 250′ TDH the $NPSH_R$ of the pump is only 10′ while the system $NPSH_A$ is 13′. At the lower system head of 243′ the pump requires 13.5′ NPSH and cavitation will probably occur.

A better selection would be a pump with a characteristic as shown in Fig. 12. The steeper characteristic will limit the flow to between 90 GPM at 257′ TDH and 110 gpm at 243′ TDH. The small increase in capacity at low head condition will mean no motor overload. Since the maximum flow is 110 gpm, the maximum $NPSH_R$ will be 12′ and the pump will not cavitate.

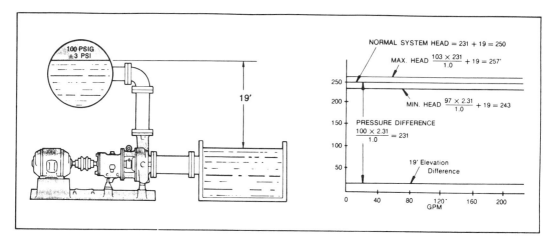

Fig. 10

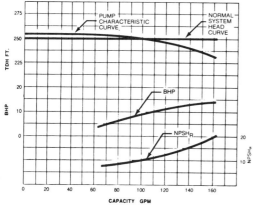

Fig. 11

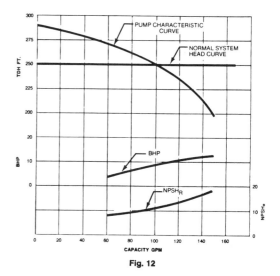

Fig. 12

DYNAMIC SYSTEM HEAD

In frictional systems where resistance to flow increases with flow, the system head characteristic becomes curved. The magnitude of the system head at each flow is the summation of the system static head plus the total friction losses at that particular flow rate. A typical example of this type of system is shown in Fig. 13.

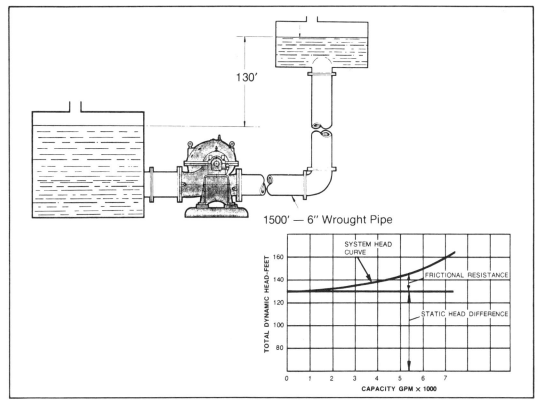

Fig. 13

Unlike the static system, the friction system is always self-correcting to some degree. Consider the above system with a flow requirement of 6000 gpm at 150' TDH. Also assume that the discharge tank level may drop 10'. The new system head curve will be parallel to the original one, but 10' lower as shown in Fig. 14. Flow under this reduced head will be 6600 gpm at 144' rather than the normal 6000 gpm at 150'. This increased flow rate will tend to raise the discharge tank level back to normal.

The frictional resistance of pipes and fittings will increase as they wear, resulting in greater curvature of the system head curve. A slight drop in the pump head curve may also result from increasing pump wear and recirculation. These changes will have less effect on the flow in a dynamic system (steep curve) than in a static system (flat curve).

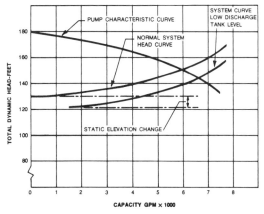

Fig. 14

PARALLEL OPERATION

It is sometimes desirable to use two or more pumps in parallel rather than a single larger pump. This is particularly advantageous when the system flow requirements vary greatly. One pump can be shut down when the flow requirement drops, allowing the remaining pump or pumps to operate closer to their peak efficiency. It also provides an opportunity for repairs or maintenance work on one unit without shutting down the entire system.

Special care must be taken in selecting pumps for parallel operation. Consideration must be given to single pump operation in the system as well as parallel operation. Consider the system shown in Fig. 15. The NPSH available is plotted along with the system head. Since entrance and line losses increase with increases in flow, the $NPSH_A$ decreases with flow increases.

The flow required is 16,000 gpm. We want to use two pumps in parallel, but each must be capable of single operation.

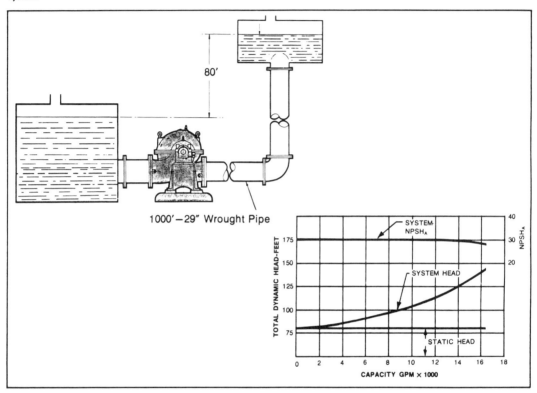

Fig. 15

The total system head at 16,000 gpm is 140'. Each pump must be sized for 8000 gpm at 140' TDH. $NPSH_R$ for each pump must be less than 28' for parallel operation. Consider applying two pumps each with characteristics as shown in Fig. 16. In order to study both parallel and single pump operation, the head-capacity curves for both single and parallel operation must be plotted with the system head curve.

The head-capacity pump curve for parallel operation is plotted by adding the capacities of each pump for several different heads and plotting the new capacity at each head. The shut-off head for the two pumps in parallel is the same as for single operation. The NPSH curve is plotted in the same manner as shown in Fig. 16. For example, the $NPSH_R$ for one pump at 8000 gpm is 14'. Therefore, in parallel operation 16,000 gpm can be pumped with 14' $NPSH_R$ by each pump.

The curve show that each pump will deliver 8000 gpm at 140' TDH when operating in parallel. Brake horsepower for each unit will be 340 HP. NPSH$_R$ is 14'. NPSH$_A$ is 28'.

With only one pump operating, the flow will be 11,000 gpm at 108' TDH. BHP will be 355 HP. NPSH$_R$ is 26' and NPSH$_A$ is 30'. A 400 HP motor would be required.

This example shows that if a 350 HP motor had been selected based on parallel operation only, the motor would have been overloaded in single pump operation. The single pump operation is also critical in terms of NPSH. For example, if the system NPSH$_A$ had been in the neighborhood of 20', parallel pump operation would have been fine, but single pump operation would result in cavitation.

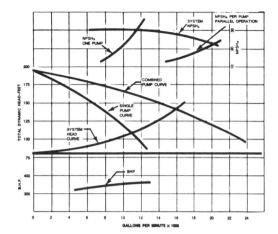

Fig. 16

Basic Formulae and Symbols

FORMULAS

$$GPM = \frac{0.002 \times Lb./Hr.}{Sp.\ Gr.}$$

$$H = \frac{2.31 \times psi}{Sp.\ Gr.}$$

$$H = \frac{1.134 \times In.\ Hg.}{Sp.\ Gr.}$$

$$h_v = \frac{V^2}{2g} = .0155\ V^2$$

$$V = \frac{GPM \times 0.321}{A} = \frac{GPM \times 0.409}{(I.D.)^2}$$

$$BHP = \frac{GPM \times H \times Sp.\ Gr.}{3960 \times Eff.} = \frac{GPM \times psi}{1715 \times Eff.}$$

$$Eff. = \frac{GPM \times H \times Sp.\ Gr.}{3960 \times BHP}$$

$$N_s = \frac{N\sqrt{GPM}}{H^{3/4}}$$

$$H = \frac{v^2}{2g}$$

$$v = \frac{N \times D}{229}$$

DEG. C = (DEG. F − 32) × 5/9

DEG. F = (Deg. C × 9/5) + 32

SYMBOLS

GPM = gallons per minute
Lb. = pounds
Hr. = hour
Sp. Gr. = specific gravity
H = head in feet
psi = pounds per square inch
In. Hg. = inches of mercury
h_v = velocity head in feet
V = velocity in feet per second
g = 32.16 ft/sec² (acceleration of gravity)

A = area in square inches
I.D. = inside diameter in inches
BHP = brake horsepower
Eff. = pump efficiency expressed as a decimal
N_s = specific speed
N = speed in revolutions per minute
v = peripheral velocity of an impeller in feet per second
D = impeller in inches

Pump Application Data
Corrosion & Materials of Construction

Corrosion is the destructive attack of a metal by chemical or electro-chemical reaction with its environment. It is important to understand the various types of corrosion and factors affecting corrosion rate to properly select materials.

Types of Corrosion

(1) Galvanic corrosion is the electro-chemical action produced when one metal is in electrical contact with another more noble metal, with both being immersed in the same corroding medium called the electrolyte. A galvanic cell is formed and current flows between the two materials. The least noble material called the anode will corrode while the more noble cathode will be protected. It is important that the smaller wearing parts in a pump be of a more noble material than the larger more massive parts, as in an iron pump with bronze or stainless steel trim.

Following is a galvanic series listing the more common metals and alloys:

Corroded End (Anodic, or	Nickel base alloy (active)
least noble)	Brasses
Magnesium	Copper
Magnesium Alloys	Bronzes
Zinc	Copper-Nickel Alloy
Aluminum 2S	Monel
Cadmium	Silver Solder
Aluminum 17ST	Nickel (Passive)
Steel or Iron	Nickel Base Alloy (Passive)
Cast Iron	Stainless Steel, 400 Series
Stainless Steel, 400 Series	(Passive)
(Active)	Stainless Steel, Type 304
Stainless Steel, Type 304	(Passive)
(Active)	Stainless Steel, Type 316
Stainless Steel, Type 316	(Passive)
(Active)	Silver
Lead-tin Solders	Graphite
Lead	Gold
Tin	Platinum
Nickel (Active)	**Protected End** (Cathodic, or
	most noble)

(2) Uniform Corrosion is the overall attack on a metal by a corroding liquid resulting in a relatively uniform metal loss over the exposed surface. This is the most common type of corrosion and it can be minimized by the selection of a material which offers resistance to the corroding liquid.

(3) Intergranular corrosion is the precipitation of chromium carbides at the grain boundaries of stainless steels. It results in the complete destruction of the mechanical properties of the steel for the depth of the attack. Solution annealing or the use of extra low carbon stainless steels will eliminate intergranular corrosion.

(4) Pitting Corrosion is a localized rather than uniform type of attack. It is caused by a breakdown of the protective film and results in rapid pit formation at random locations on the surface.

(5) Crevice or Concentration Cell Corrosion occurs in joints or small surface imperfections. Portions of the liquid become trapped and a difference in potential is established due to the oxygen concentration difference in these cells. The resulting corrosion may progress rapidly leaving the surrounding area unaffected.

(6) Stress Corrosion is the failure of a material due to a combination of stress and a corrosive environment, whereas the material would not be affected by the environment alone.

(7) Erosion-Corrosion is the corrosion resulting when a metal's protective film is destroyed by high velocity fluids. It is distinguished from abrasion which is destruction by fluids containing abrasive solid particles.

pH Values

The pH of a liquid is an indication of its corrosive qualities, either acidic or alkaline. It is a measure of the hydrogen or hydroxide ion concentration in gram equivalents per liter. pH value is expressed as the logarithm to the base 10 of the reciprocal of the hydrogen ion concentration. The scale of pH values is from zero to 14, with 7 as a neutral point. From 6 to zero denotes increasing hydrogen ion concentration and thus increasing acidity; and from 8 to 14 denotes increasing hydroxide ion concentration and thus increasing alkalinity.

The table below outlines materials of construction usually recommended for pumps handling liquids of known pH value.

pH Value	Material of Construction
10 to 14	Corrosion Resistant Alloys
8 to 10	All Iron
6 to 8	Bronze fitted or Standard fitted
4 to 6	All Bronze
0 to 4	Corrosion Resistant Alloy Steels

The pH value should only be used as a guide with weak aqueous solutions. For more corrosive solutions, temperature and chemical composition should be carefully evaluated in the selection of materials of construction.

Materials Selection Chart

This chart is intended as a guide in the selection of economical materials. It must be kept in mind that corrosion rates may vary widely with temperature, concentration, and the presence of trace elements or abrasive solids. Blank spaces indicate a lack of accurate corrosion information for those specific conditions.

C.I. — Cast Iron, ASTM A48.
D.I. — Ductile Iron, ASTM A536.
Steel — Carbon Steel, ASTM A216-WCA or WCB.
Brz. — Anti-Acid Bronze, Similar to ASTM B143A2.
316SS — Stainless Steel, ASTM A744 Gr. CF-8M, AISI 316.
GA-20 — Carpenter Stainless No. 20, ASTM A744 Gr. CN-7M.
CD4MCu — Stainless Steel, ACI CD-4MCu.
Mon — Monel Grade E, ASTM A744 Gr. M-35.
Ni — Nickel, ASTM A744 Gr. CZ-100.
H-B — Hastelloy Alloy-B, ASTM A494.
H-C — Hastelloy Alloy-C, ASTM A494.
Ti — Titanium Unalloyed, ASTM B367 Gr. C-1.
Zl — Zirconium

Code

A—Fully Satisfactory. C—Limited Use.
B—Useful Resistance. X—Unsuitable.

Corrosive	Steel C.I. D.I.	Brz.	316SS	GA-20	CD4MCu	Mon	Ni	H-B	H-C	Ti	Zi
Acetaldehyde, 70° F.	B	A	A	A	A	A	A		A	A	A
Acetic Acid, 70° F.	X	A	A	A	A	B	B	A	A	A	A
Acetic Acid, < 50%, To Boiling	X	B	A	A	B	B	B	C	A	A	A
Acetic Acid, > 50%, To Boiling	X	X	B	B	C	B	B	X	A	A	A
Acetone, To Boiling	A	A	A	A	A	A	A	A	A	A	A
Aluminum Chloride, < 10%, 70° F.	X	B	C	B	C	B	C	A		B	A
Aluminum Chloride, > 10%, 70° F.	X	X	C	B	C	C	X	A		B	A
Aluminum Chloride, < 10%, To Boiling	X	X	X	C	X	X	X	A		X	A
Aluminum Chloride, > 10%, To Boiling	X	X	X	X	X	X	X	A	X	X	A
Aluminum Sulphate, 70° F.	X	B	A	A	A	B	B	B	B	A	A
Aluminum Sulphate, < 10%, To Boiling	X	B	B	A	B	X	X	A	A	A	A
Aluminum Sulphate, > 10%, To Boiling	X	C	C	B	C	X	X	B	B	C	B
Ammonium Chloride, 70° F.	X	X	B	B	B	B	B		A	A	A
Ammonium Chloride, < 10%, To Boiling	X	X	B	B	C	B	B	B	A	A	A
Ammonium Chloride, > 10%, To Boiling	X	X	X	C	X	C	C		C	C	C
Ammonium Fluosilicate, 70° F.	X	X	C	B	C	X	X		C	X	X
Ammonium Sulphate, < 40%, To Boiling	X	X	B	B	C	B	B	X	B	A	A
Arsenic Acid, to 225° F.	X	X	C	B	C	X	X				
Barium Chloride, 70° F. < 30%	X	B	C	B	C	B	B	B	B	B	B
Barium Chloride, < 5%, To Boiling	X	B	C	B	C	B	B	B	A	A	A
Barium Chloride, > 5%, To Boiling	X	C	X	C	X	C	C	C	C	C	C
Barium Hydroxide, 70° F.	B	X	A	A	A	B	A	B	B	A	A
Barium Nitrate, To Boiling	C	X	B	B	B		B	B		B	B
Barium Sulphide, 70° F.	C	X	B	B	B	X	X		A	A	A
Benzoic Acid	X	C	B	B	B	B	B	A	A	A	A
Boric Acid, To Boiling	X	C	B	B	B	C	C	A	A	B	B
Boron Trichloride, 70° F. Dry	B	B	B	B	B	B	B	B	B		
Boron Trifluoride, 70° F. 10%, Dry	B	B	B	A	B	A	A		A		
Brine (acid), 70° F.	X	X	X	X	X				B	B	
Bromine (dry), 70° F.	X	X	X	X	X	X	C	B	B	X	X
Bromine (wet), 70° F.	X	X	X	X	X	X	C		B	X	X
Calcium Bisulphite, 70° F.	X	X	B	B	B	X	X		B	A	A
Calcium Bisulphite, To Hot	X	X	C	B	C	X	X		C	A	A
Calcium Chloride, 70° F.	B	C	B	B	B	B	B	A	A	A	A
Calcium Chloride, < 5%, To Boiling	C	C	B	B	B	A	A	A	A	A	A
Calcium Chloride, > 5%, To Boiling	X	C	C	B	C	C	C	A	A	B	B
Calcium Hydroxide, 70° F.	B	B	B	B	B	B	B		A	A	
Calcium Hydroxide, < 30%, To Boiling	C	B	B	B	B	B	B		A	A	
Calcium Hydroxide, > 30%, To Boiling	X	X	C	C	C	C	C		B	A	
Calcium Hypochlorite, < 2%, 70° F.	X	X	X	C	X	X	X		A	A	A
Calcium Hypochlorite, > 2%, 70° F.	X	X	X	C	X	X	X		B	A	B
Carbolic Acid, 70° F. (phenol)	C	B	A	A	A	A	A	A	A	A	A
Carbon Bisulphide, 70° F.	B	B	A	A	A	B	B		A		
Carbonic Acid, 70° F.	B	C	A	A	A	C	B	A	A	A	A
Carbon Tetrachloride, Dry to Boiling	B	B	A	A	A	A	A	B	B	A	A
Chloric Acid, 70° F.	X	X	X	B	C	X	X	X	C		
Chlorinated Water, 70° F.	C	C	B	B	B				A	A	A
Chloroacetic Acid, 70° F.	X		X	X						A	B
Chlorosulphonic Acid, 70° F.	X	X	X	C	X	X	X	A	A	B	X
Chromic Acid, < 30%	X	X	C	B	C	X	X		B	A	A
Citric Acid	X	C	A	A	A	C	C	A	A	A	A
Copper Nitrate, to 175° F.	X	X	B	B	B	X	X	X	X	B	
Copper Sulphate, To Boiling	X	C	C	B	C	X	X		A	A	A
Cresylic Acid	C	C	B	B	B	C	C	B	B		
Cupric Chloride	X	C	X	X	X	C	X		C	B	X
Cyanohydrin, 70° F.	C		B	B	B						

Corrosive	Steel C.I. D.I.	Brz.	316SS	GA-20	CD4MCu	Mon	NI	H-B	H-C	TI	ZI
Dichloroethane	C	B	B	B	B	C	B	B	B	A	B
Diethylene Glycol, 70° F.	A	B	A	A	A	B	B	B	B	A	A
Dinitrochlorobenzene, 70° F. (dry)	C	B	A	A	A	A	A	A	A	A	A
Ethanolamine, 70° F.	B	X	B	B	B	C	X			A	A
Ethers, 70° F.	B	B	B	A	A	B	B	B	B	A	A
Ethyl Alcohol, To Boiling	A	A	A	A	A	A	A	A	A	A	A
Ethyl Cellulose, 70° F.	A	B	B	B	B	B	B	B	B	A	A
Ethyl Chloride, 70° F.	C	B	B	A	B	B	B	B	B	A	A
Ethyl Mercaptan, 70° F.	C	X	B	A	B			B	B		
Ethyl Sulphate, 70° F.	C	B	B	A	B	B					
Ethylene Chlorohydrin, 70° F.	C	B	B	B	B	B	B	B	B	A	A
Ethylene Dichloride, 70° F.	C	B	B	B	B	B	B	B	C	A	A
Ethylene Glycol, 70° F.	B	B	B	B	B	B	B	A	A	A	A
Ethylene Oxide, 70° F.	C	X	B	B	B	B	B	A	A	A	A
Ferric Chloride, < 5%, 70° F.	X	X	X	X	X	X	X	X	A	A	B
Ferric Chloride, > 5%, 70° F.	X	X	X	X	X	X	X	X	B	B	X
Ferric Nitrate, 70° F.	X	X	B	A	B	X	X		B		
Ferric Sulphate, 70° F.	X	X	C	B	C	C	C		B	B	B
Ferrous Sulphate, 70° F.	X	C	C	B	C	C	C	B	B	A	A
Formaldehyde, To Boiling	B	B	A	A	A	B	B	B	B	A	A
Formic Acid, to 212° F.	X	C	X	A	B	C	C	A	A	C	A
Freon, 70° F.	A	A	A	A	A	A	A	A	A	A	A
Hydrochloric Acid, < 1%, 70° F.	X	X	C	B	C	B	B	B	A	B	A
Hydrochloric Acid, 1-20%, 70° F.	X	X	X	X	X	X	X	B	C	X	A
Hydrochloric Acid, > 20%, 70° F.	X	X	X	X		X	X	B	C	X	B
Hydrochloric Acid, < ½%, 175° F.	X	X	C	C	C	X	X	A	C	X	A
Hydrochloric Acid, ½-2%, 175° F.	X	X	X		X	X	X	B	C	X	A
Hydrocyanic Acid, 70° F.	X	X	C	B	C	C	C	C	C		
Hydrogen Peroxide, < 30% < 150° F.	C	X	B	B	B	B	B	B	B	A	A
Hydrofluoric Acid, < 20%, 70° F.	X	B	X	B	C	C	C	C	B	X	X
Hydrofluoric Acid, > 20%, 50° F.	X	C	X	C	X	C	C	C	B	X	X
Hydrofluoric Acid, To Boiling	X	X	X	X	X	C	X		C	X	X
Hydrofluorsilicic Acid, 70° F.	X		C	B	C				B		
Lactic Acid, < 50%, 70° F.	X	B	A	A	A	X	C	B	B	A	A
Lactic Acid, > 50%, 70° F.	X	B	B	B	B	C	C	B	B	A	A
Lactic Acid, < 5%, To Boiling	X	X	C	B	C	X	X	B	B	A	A
Lime Slurries, 70° F.	B	B	B	B	A	B	B	B	B	B	B
Magnesium Chloride, 70° F.	C	C	B	A	B	C	C	A	A	A	A
Magnesium Chloride, < 5%, To Boiling	X	C	C	B	C	C	C	A	A	A	A
Magnesium Chloride, > 5%, To Boiling	X	C	X	C	X	C	C	B	B	B	B
Magnesium Hydroxide, 70° F.	B	A	B	B	A	B	A	B	B	A	
Magnesium Sulphate	C	C	B	A	B	B	B	C	C	B	B
Maleic Acid	C	C	B	B	B	C	C	B	B	A	
Mercaptans	A	X	A	A	A	X	X				
Mercuric Chloride, < 2%, 70° F.	X	X	X	X	X	X	C		B	A	A
Mercurous Nitrate, 70° F.	C	X	B	B	B	C			C		
Methyl Alcohol, 70° F.	A	A	A	A	A	A	A	A	A	A	A
Naphthalene Sulphonic Acid, 70° F.	X	C	B	B	B	C	C	B	B		
Napthalenic Acid, To Hot	C	B	B	B	B	C	C	B	B		
Nickel Chloride, 70° F.	X	X	C	B	C	C	X	A		B	B
Nickel Sulphate	X	C	B	B	B	C	C		B		A
Nitric Acid	X	X	B	B	B	X	X			B	B
Nitrobenzene, 70° F.	A	C	A	A	A	B	B	B	B	A	
Nitroethane, 70° F.	A	A	A	A	A	A	A	A	A	A	A
Nitropropane, 70° F.	A	A	A	A	A	A	A	A	A	A	A
Nitrous Acid, 70° F.	X	X	X	C	X	X	X				
Nitrous Oxide, 70° F.	C	C	C	C	C	X	X		C		
Oleic Acid	C	C	B	B	B	C	C	C	C	C	C
Oleum, 70° F.	B	X	B	B	B	X	X	B	B	B	
Oxalic Acid	X	C	C	B	C	C	C	B	B	X	A
Palmitic Acid	B	B	B	A	B	B	B				
Phenol (see carbolic acid)											
Phosgene, 70° F.	C	C	B	B	B	C	C	B	B		
Phosphoric Acid, < 10%, 70° F.	X	C	A	A	A	C	C	A	A	A	A
Phosphoric Acid, > 10-70%, 70° F.	X	C	A	A	A	C	C	B	C	B	B
Phosphoric Acid, < 20%, 175° F.	X	C	B	B	B	C	C	A	A	C	B
Phosphoric Acid, > 20%, 175° F. < 85%	X	C	C	B	C	C	C	B	C	C	C
Phosphoric Acid, > 10%, Boil, < 85%	X	C	X	C	C	C	C	C	C	C	C
Phthalic Acid, 70° F.	C	B	B	A	B	B	B	B	B	A	A
Phthalic Anhydride, 70° F.	B	C	A	A	A	A	A	A	A		
Picric Acid, 70° F.	X	X	C	B	C	C	X		B		
Potassium Carbonate	B	B	A	A	A	B	B	B	B	A	A
Potassium Chlorate	B	C	A	A	A	C	C		B	A	A
Potassium Chloride, 70° F.	C	C	B	A	B	B	B	B	B	A	A
Potassium Cyanide, 70° F.	B	X	B	B	B	C	C	B	B		
Potassium Dichromate	B	B	A	A	A	B	B		B	A	A
Potassium Ferricyanide	C	B	B	B	B	B	B	B	B	A	A
Potassium Ferrocyanide, 70° F.	X	B	B	B	B	B	B	B	B		B
Potassium Hydroxide, 70° F.	C	C	B	A	B	A	A	B	C	B	A

Corrosive	Steel C.I. D.I.	Brz.	316SS	GA-20	CD4MCu	Mon	Ni	H–B	H–C	Ti	Zi
Potassium Hypochlorite	X	C	C	B	C	X	X		B	A	
Potassium Iodide, 70° F.	C	B	B	B	B	B	B	B	B	A	A
Potassium Permanganate	B	B	B	B	B	C	B		B		
Potassium Phosphate	C	C	B	B	B					B	B
Sea Water, 70° F.	C	B	B	A	B	A	A	A	A	A	A
Sodium Bisulphate, 70° F.	X	C	C	B	C	C	C	B	B	B	A
Sodium Bromide, 70° F.	B	C	B	B	B	B	B	B	B		
Sodium Carbonate	B	B	B	A	B	B	B	B	B	A	A
Sodium Chloride, 70° F.	C	B	B	B	B	A	A	B	B	A	A
Sodium Cyanide	B	X	B	B	B	X	X			B	
Sodium Dichromate	B	X	B	B	B					B	
Sodium Ethylate	B	A	A	A	A	A	A				
Sodium Fluoride	C	C	B	B	B	B	B	C	C	B	B
Sodium Hydroxide, 70° F.	B	B	B	A	B	A	A	A	A	A	A
Sodium Hypochlorite	X	X	C	C	C	X	X		B	A	B
Sodium Lactate, 70° F.	B	C	C	C	C	C		C	C		
Stannic Chloride, < 5%, 70° F.	X	C	X	C	X	C	C	B	B	A	A
Stannic Chloride, > 5%, 70° F.	X	X	X	X	X	X	X	B	C	B	B
Sulphite Liquors, To 175° F.	X	C	B	B	B	C	C		B	A	
Sulphur (molten)	B	X	A	A	A	C	C	C	A	A	
Sulphur Dioxide (spray), 70° F.	C	C	B	B	B	C	C		B	C	
Sulphuric Acid, < 2%, 70° F.	X	C	B	A	B	C	C	A	A	B	A
Sulphuric Acid, 2–40%, 70° F.	X	C	C	B	C	C	C	A	A	X	A
Sulphuric Acid, 40%, < 90%, 70° F.	X	X	X	B	X	X	X	A	A	X	C
Sulphuric Acid, 93–98%, 70° F.	B	X	B	B	B	X	X	B	B	X	C
Sulphuric Acid, < 10%, 175° F.	X	C	X	B	X	X	X	A	C	X	B
Sulphuric Acid, 10–60% & > 80%, 175° F.	X	X	X	B	X	X	X	B	C	X	C
Sulphuric Acid, 60–80%, 175° F.	X	X	X	X	X	X	X	B	C	X	C
Sulphuric Acid, < ¾%, Boiling	X	X	C	B	C	X	X	B	B	X	B
Sulphuric Acid, ¾–40%, Boiling	X	X	X	C	X	X	X	B	C	X	B
Sulphuric Acid, 40–65% & > 85%, Boil	X	X	X	X	X	X	X	X	X	X	X
Sulphuric Acid, 65–85%, Boiling	X	X	X	X	X	X	X	X	X	X	X
Sulphurous Acid, 70° F.	X	C	C	B	C	X	X	B	B	A	B
Titanium Tetrachloride, 70° F.	C		C	B	C	C			C		
Tirchlorethylene, To Boiling	B	C	B	B	B	B	B	B	B	A	A
Urea, 70° F.	C	C	B	B	B	C	C	C	C	B	B
Vinyl Acetate	B	B	B	B	B				B		
Vinyl Chloride	B	C	B	B	B	C	C	C	B	A	
Water, To Boiling	B	A	A	A	A	A	A	A	A	A	A
Zinc Chloride	C	C	B	A	B	B	B	B		A	A
Zinc Cyanide, 70° F.	X	B	B	B	B	B	B	B	B	B	B
Zinc Sulphate	X	C	A	A	A	C	C	C	C	A	

Piping Design

The design of a piping system can have an important effect on the successful operation of a centrifugal pump. Such items as sump design, suction piping design, suction and discharge pipe size, and pipe supports must all be carefully considered.

Selection of the discharge pipe size is primarily a matter of economics. The cost of the various pipe sizes must be compared to the pump size and power cost required to overcome the resulting friction head.

The suction piping size and design is far more important. Many centrifugal pump troubles are caused by poor suction conditions.

The suction pipe should never be smaller than the suction connection of the pump, and in most cases should be at least one size larger. Suction pipes should be as short and as straight as possible. Suction pipe velocities should be in the 5 to 8 feet per second range unless suction conditions are unusually good.

Higher velocities will increase the friction loss and can result in troublesome air or vapor separation. This is further complicated when elbows or tees are located adjacent to the pump suction nozzle, in that uneven flow patterns or vapor separation keeps the liquid from evenly filling the impeller. This upsets hydraulic balance leading to vibration, possible cavitation, and excessive shaft deflection. Shaft breakage or premature bearing failure may result.

On pump installations involving suction lift, air pockets in the suction line can be a source of trouble. The suction pipe should be exactly horizontal, or with a uniform slope upward from the sump to the pump as shown in Fig. 1. There should be no high spots where air can collect and cause the pump to lose its prime. Eccentric rather than concentric reducers should always be used.

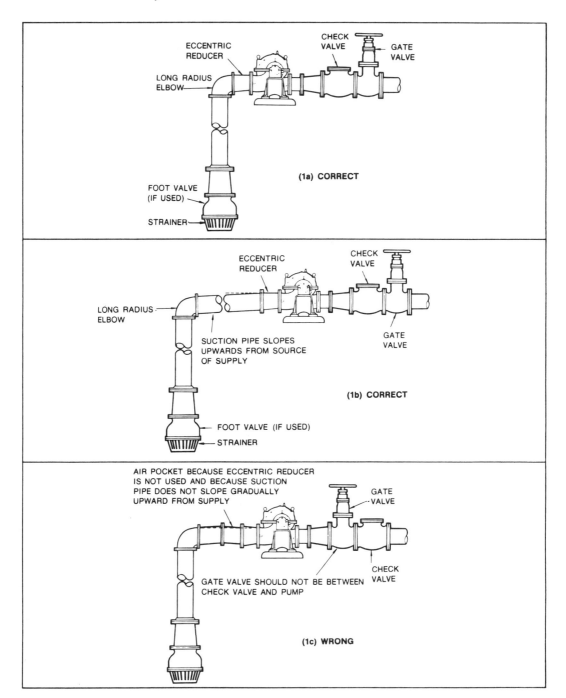

Fig. 1 Air Pockets in Suction Piping

If an elbow is required at the suction of a double suction pump, it should be in a vertical position if at all possible. Where it is necessary for some reason to use a horizontal elbow, it should be a long radius elbow and there should be a minimum of two diameters of straight pipe between the elbow and the pump as shown in Fig. 2. Fig. 3 shows the effect of an elbow directly on the suction. The liquid will flow toward the outside of the elbow and result in an uneven flow distribution into the two inlets of the double suction impeller. Noise and excessive axial thrust will result.

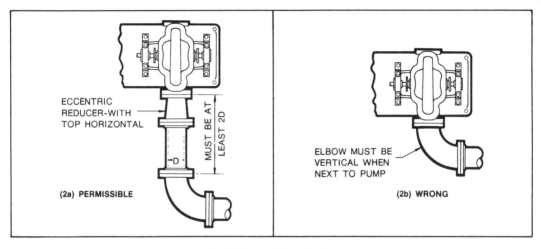

Fig. 2 Elbows At Pump Suction

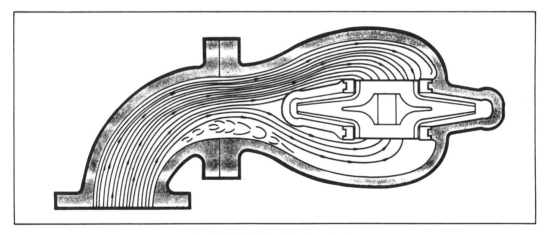

Fig. 3 Effect of Elbow Directly on Suction

There are several important considerations in the design of a suction supply tank or sump. It is imperative that the amount of turbulence and entrained air be kept to a minimum. Entrained air will cause reduced capacity and efficiency as well as vibration, noise, shaft breakage, loss of prime, and/or accelerated corrosion.

The free discharge of liquid above the surface of the supply tank at or near the pump suction can cause entrained air to enter the pump. All lines should be submerged in the tank, and baffles should be used in extreme cases as shown in Fig. 4.

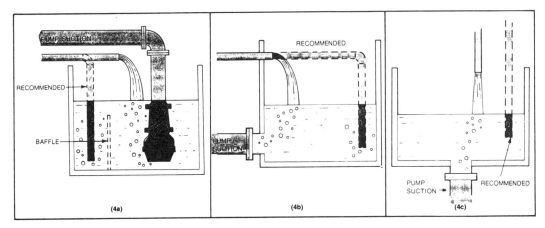

Fig. 4   Keeping Air Out of Pump

Improper submergence of the pump suction line can cause a vortex which is a swirling funnel of air from the surface directly into the pump suction pipe. In addition to submergence, the location of the pipe in the sump and the actual dimensions of the sump are also important in preventing vortexing and/or excess turbulence.

For horizontal pumps, Fig. 5 can be used as a guide for minimum submergence and sump dimensions for flows up to approximately 3000 gpm. Baffles can be used to help prevent vortexing in cases where it is impractical or impossible to maintain the required submergence. Fig. 6 shows three such baffling arrangements.

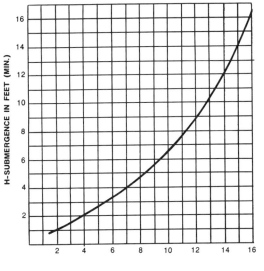

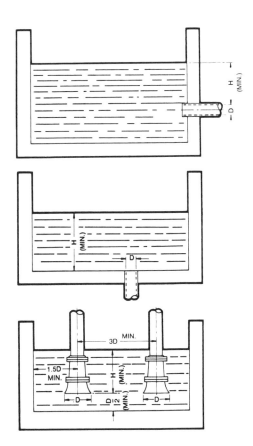

$$\text{VELOCITY IN FEET PER SEC.} = \frac{\text{QUAN. (G.P.M.) x .321}}{\text{AREA (inches)}^2} \text{ OR } \frac{\text{G.P.M. x.4085}}{D^2}$$

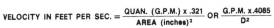

Fig. 5   Minimum Suction Pipe Submergence and Sump Dimensions

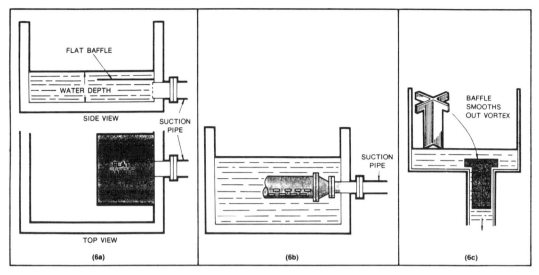

Fig. 6   Baffle Arrangements for Vortex Prevention

Large units (over 3000 gpm) taking their suction supply from sumps, especially vertical submerged pumps, require special attention. The larger the unit, the more important the sump design becomes.

Fig. 7 illustrates several preferred piping arrangements within a multiple pump pit. Note that the pipe should always be located near the back wall and should not be subjected to rapid changes in direction of the flow pattern. The velocity of the water in the area of the suction pipes should be kept below one foot per second to avoid air being drawn into the pump.

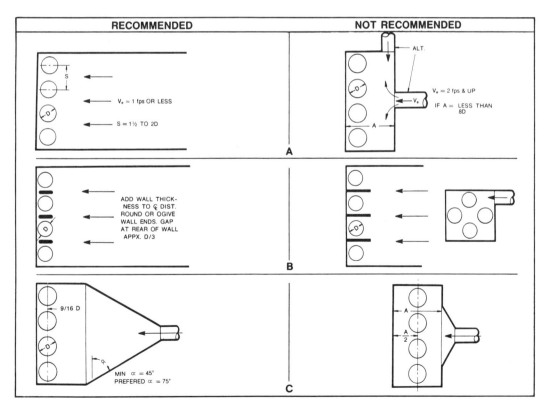

Fig. 7   Piping Arrangements Within Multiple Pump Pits

On horizontal pumps, a bell should be used on the end of the suction pipe to limit the entrance velocity to 3.5 feet per second. Also, a reducer at the pump suction flange to smoothly accelerate and stabilize the flow into the pump is desirable.

The submergence of the suction pipe must also be carefully considered. The amount of submergence required depends upon the size and capacity of the individual pumps as well as on the sump design. Past experience is the best guide for determining the submergence. The pump manufacturer should be consulted for recommendations in the absence of other reliable data.

Stuffing Box Sealing

The stuffing box of a pump provides an area in which to seal against leakage out of the pump along the shaft. Packing and mechanical seals are the two devices used to accomplish this seal.

Packing

A typical packed stuffing box arrangement is shown in Fig. 1. It consists of: A) Five rings of packing, B) A lantern ring used for the injection of a lubricating and/or flushing liquid, and C) A gland to hold the packing and maintain the desired compression for a proper seal.

The function of packing is to control leakage and not to eliminate it completely. The packing must be lubricated, and a flow of from 40 to 60 drops per minute out of the stuffing box must be maintained for proper lubrication.

The method of lubricating the packing depends on the nature of the liquid being pumped as well as on the pressure in the stuffing box. When the pump stuffing box pressure is above atmospheric pressure and the liquid is clean and nonabrasive, the pumped liquid itself will lubricate the packing (Fig. 2.) When the stuffing box pressure is below atmospheric pressure, a lantern ring is employed and lubrication is injected into the stuffing box (Fig. 3.) A bypass line from the pump discharge to the lantern ring connection is normally used providing the pumped liquid is clean.

When pumping slurries or abrasive liquids, it is necessary to inject a clean lubricating liquid from an external source into the lantern ring (Fig. 4.) A flow of from .2 to .5 gpm is desirable and a valve and flowmeter should be used for accurate control. The seal water pressure should be from 10 to 15 psi above the stuffing box pressure, and anything above this will only add to packing wear. The lantern ring is normally located in the center of the stuffing box. However, for extremely thick slurries like paper stock, it is recommended that the lantern ring be located at the stuffing box throat to prevent stock from contaminating the packing.

The gland shown in Figs. 1-4 is a quench type gland. Water, oil, or other fluids can be injected into the gland to remove heat from the shaft, thus limiting heat transfer to the bearing frame. This permits the operating temperature of the pump to be higher than the limits of the bearing and lubricant design. The same quench gland can be used to prevent the escape of a toxic or volatile liquid into the air around the pump. This is called a smothering gland, with an external liquid simply flushing away the undesirable leakage to a sewer or waste receiver.

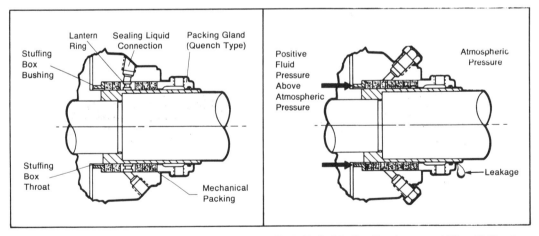

Fig. 1   Typical Stuffing Box Arrangement (Description of Parts)

Fig. 2   Typical Stuffing Box Arrangement When Stuffing Box Pressure is Above Atmospheric Pressure

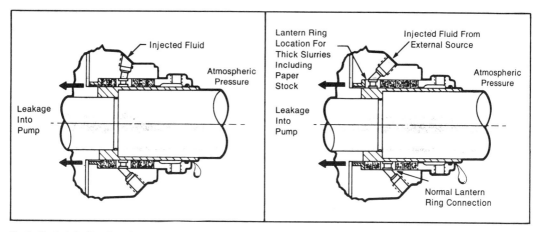

Fig. 3 Typical Stuffing Box Arrangement When Stuffing Box Pressure is Below Atmospheric Pressure

Fig. 4 Typical Stuffing Box Arrangement When Pumping Slurries

Mechanical Seals

The Basic Seal

A mechanical seal is a sealing device which forms a running seal between rotating and stationary parts. The design of liquid handling equipment with rotating parts today would include the consideration for the use of mechanical seals. Advantages over conventional packing are as follows:

1. Reduced friction and power losses.

2. Zero or limited leakage of product.

3. Elimination of shaft or sleeve wear.

4. Reduced maintenance.

5. Ability to seal higher pressures and more corrosive environments.

The wide variety of styles and designs together with extensive experience allows the use of seals on most pump applications.

A mechanical seal must seal at three points:

1. Static seal between the stationary part and the housing.

2. Static seal between the rotary part and the shaft.

3. Dynamic seal between the rotating seal face and the stationary seal face.

Fig. 5 shows a basic seal with these components:

1. Stationary seal part positioned in the housing with preload on the "O" ring to effect sealing and prevent rotation.

2. Rotating seal part positioned on the shaft by the "O" ring. The "O" ring seals between it and the shaft and provides resiliency.

3. The mating faces. The faces are precision lapped for a flatness of 3 light bands and a surface finish of 5 microinches.

4. Spring assembly, rotates with the shaft and provides pressure to keep the mating faces together during periods of shut down or lack of hydraulic pressure.

5. Driving member, positions the spring assembly and the rotating face. It also provides the positive drive between shaft and the other rotating parts.

As wear takes place between the mating faces, the rotating face must move along the shaft to maintain contact with stationary face. The "O" ring must be free to move.

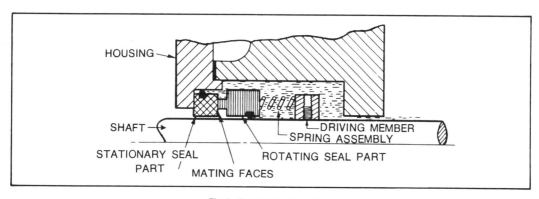

Fig. 5 Basic Mechanical Seal

A mechanical seal operates as each basic component performs its duty. Liquid pressure in the seal chamber forces the faces together and provides a thin film of lubricant between them. The faces, selected for low frictional qualities, are the only rubbing parts. These basic components are a part of every seal. The form, shape, style and design will vary greatly depending on service and manufacture. The basic theory, however, remains the same.

Types

Mechanical seals can be classified into the general types and arrangements shown below. Understanding these classes provides the first step in proper seal selection.

       (a) Single seals—
          Inside, outside, unbalanced, balanced
       (b) Double seals—
          Unbalanced or balanced

Single Seal, Inside Unbalanced

The single inside seal mounts on the shaft or sleeve within the stuffing box housing. The pumpage is in direct contact with all parts of the seal and provides the lubrication for the faces. The full force of pressure in the box acts on the faces providing good sealing to approximately 100 P.S.I.G. This is the most widely used type for services handling clear liquids. A circulation or by-pass line connected from the volute to the stuffing box provides continual flushing of the seal chamber.

Single Seals, Outside Unbalanced

This type mounts with the rotary part outside of the stuffing box. The springs and drive element are not in contact with the pumpage, thus reducing corrosion problems and preventing product accumulation in the springs. Pressures are limited to the spring rating, usually 35 P.S.I.G. Usually the same style seal can be mounted inside or outside. The outside seal is easier to install, adjust and maintain. A restricting bushing can be used to control leakage of an external sealing liquid into the pumpage.

Single Seals, Balanced

Balancing a seal varies the face loading exerted by the box pressure, thus extending the pressure limits of the seal. A balanced rotating part utilizes a stepped face and a sleeve. Balanced seals are used to pressures of 2000 P.S.I.G. Their use is also extensive on light hydrocarbons which tend to vaporize easily.

Balanced outside seals allow box pressure to be exerted toward the seal faces, thus allowing pressure ranges to above 150 P.S.I. as compared to the 35 P.S.I.G. limit for the unbalanced outside seal.

Double Seals

Double seals use two seals mounted back to back in the stuffing box. The stuffing box is pressurized with a clear liquid from an external source. This liquid is circulated thru the double seal chamber at ¼-1 GPM to cool and lubricate the mechanical seals. Double seals are used on solutions that contain solids, are toxic or extremely corrosive. The external source fluid should be compatible with the pumpage.

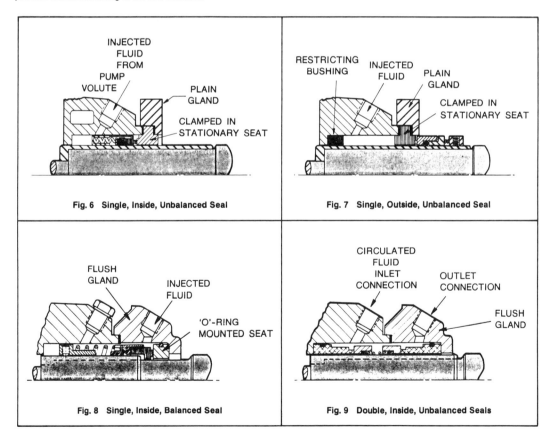

Fig. 6   Single, Inside, Unbalanced Seal

Fig. 7   Single, Outside, Unbalanced Seal

Fig. 8   Single, Inside, Balanced Seal

Fig. 9   Double, Inside, Unbalanced Seals

TANDEM SEALS

A variation of the double seal arrangement. The purpose of this seal is to provide a backup seal in the event the primary seal fails. The primary mechanical seal functions in a manner identical to that of a conventional single inside seal. The cavity between the primary seal and the backup seal is flooded with liquid to provide lubrication for the backup seal. The seal arrangement is used on toxic or hazardous chemical, and transfer and pipeline services to provide an extra measure of safety and allow equipment to operate until time to shut down.

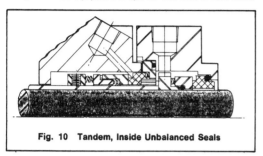

Fig. 10   Tandem, Inside Unbalanced Seals

Selection

The proper selection of a mechanical seal can be made only if the full operating conditions are known. These conditions are as follows:

1. Liquid
2. Pressure
3. Temperature
4. Characteristics of Liquid

1. Liquid   Identification of the exact liquid to be handled provides the first step in seal selection. The metal parts must be corrosion resistant. These parts, usually available in steel, bronze, stainless steel, or Hastelloy, provide a wide choice to meet specific services.

The mating faces must also resist both corrosion and wear Carbon, ceramic glass-filled Teflon, Stellite or tungsten carbide are available and offer both excellent wear properties and corrosion resistance.

Stationary sealing members of synthetic rubber, asbestos and Teflon complete the proper material selection.

2. Pressure   The proper type of seal, unbalanced or balanced, is based on the pressure on the seal and on the seal size. Fig. 11 shows the normal limits for unbalanced seals of various types.

3. Temperatures   The temperature will in part determine the use of the sealing members Synthetic rubbers are used to approximately 400 F., Teflon to 500 F. and asbestos to 750 F. Cooling the liquid in the seal chamber by water cooling jackets or cool liquid flushing, often extends seal life and allows wider selection of materials.

4. Characteristics of Liquid   Abrasive liquids create excessive wear and short seal life. Double seals or clear liquid flushing from an external source allows the use of mechanical seals on these difficult liquids. On light hydrocarbons balanced seals are often used to promote longer seal life, even though pressures are low.

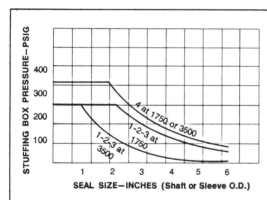

Maximum stuffing box pressures for Unbalanced Mechanical Seals on water solutions 160° max. Unbalanced seals generally limited to 200 PSIG maximum stuffing box pressure. All ratings based on one carbon face against hard face:

1—CERAMIC
2—NI-RESIST
3—STELLITE (not generally recommended on water services)
4—TUNGSTEN CARBIDE

Fig. 11   Pressure-Velocity Limits, Unbalanced Seals

Environmental Controls

Environmental controls are necessary for reliable performance of a mechanical seal on many applications. Pump manufacturers and the seal vendors offer a variety of arrangements to combat these problems.

1. Corrosion
2. Temperature Control
3. Dirty or incompatible environments

CORROSION

Corrosion can be controlled by selecting seal materials that are not attacked by the pumpage. When this is difficult, external fluid injection of a non-corrosive chemical to lubricate the seal is possible. Single or double seals could be used, depending on if the customer can stand delusion of his product.

TEMPERATURE CONTROL

As the seal rotates, the faces are in contact. This generates heat and if this heat is not removed, the temperature in the stuffing box can increase and cause sealing problems. A simple by-pass flush of the product over the seal faces will remove the heat generated by the seal (Figure 12). For higher temperature services, by-pass of product through a cooler may be required to cool the seal sufficiently (Figure 13). External cooling fluid injection can also be used.

Jacketed stuffing boxes are used on many pumps to cool the environment around the mechanical seal (Figure 6). This will also allow the use of a mechanical seal on services where it would not normally function (hot heat transfer oil). For other services, heat is provided to the jacket to melt or prevent the liquid from freezing (liquid sulfur).

DIRTY or INCOMPATIBLE ENVIRONMENTS

Mechanical seals do not normally function well on liquids which contain solids or can solidify on contact with the atmosphere. Here, by-pass flush through a filter, a cyclone separator or a strainer are methods of providing a clean fluid to lubricate the stuffing box.

Strainers are effective for particles larger than the openings on a 40 mesh screen.

Cyclone separators are effective on solids 10 micron or more in diameter, if they have a specific gravity of 2.7 and the pump develops a differential pressure of 30-40 psi. Filters are available to remove solids 2 micron and larger.

If external flush with clean liquid is available, this is the most fail proof system. Lip seal or restricting bushings are available to control flow of injected fluid to flows as low as 1/8 GPM (Figure 7).

Quench type glands are used on fluids which tend to crystalize on exposure to air. Water or steam is put through this gland to wash away any build up. Other systems are available as required by the service.

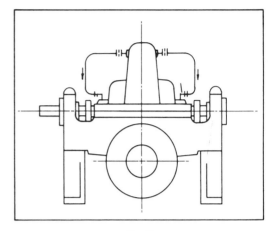

Fig. 12

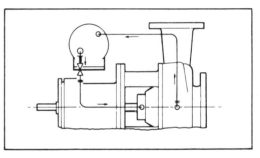

Fig. 13

Field Testing Methods

A. Determination of total head

The total head of a pump can be determined by gauge readings as illustrated in Fig. 1.

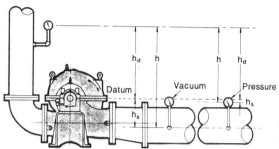

Fig. 1 Determination of Total Head From Gauge Readings

Negative Suction Pressure:

TDH = Discharge gauge reading converted to feet of liquid + vacuum gauge reading converted to feet of liquid + distance between point of attachment of vacuum gauge and the centerline of the discharge gauge, h, in feet $+ \left(\dfrac{Vd^2}{2g} - \dfrac{Vs^2}{2g} \right)$

Positive Suction Pressure:

or TDH=Discharge gauge reading converted to feet of liquid−pressure gauge reading in suction line converted to ft. of liquid + distance between center of discharge and suction gauges, h, in feet $+ \left(\dfrac{Vd^2}{2g} - \dfrac{Vs^2}{2g} \right)$

In using gauges when the pressure is positive or above atmospheric pressure, any air in the gauge line should be vented off by loosening the gauge until liquid appears. This assures that the entire gauge line is filled with liquid and thus the gauge will read the pressure at the elevation of the centerline of the gauge. However, the gauge line will be empty of liquid when measuring vacuum and the gauge will read the vacuum at the elevation of the point of attachment of the gauge line to the pipe line. These assumptions are reflected in the above definitions.

The final term in the above definitions accounts for a difference in size between the suction and discharge lines. The discharge line is normally smaller than the suction line and thus the discharge velocity is higher. A higher velocity results in a lower pressure since the sum of the pressure head and velocity head in any flowing liquid remains constant. Thus, when the suction and discharge line sizes at the gauge attachment points are different, the resulting difference in velocity head must be included in the total head calculation.

Manometers can also be used to measure pressure. The liquid used in a manometer is normally water or mercury, but any liquid of known specific gravity can be used. Manometers are extremely accurate for determining low pressures or vacuums and no calibration is needed. They are also easily fabricated in the field to suit any particular application. Figs. 2 & 3 illustrate typical manometer set ups.

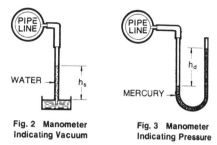

Fig. 2 Manometer Indicating Vacuum   **Fig. 3 Manometer Indicating Pressure**

B. Measurement of capacity

a.) Magnetic Flow Meter

A calibrated magnetic flow meter is an accurate means of measuring flow in a pumping system. However, due to the expense involved, magnetic flow meters are only practical in small factory test loops and in certain process pumping systems where flow is critical.

b.) Volumetric measurement

Pump capacity can be determined by weighing the liquid pumped or measuring its volume in a calibrated vessel. This is often practical when pumping into an accurately measured reservoir or tank, or when it is possible to use small containers which can be accurately weighed. These methods, however, are normally suited only to relatively small capacity systems.

c.) Venturi meter

A venturi meter consists of a converging section, a short constricting throat section and then a diverging section. The object is to accelerate the fluid and temporarily lower its static pressure. The flow is then a function of the pressure differential between the full diameter line and the throat. Fig. 4 shows the general shape and flow equation. The meter coefficient is determined by actual calibration by the manufacturer and when properly installed the Venturi meter is accurate to within plus or minus 1%.

$$Q(GPM) = 5.67 \ C D_2^2 \ \sqrt{\frac{H}{1 - R^4}}$$

C = Instrument Coefficient
D_1 = Entrance Diameter in Inches
D_2 = Throat Diameter in Inches
$R = D_2/D_1$
H = Differential Head in Inches = $h_1 - h_2$

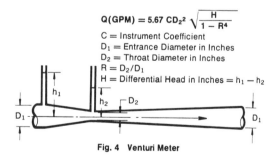

Fig. 4 Venturi Meter

d.) Nozzle

A nozzle is simply the converging portion of a venturi tube with the liquid exiting to the atmosphere. Therefore, the same formula can be used with the differential head equal to the gauge reading ahead of the nozzle. Fig. 5 lists theoretical nozzle discharge flows.

Theoretical Discharge of Nozzles in U.S. GPM

Head		Veloc'y of Disch. Feet per Sec.	Diameter of Nozzle in Inches												
Lbs.	Feet		1/16	1/8	3/16	1/4	3/8	1/2	5/8	3/4	7/8	1	1 1/8	1 1/4	1 3/8
10	23.1	38.6	0.37	1.48	3.32	5.91	13.3	23.6	36.9	53.1	72.4	94.5	120	148	179
15	34.6	47.25	0.45	1.81	4.06	7.24	16.3	28.9	45.2	65.0	88.5	116.	147	181	219
20	46.2	54.55	0.52	2.09	4.69	8.35	18.8	33.4	52.2	75.1	102.	134.	169	209	253
25	57.7	61.0	0.58	2.34	5.25	9.34	21.0	37.3	58.3	84.0	114.	149.	189	234	283
30	69.3	66.85	0.64	2.56	5.75	10.2	23.0	40.9	63.9	92.0	125.	164.	207	256	309
35	80.8	72.2	0.69	2.77	6.21	11.1	24.8	44.2	69.0	99.5	135.	177.	224	277	334
40	92.4	77.2	0.74	2.96	6.64	11.8	26.6	47.3	73.8	106.	145.	189.	239	296	357
45	103.9	81.8	0.78	3.13	7.03	12.5	28.2	50.1	78.2	113.	153.	200.	253	313	379
50	115.5	86.25	0.83	3.30	7.41	13.2	29.7	52.8	82.5	119.	162.	211.	267	330	399
55	127.0	90.4	0.87	3.46	7.77	13.8	31.1	55.3	86.4	125.	169.	221.	280	346	418
60	138.6	94.5	0.90	3.62	8.12	14.5	32.5	57.8	90.4	130.	177.	231.	293	362	438
65	150.1	98.3	0.94	3.77	8.45	15.1	33.8	60.2	94.0	136.	184.	241.	305	376	455
70	161.7	102.1	0.98	3.91	8.78	15.7	35.2	62.5	97.7	141.	191.	250.	317	391	473
75	173.2	105.7	1.01	4.05	9.08	16.2	36.4	64.7	101.	146.	198.	259.	327	404	489
80	184.8	109.1	1.05	4.18	9.39	16.7	37.6	66.8	104.	150.	205.	267.	338	418	505
85	196.3	112.5	1.08	4.31	9.67	17.3	38.8	68.9	108.	155.	211.	276.	349	431	521
90	207.9	115.8	1.11	4.43	9.95	17.7	39.9	70.8	111.	160.	217.	284.	359	443	536
95	219.4	119.0	1.14	4.56	10.2	18.2	41.0	72.8	114.	164.	223.	292.	369	456	551
100	230.9	122.0	1.17	4.67	10.5	18.7	42.1	74.7	117.	168.	229.	299.	378	467	565
105	242.4	125.0	1.20	4.79	10.8	19.2	43.1	76.5	120.	172.	234.	306.	388	479	579
110	254.0	128.0	1.23	4.90	11.0	19.6	44.1	78.4	122.	176.	240.	314.	397	490	593
115	265.5	130.9	1.25	5.01	11.2	20.0	45.1	80.1	125.	180.	245.	320.	406	501	606
120	277.1	133.7	1.28	5.12	11.5	20.5	46.0	81.8	128.	184.	251.	327.	414	512	619
125	288.6	136.4	1.31	5.22	11.7	20.9	47.0	83.5	130.	188.	256.	334.	423	522	632
130	300.2	139.1	1.33	5.33	12.0	21.3	48.0	85.2	133.	192.	261.	341.	432	533	645
135	311.7	141.8	1.36	5.43	12.2	21.7	48.9	86.7	136.	195.	266.	347.	439	543	656
140	323.3	144.3	1.38	5.53	12.4	22.1	49.8	88.4	138.	199.	271.	354.	448	553	668
145	334.8	146.9	1.41	5.62	12.6	22.5	50.6	89.9	140.	202.	275.	360.	455	562	680
150	346.4	149.5	1.43	5.72	12.9	22.9	51.5	91.5	143.	206.	280.	366.	463	572	692
175	404.1	161.4	1.55	6.18	13.9	24.7	55.6	98.8	154.	222.	302.	395.	500	618	747
200	461.9	172.6	1.65	6.61	14.8	26.4	59.5	106.	165.	238.	323.	423.	535	660	799
250	577.4	193.0	1.85	7.39	16.6	29.6	66.5	118.	185.	266.	362.	473.	598	739	894
300	692.8	211.2	2.02	8.08	18.2	32.4	72.8	129.	202.	291.	396.	517.	655	808	977

			1 1/2	1 3/4	2	2 1/4	2 1/2	2 3/4	3	3 1/2	4	4 1/2	5	5 1/2	6
10	23.1	38.6	213	289	378	479	591	714	851	1158	1510	1915	2365	2855	3405
15	34.6	47.25	260	354	463	585	723	874	1041	1418	1850	2345	2890	3490	4165
20	46.2	54.55	301	409	535	676	835	1009	1203	1638	2135	2710	3340	4040	4810
25	57.7	61.0	336	458	598	756	934	1128	1345	1830	2385	3025	3730	4510	5380
30	69.3	66.85	368	501	655	828	1023	1236	1473	2005	2615	3315	4090	4940	5895
35	80.8	72.2	398	541	708	895	1106	1335	1591	2168	2825	3580	4415	5340	6370
40	92.4	77.2	425	578	756	957	1182	1428	1701	2315	3020	3830	4725	5710	6810
45	103.9	81.8	451	613	801	1015	1252	1512	1802	2455	3200	4055	5000	6050	7210
50	115.5	86.25	475	647	845	1070	1320	1595	1900	2590	3375	4275	5280	6380	7600
55	127.0	90.4	498	678	886	1121	1385	1671	1991	2710	3540	4480	5530	6690	7970
60	138.6	94.5	521	708	926	1172	1447	1748	2085	2835	3700	4685	5790	6980	8330
65	150.1	98.3	542	737	964	1220	1506	1819	2165	2950	3850	4875	6020	7270	8670
70	161.7	102.1	563	765	1001	1267	1565	1888	2250	3065	4000	5060	6250	7560	9000
75	173.2	105.7	582	792	1037	1310	1619	1955	2330	3170	4135	5240	6475	7820	9320
80	184.8	109.1	602	818	1070	1354	1672	2020	2405	3280	4270	5410	6690	8080	9630
85	196.3	112.5	620	844	1103	1395	1723	2080	2480	3375	4400	5575	6890	8320	9920
90	207.9	115.8	638	868	1136	1436	1773	2140	2550	3475	4530	5740	7090	8560	10210
95	219.4	119.0	656	892	1168	1476	1824	2200	2625	3570	4655	5900	7290	8800	10500
100	230.9	122.0	672	915	1196	1512	1870	2255	2690	3660	4775	6050	7470	9030	10770
105	242.4	125.0	689	937	1226	1550	1916	2312	2755	3750	4890	6200	7650	9250	11020
110	254.0	128.0	705	960	1255	1588	1961	2366	2820	3840	5010	6350	7840	9470	11300
115	265.5	130.9	720	980	1282	1621	2005	2420	2885	3930	5120	6490	8010	9680	11550
120	277.1	133.7	736	1002	1310	1659	2050	2470	2945	4015	5225	6630	8180	9900	11800
125	288.6	136.4	751	1022	1338	1690	2090	2520	3005	4090	5340	6760	8350	10100	12030
130	300.2	139.1	767	1043	1365	1726	2132	2575	3070	4175	5450	6900	8530	10300	12290
135	311.7	141.8	780	1063	1390	1759	2173	2620	3125	4250	5550	7030	8680	10490	12510
140	323.3	144.3	795	1082	1415	1790	2212	2670	3180	4330	5650	7160	8850	10690	12730
145	334.8	146.9	809	1100	1440	1820	2250	2715	3235	4410	5740	7280	8990	10880	12960
150	346.4	149.5	824	1120	1466	1853	2290	2760	3295	4485	5850	7410	9150	11070	13200
175	404.1	161.4	890	1210	1582	2000	2473	2985	3560	4840	6310	8000	9890	11940	14250
200	461.9	172.6	950	1294	1691	2140	2645	3190	3800	5175	6750	8550	10580	12770	15220
250	577.4	193.0	1063	1447	1891	2392	2955	3570	4250	5795	7550	9570	11820	14290	17020
300	692.8	211.2	1163	1582	2070	2615	3235	3900	4650	6330	8260	10480	12940	15620	18610

Note:—The actual quantities will vary from these figures, the amount of variation depending upon the shape of nozzle and size of pipe at the point where the pressure is determined. With smooth taper nozzles the actual discharge is about 94% of the figures given in the tables.

Fig. 5

e.) Orifice

An orifice is a thin plate containing an opening of specific shape and dimensions. The plate is installed in a pipe and the flow is a function of the pressure upstream of the orifice. There are numerous types of orifices available and their descriptions and applications are covered in the Hydraulic Institute Standards and the ASME Fluid Meters Report. Orifices are not recommended for permanent installations due to the inherent high head loss across the plate.

f.) Weir

A weir is particularly well suited to measuring flows in open conduits and can be adapted to extremely large capacity systems. For best accuracy, a weir should be calibrated in place. However, when this is impractical, there are formulas which can be used for the various weir configurations. The most common types are the rectangular contracted weir and the 90° V-notch weir. These are shown in Fig. 6 with the applicable flow formulas.

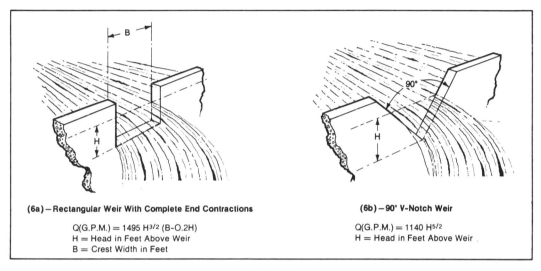

(6a) — Rectangular Weir With Complete End Contractions

$Q(G.P.M.) = 1495\ H^{3/2}\ (B-0.2H)$
H = Head in Feet Above Weir
B = Crest Width in Feet

(6b) — 90° V-Notch Weir

$Q(G.P.M.) = 1140\ H^{5/2}$
H = Head in Feet Above Weir

Fig. 6 Weirs

g.) Pitot tube

A pitot tube measures fluid velocity. A small tube placed in the flow stream gives two pressure readings; one receiving the full impact of the flowing stream reads static head + velocity head, and the other reads the static head only (Fig. 7). The difference between the two readings is the velocity head. The velocity and the flow are then determined from the following well known formulas.

$V = C\ \sqrt{2gh_v}$ where C is a coefficient for the meter determined by calibration, and h_v = velocity head,

Capacity = Area x Average Velocity

Since the velocity varies across the pipe, it is necessary to obtain a velocity profile to determine the average velocity. This involves some error, but when properly applied a calibrated pitot tube is within plus or minus 2% accuracy.

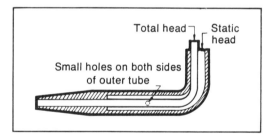

Fig. 7 Pitot Tube

Vertical Turbine Pumps

Turbine Nomenclature

1. **DATUM OR GRADE** — The elevation of the surface from which the pump is supported.

2. **STATIC LIQUID LEVEL** — The vertical distance from grade to the liquid level when no liquid is being drawn from the well or source.

3. **DRAWDOWN** — The distance between the static liquid level and the liquid level when pumping at required capacity.

4. **PUMPING LIQUID LEVEL** — The vertical distance from grade to liquid level when pumping at rated capacity. Pumping liquid level equals static water level plus drawdown.

5. **SETTING** — The distance from grade to the top of the pump bowl assembly.

6. **TPL (TOTAL PUMP LENGTH)** — The distance from grade to lowest point of pump.

7. **RATED PUMP HEAD** — Lift below discharge plus head above discharge plus friction losses in discharge line. This is the head for which the customer is responsible and does not include any losses within the pump.

8. **COLUMN AND DISCHARGE HEAD FRICTION LOSS** — Head loss in the pump due to friction in the column assembly and discharge head. Friction loss is measured in feet and is dependent upon column size, shaft size, setting, and discharge head size. Values given in appropriate charts in Data Section.

9. **BOWL HEAD** — Total head which the pump bowl assembly will deliver at the rated capacity. This is curve performance.

10. **BOWL EFFICIENCY** ——— The efficiency of the bowl unit only. This value is read directly from the performance curve.

11. **BOWL HORSEPOWER** ——— The horsepower required by the bowls only to deliver a specified capacity against bowl head.

$$\text{BOWL HP} = \frac{\text{Bowl Head} \times \text{Capacity}}{3960 \times \text{Bowl Efficiency}}$$

12. **TOTAL PUMP HEAD** — Rated pump head plus column and discharge head loss.
Note: This is new or final bowl head.

13. **SHAFT FRICTION LOSS** — The horsepower required to turn the lineshaft in the bearings. These values are given in appropriate table in Data Section.

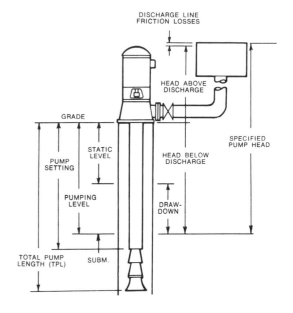

14. **PUMP BRAKE HORSEPOWER** — Sum of bowl horsepower plus shaft loss (and the driver thrust bearing loss under certain conditions).

15. **TOTAL PUMP EFFICIENCY (WATER TO WATER)** — The efficiency of the complete pump less the driver, with all pump losses taken into account.

$$\text{Efficiency} = \frac{\text{Specified Pump Head} \times \text{Capacity}}{3960 \times \text{Brake Horsepower}}$$

16. **OVERALL EFFICIENCY (WIRE TO WATER)** — The efficiency of the pump and motor complete. Overall efficiency = total pump efficiency × motor efficiency.

17. **SUBMERGENCE** — Distance from liquid level to suction bell.

Vertical Turbine Pumps
Calculating Axial Thrust

Under normal circumstances Vertical Turbine Pumps have a thrust load acting parallel to the pump shaft. This load is due to unbalanced discharge pressure, dead weight and liquid direction change. Optimum selection of the motor bearing and correct determination of required bowl lateral for deep setting pumps require accurate knowledge of both the magnitude and direction (usually down) of the resultant of these forces. In addition, but with a less significant role, thrust influences shaft H.P. rating and shaft critical speeds.

IMPELLER THRUST
Impeller Thrust in the downward direction is due to the unbalanced discharge pressure across the eye area of the impeller. See diagram A.

Counteracting this load is an upward force primarily due to the change in direction of the liquid passing through the impeller. The resultant of these two forces constitutes impeller thrust. Calculating this thrust using a thrust constant (K) will often produce only an approximate thrust value because a single constant cannot express the upthrust component which varies with capacity.

To accurately determine impeller thrust, thrust-capacity curves based on actual tests are required. Such curves now exist for the "A" Line. To determine thrust, the thrust factor "K" is read from the thrust-capacity curve at the required capacity and given RPM. "K" is then multiplied by the Total Pump Head (Final Lab Head) times Specific Gravity of the pumped liquid.

If impeller thrust is excessively high, the impeller can usually be hydraulically balanced. This reduces the value of "K". Balancing is achieved by reducing the discharge pressure above the impeller eye by use of balancing holes and rings. See diagram B.

Although hydraulic balancing reduces impeller thrust, it also decreases efficiency by 1 to 5 points by providing an additional path for liquid recirculation.

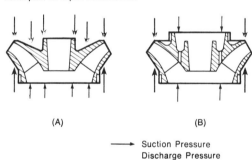

(A) (B)

$\longrightarrow$ Suction Pressure
Discharge Pressure

NOTE:
Although hydraulic balancing reduces impeller thrust, it also decreases efficiency by one to five points by providing an additional path for liquid recirculation. Of even greater concern is that should the hydraulic balancing holes become clogged, (unclean fluids, fluids with solid content, intermittent services, etc.), the impeller thrust will increase and possibly cause the driver to fail. Hydraulically balanced impellers cannot be used in applications requiring rubber bowl bearings because the flutes on the inside diameter of the bearings provide an additional path to the top side of the impeller, thus creating an additional down thrust.

Hydraulically balanced impellers should be used as a "last resort" for those situations where the pump thrust exceeds the motor thrust bearing capabilities.

DEAD WEIGHT
In addition to the impeller force, dead weight (shaft plus impeller weight less the weight of the liquid displaced) acts downward. On pumps with settings less than 50 feet, dead weight may be neglected on all but the most critical applications as it represents only a small part of the total force. On deeper setting pumps, dead weight becomes significant and must be taken into account.

NOTE:
We normally only take shaft weight into consideration as dead weight, the reason being that impeller weight less its liquid displacement weight is usually a small part of the total.

SHAFT SLEEVES
Finally, there can be an upward force across a head shaft sleeve or mechanical seal sleeve. In the case of can pumps with suction pressure there can be an additional upward force across the impeller shaft area. Again for most applications, these forces are small and can be neglected; however, when there is a danger of upthrusts or when there is high discharge pressure (above 600 psi) or high suction pressure (above 400 psi) these forces should be considered.

MOTOR BEARING SIZING
Generally speaking a motor for a normal thrust application has as standard, a bearing adequate for shutoff thrust. When practical, motor bearings rated for shutoff conditions are preferred.

For high thrust applications (when shutoff thrust exceeds the standard motor bearing rating) the motor bearing may be sized for the maximum anticipated operating range of the pump.

Should the pump operate to the left of this range for a short period of time, anti-fraction bearings such as angular contact or spherical roller can handle the overload. It should be remembered, however, that bearing life is approximately inversely proportional to the cube of the load. Should the load double, motor bearing life will be cut to $\frac{1}{8}$ of its original value. Although down thrust overloading is possible, the pump must never be allowed to operate in a continuous up thrust condition even for a short interval without a special motor bearing equipped to handle it. Such upthrust will fail the motor bearing.

CALCULATING MOTOR BEARING LOAD
As previously stated, for short setting non-hydraulic balanced pumps below 50 feet with discharge pressures below 600 psi and can pumps with suction pressures below 100 psi, only impeller thrust need be considered.

Under these conditions:	Where:
Motor Bearing Load (lbs)	Impeller Thrust (lbs)
$T_{imp} = KH_L \times SG$	K = Thrust factors (lbs./ft.)
	H_L = Lab Head (ft.)
	SG = Specific Gravity

For more demanding applications, the forces which should be considered are impeller thrust plus dead weight minus any sleeve or shaft area force.

In equation form:

Motor Bearing Load = T_{imp} + Wt[1]—sleeve force[2]—shaft area force[3] = T_t

(1) Wt. = Shaft Dead Wt. x Setting In Ft.
(2) Sleeve Force = Sleeve area x Discharge pressure
(3) Shaft Area Force = Shaft area x Suction pressure
*Oil Lube shaft does not displace liquid above the pumping water level and therefore has a greater net weight.

CALCULATING AXIAL THRUST — CONTINUED

| Shaft Dia (in) | Shaft Dead Wt. (lbs/ft.) | | Shaft Area (in²) | Sleeve Area (in) |
	Open Lineshaft	Closed Lineshaft		
1	2.3	2.6	.78	1.0
1³⁄₁₆	3.3	3.8	1.1	1.1
1½	5.3	6.0	1.8	1.1
1¹¹⁄₁₆	6.7	7.6	2.2	1.5
1¹⁵⁄₁₆	8.8	10.0	2.9	1.8
2³⁄₁₆	11.2	12.8	3.7	2.0

THRUST BEARING LOSS

Thrust bearing loss is the loss of horsepower delivered to the pump at the thrust bearings due to thrust. In equation form:

$$L_{TB} = .0075 \left(\frac{BHP}{100} \right) \left(\frac{T_t}{1000} \right)$$

where:

L_{TB} = Thrust bearing loss (HP)
BHP = Brake horsepower
T_t = Motor Bearing Load (Lbs.)
 = T_{imp} + Wt[1]—sleeve force[2]—shaft area force[3]

Vertical Turbine Bearing Material Data

Material Description	Temp. and S.G. Limits	Remarks
1. Bronze-SAE 660 (Standard) #1104 ASTM-B-584-932	-50 to 250°F Min. S.G. of 0.6	General purpose material for non-abrasive, neutral pH service. 7% Tin/7% Lead/3% Zinc/83% Cu.
2. Bronze-SAE 64 (Zincless) #1107 ASTM-B-584-937	-50 to 180°F Min. S.G. of 0.6	Similar to std. bronze. Used for salt water services. 10% Tin/ 10% Lead/80% Cu.
3. Carbon Graphite Impregnated with Babbitt	-450 to 300°F All Gravities	Corrosion resistant material not suitable for abrasive services. Special materials available for severe acid services and for temp. as high as 650°. Good for low specific gravity fluids because the carbon is self-lubricating.
4. Teflon 25% Graphite with 75% Teflon	-50 to 250°F All Gravities	Corrosion resistant except for highly oxidizing solutions. Not suitable for abrasive services. Glass filled Teflon also available.
5. Cast Iron ASTM-A-48 CL30 Flash Chrome Coated	32 to 180°F Min. S.G. of 0.6	Used on non-abrasive caustic services and some oil products. Avoid water services as bearings can rust to shaft when idle. Test with bronze Bearings.
6. Lead Babbitt	32 to 300°F	Excellent corrosion resistance to a pH of 2. Good in mildly abrasive services. 80% Lead/3% Tin/17% Antimony.
7. Rubber w/Phenolic backing (Nitrile Butadiene or Neoprene)	32 to 150°F	Use in abrasive water services. Bearings must be wet prior to start-up for TPL 50'. Do not use: For oily services, for stuffing box bushing, or with hydraulically balanced impeller. For services that are corrosive, backing material other than Phenolic must be specified.
8. Hardened Metals: Sprayed on stainless steel shell (Tungsten Carbide)	All Temperatures All Gravities	Expensive alternate for abrasive services. Hardfaced surfaces typically in the range of Rc72. Other coatings are chromium oxide, tungsten carbide, colmonoy, etc. Consult factory for pricing and specific recommendation.

APPENDIX 7B

Change of Performance*

Different industries with many different processes will have requirements for the same pump to operate at different capacities and different heads, and to have a different shape of the head-capacity curves. To ideally satisfy these requirements, one should have a variable-speed pump with adjustable vanes in the impellers. But because most of the drivers in the process industries operate at constant speed, and because the adjustable vanes cannot be produced economically, variable pump performance must be achieved by mechanical means without sacrificing efficiency.

In order to provide this flexibility at minimum cost, studies were made to change pump performance within a given pump casing. This can be accomplished by varying the impeller design, cutting impellers, changing the running speed, modifying the impeller vane tips, filing the volute cutwater tip, or orificing the pump discharge.

Pump users would prefer to use the same casing for a wide variation of pump performance. The pump casing is usually the most costly part of the pump. To replace a pump casing means extensive and costly work on base plate and piping.

The prediction of pump performance by modifying parts other than the casing is based largely on experimentation. Many tests have been conducted by the various pump companies in such areas as:

1. Trimming the pump impellers
2. Removing metal from the tips of impeller vanes at the impeller periphery
3. Removing metal from the volute tip in the pump casing
4. Providing impeller vanes of the same angularity, but different width
5. Providing impellers with different numbers of vanes and different discharge angles
6. Orificing the pump discharge in the pump casing

We will consider each of these means, but before we do so, we should review the so-called laws of affinity relating to centrifugal pumps. These are theoretical laws or rules that apply to the change in performance of a centrifugal pump by a change in the speed of rotation or a change in the impeller diameter of a particular pump. It should always be remembered in using these laws of affinity that they are theoretical and do not always give exact results as compared with tests. However, they are a good guide for predicting the hydraulic performance characteristic of a pump from a known characteristic. A performance change can be obtained by either altering the speed of rotation or the outside diameter of the impeller.

---

*Source: Goulds Pumps, Inc., Seneca Falls, NY. Adapted by permission.

I.   Constant impeller diameter
     A.   The capacity varies directly as the speed

$$\frac{GPM_1}{GPM_2} = \frac{RPM_1}{RPM_2}$$

B.   The head varies as the square of the speed

$$\frac{Head_1}{Head_2} = \left[\frac{RPM_1}{RPM_2} \right]^2$$

C.   The horsepower varies as the cube of the speed

$$\frac{BHP_1}{BHP_2} = \left[\frac{RPM_1}{RPM_2} \right]^3$$

II. Constant speed
     A.   The capacity varies directly with the impeller diameter.
     B.   The head varies as the square of the impeller diameter.
     C.   The horsepower varies as the cube of the impeller diameter.

These relationships can be expressed in a simple formula:

$$\frac{Impeller\ diameter_1}{Impeller\ diameter_2} = \frac{GPM_1}{GPM_2} = \left[\frac{Head_1}{Head_2} \right]^2 = \left[\frac{BHP_1}{BHP_2} \right]^3$$

IMPELLER CUTS

Assuming that the impeller represents a standard design and that the impeller profile is typically of average layout and not specifically designed for high NPSH, pump performance with trimmed impellers will follow the affinity laws as some vane overlap is maintained. To compensate for casting and mechanical imperfections, correction factors are normally applied to the impeller cuts (Figure 7B-1). The efficiency of the cut impellers (within a 25 percent cut) will usually drop about two points at the maximum cut. On high specific speed pumps, the performance of the cut impellers should be determined by shop tests.

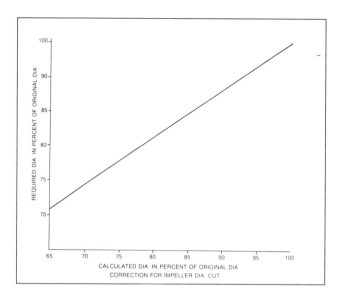

FIGURE 7B–1

REMOVING METAL FROM VANE TIPS

The pump performance can be changed by removing metal from the vane tips at the impeller periphery. Removing metal from the underside of the vane is known as *underfiling*. Removing metal from the working side of the vane is known as *overfiling*. The effect of overfiling on pump performance is very difficult to predict and to duplicate. This is because filing vanes by hand on the working side changes the discharge angle of the impeller, and nonuniformity exists between each vane.

Underfiling, however, is more consistent, more predictable, and easier to apply. Underfiling is most effective at peak efficiency and to the right of peak efficiency. Also, underfiling will be more effective where vanes are thick and specific speeds are high. Underfiling increases the area at the impeller discharge, thereby increasing the capacity at peak. This increase is directly proportional to the increase in an area due to filing, or it can be said to equal dimension ''A'' over ''B'' in Figure 7B-2. The head rating will move to the right of peak efficiency in a straight line toward the new capacity.

With underfiling, the shutoff head does not change; therefore, the change of the performance by impeller underfiling is less effective to the left of peak efficiency.

REMOVING METAL FROM THE VOLUTE TIP

Capacity increase in a given pump can also be achieved by trimming of the volute tip in the pump casing. This is illustrated in Figure 7B–3. Removing metal at this point increases the total volute area. The peak efficiency and peak capacity will move to the right as the square root ratio of the new area divided by the original area. Pump peak efficiency will normally drop one or two points.

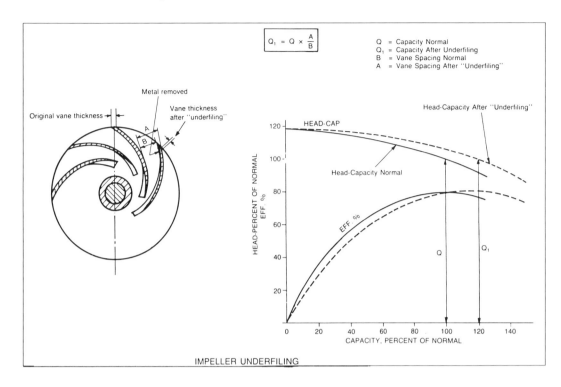

IMPELLER UNDERFILING

FIGURE 7B-2

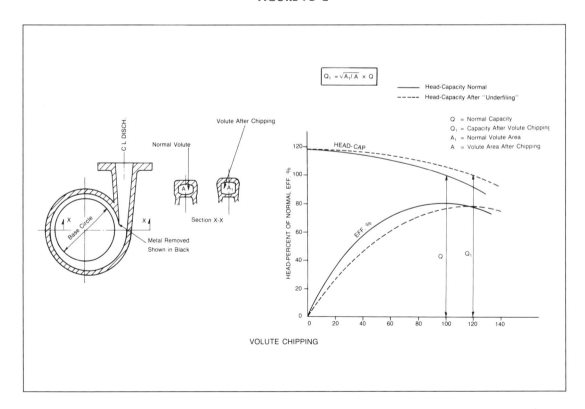

VOLUTE CHIPPING

FIGURE 7B-3

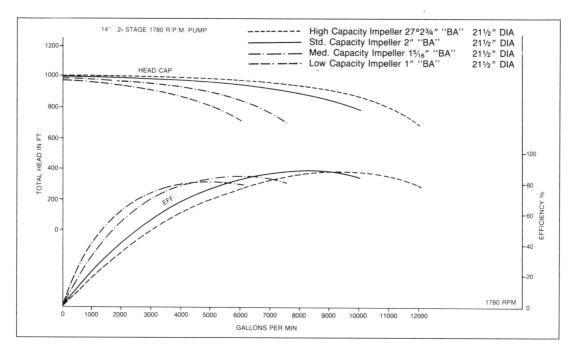

FIGURE 7B–4

LOW- AND HIGH-CAPACITY IMPELLERS

In the majority of pump casings, we can install impellers of different widths for low-
or high-capacity performance. Because of the variations in the design of the impeller
vanes (angularity and number of vanes), it is very difficult to predict their performance.
However, if we take a given impeller with a given angularity and number of vanes,
we can reasonably predict the performance of the narrow, medium, and wide impellers.
Figure 7B–4 shows actual test data of a two-stage fourteen-inch pipeline pump with a
specific speed of 1600. In this pump, the peripheral width of the normal impeller was
two inches, whereas the high, medium, and low capacity impellers were 2 3/4 inches,
1 15/16 inches, and 1 inch wide, respectively. Capacities ranged from 5000 GPM to
9000 GPM, and efficiencies bracketed 82 percent to 88 percent. The performance of
the different impellers in the same casing is to some degree related to specific speed
and running speed. Slow running speed pumps respond better to low-capacity impellers; also, pumps of higher specific speed respond with higher efficiency to a low-
capacity impeller. Figure 7B–4 shows the performance of different impeller widths and
Figure 7B–5 shows the loss of efficiency for different specific speeds.

IMPELLERS OF DIFFERENT NUMBER OF VANES

Certain pump applications require the pump performance curves to have differently
shaped head capacity curves. For instance, to overcome friction only, as in pipeline
service, the highest head per stage, or a very flat curve, is desirable. To overcome static
head or to have pumps run in parallel as is customary in process or boiler feed services,
a continuously rising or steep curve must be developed. A medium rising head capacity
curve is usually needed for highest possible efficiency.

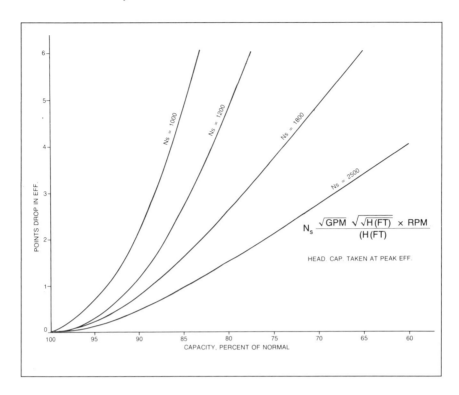

FIGURE 7B–5

There are two ways to vary the shape of a head capacity curve:

1. If the existing impeller has six or more vanes, removing some of the vanes and equally spacing the remaining vanes will produce a steeper head capacity curve. The fewer the number of vanes, the steeper the curve. When vanes are removed, the total discharge area of the impeller is reduced and the peak efficiency point will move to the left, as shown in Figure 7B–6. The efficiency will also drop, the lowest efficiency occurring at the least number of vanes. In a seven-vane impeller reduced to four vanes, the efficiency will drop about four points.
2. If a different head capacity curve shape is required in a given casing and the same peak capacity must be maintained, a new impeller must be designed for each head capacity shape. The steeper the head capacity curve, the wider the impeller, and the fewer will be the number of vanes. For example, a seven-vane 27-degree exit angle will have a flat curve and a narrow impeller, whereas a three-vane 15-degree exit angle will have a steep curve and the widest impeller (refer to Figure 7B–7). In other words, to peak at the same capacity, the impeller discharge area must be the same, regardless of head capacity relationships. Also, for a given impeller diameter, the head coefficient will be the highest for the flattest curve. The efficiency of the above impeller can be maintained within one point.

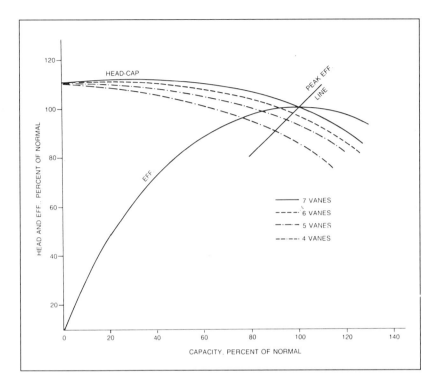

FIGURE 7B–6

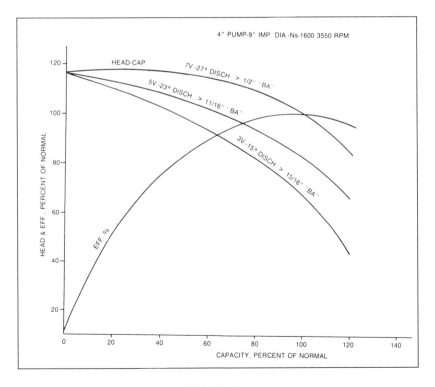

FIGURE 7B–7

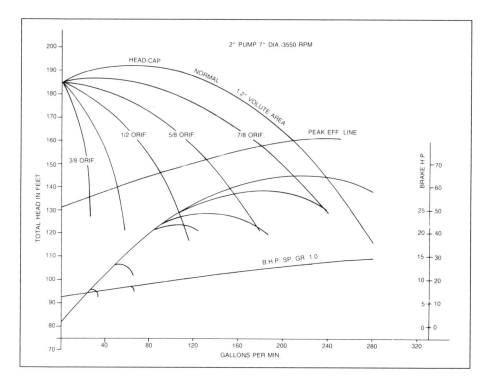

FIGURE 7B–8

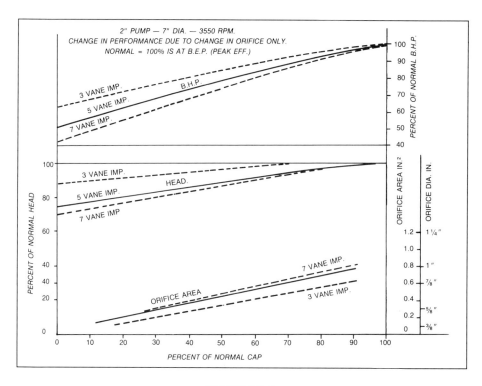

FIGURE 7B–9

ORIFICING PUMP DISCHARGE

In low specific speed pumps, where impellers are already very narrow and low capacity or narrower impellers cannot be cast, capacity reductions can be obtained by using restriction orifices in the pump discharge nozzle. Figures 7B-8 and 7B-9 illustrate these points. Figure 7B-8 shows the performance of a two-inch pump where the discharge was throttled with different size orifices. Figure 7B-9 shows the predicted performance of a throttled pump and illustrates how orifice size selection and changes in the number of impeller vanes can influence absorbed power and head developed by the pump.

Chapter 8

Positive Displacement Pumps

Positive displacement pumps can be divided into two major categories: reciprocating and rotating. Reciprocating pumps include steam pumps and power pumps, as defined later. Many reciprocating pumps use a flexible membrane or diaphragm and are collectively called diaphragm pumps. Every one of the various types comes in a wide range of sizes, or with modifications, additions, and perhaps auxiliary support equipment.

The same is true for the many different types of rotating positive displacement pumps. They include gear pumps, screw pumps, and peristaltic pumps, to name just a few.

Each pump category, reciprocating and rotating, can be found in virtually every process plant we would typically encounter in the industrialized world. Not surprising, each has a definite application range, and the vast majority of these application ranges overlap each other.

RECIPROCATING POSITIVE DISPLACEMENT PUMPS*

Reciprocating positive displacement pumps incorporate a plunger or piston that displaces, or feeds forward, a given volume of fluid per stroke. The basic principle of a reciprocating positive displacement pump is that moving a solid component into the space occupied by a liquid will result in an equal volume of liquid being moved out of that space.

To better understand reciprocating positive displacement pumps and their subgroup metering pumps, we must investigate their place within the universe of pumps. The pump universe could be organized in a variety of ways, such as by design, materials of construction, or the liquids pumped. For the purpose of this discussion, it is appropriate to organize the pump universe by classifying pumps based on the method by which the pump imparts energy to the liquid being pumped. This results in two basic classes of pumps: dynamic and displacement. Dynamic pumps encompass those shown on the left-hand side of Figure 8-1, and these impart energy to the liquid in a steady fashion. Displacement pumps encompass the remaining pumps in Figure 8-1, and these impart energy to the liquid in a pulsating fashion.

The usual basic characteristics of the dynamic and displacement pumps are shown in Table 8.1. By examining this table, it is possible to identify the class of pump required for the job from its characteristics.

This segment of our text is primarily concerned with the world of metering pumps, which is within the positive displacement, reciprocating class of pumps. Reciprocating pumps can be divided into two general categories: steam pumps and power pumps.

---

*Source: *Metering Pump Handbook,* Industrial Press, Inc., New York, NY, 1984. Adapted by permission.

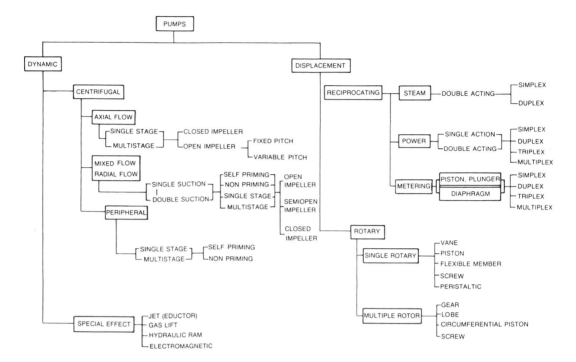

FIGURE 8–1   *Classification of the pump universe. (Source:* Pump Handbook, *edited by Igor J. Karassik, William J. Krutzsch, Warrent H. Fraser, and Joseph P. Messina. McGraw Hill, New York, NY, 1985).*

Steam Pumps

Steam pumps consist of a liquid and steam cylinder joined together by a spacer cradle (Figure 8–2). These pumps may be steam or air driven. The liquid end consists of liquid inlet and outlet ports, valves, and a piston or plunger. The steam or air end consists of valve mechanisms and pistons. Normally, steam pumps are not designed to allow adjustment of the output flow while operating. This precludes their use as a metering pump in a system requiring frequent adjustment of flow rate.

Power Pumps

Power pumps consist of a liquid end and a power end (Figure 8–3). These pumps are generally driven by electric motors, air- or gasoline-driven motors, or any device imparting a rotary or reciprocating motion to the pump. The liquid end consists of inlet and outlet ports, valves, and pistons or plungers. The power end consists of the frame, crankshaft, bearings, connecting rods, crossheads, and, sometimes, reduction gears.

Falling within this category of pumps are those designed with adjustment of output flow as well as those lacking this feature. Because a metering pump requires the adjustment of output flow in a typical process system, we will limit the remainder of this chapter to those power pumps with adjustable output flow.

Metering Pumps

Figure 8–4 subdivides the metering pump class of power pumps into various types. These types delineate the methods and geometries commercially used to produce a metering pump.

Table 8.1   Basic Characteristics of Modern Pumps

	Dynamic		Displacement				
	Centrifugal		Rotary	Reciprocating			
						Metering	
						Piston or Plunger	Diaphragm
				Steam	Power		
Discharge flow	Steady	Steady	Steady	Pulsating	Pulsating	Pulsating	
Usual max suction lift (ft)	15	15	22	22	22	22	
Liquids handled	Clean, clear; dirty, abrasive; slurries		Viscous, non-abrasive	Clean and clear			Clean, clear; dirty, abrasive, slurries
Discharge pressure range	Low to high		Medium	Low to highest produced		Low to highest produced	
Usual capacity range	Small to largest available		Small to medium	Relatively small		Relatively small	
How increased head affects:							
Capacity	Decrease		Almost none	Decrease	None	None	
Power input	Depends on specific speed		Increase	Increase	Increase	Increase	
How decreased head affects:							
Capacity	Increase		Almost none	Small increase	None	None	
Power input	Depends on specific speed		Decrease	Decrease	Decrease	Decrease	
External leakage	Some to none		Some	Some	Some	Some	None
Volume control	Possible with added equipment					Inherent in design	
Remote pumping chamber	Not available					Possible	

Source: *Metering Pump Handbook,* Industrial Press, New York, 1984. Reprinted by permission.

Metering pumps should be considered, first, as precision instruments used to feed accurately a predetermined volume of liquid into a process or system. Their secondary function is to pump, or move, a liquid from one point to another. They contain special adaptations of the conventional positive displacement reciprocating class of pumps, which are designed primarily to transfer liquid at an accurately controlled rate. They differ in that the pumping rates of metering pumps can be varied by changing the effective stroke length and, perhaps, by changing the speed. More importantly, the flow rates of metering pumps can be accurately predetermined, with repeatable flows maintained consistently to within ± 1 percent.

Metering pumps come in an extremely large variety of sizes and configurations. Figures 8–5 and 8–6 give a glimpse of this variety.

Ideally, a metering pump should be capable of handling a wide range of liquids, including those that are toxic, corrosive, dangerous, volatile, and abrasive, as well as those containing concentrations of suspended solids (slurries). In addition, a metering pump should be able to generate sufficiently high discharge pressures to permit injection of liquids into processes. To accomplish this wide range of requirements, many options in design must be available, including the following:

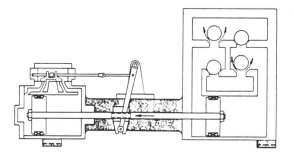

FIGURE 8–2 *Steam pump. (Source:* Hydraulic Institute Standards, *13th Edition, Cleveland, OH, 1975.)*

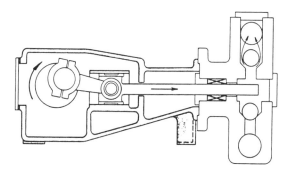

FIGURE 8–3 *Power pump. (Source:* Hydraulic Institute Standards, *13th Edition, Cleveland, OH, 1975.)*

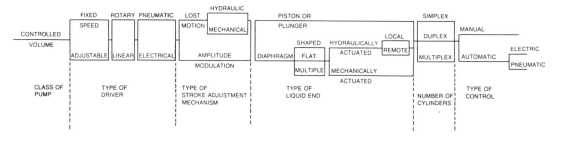

FIGURE 8–4 *Classes of metering pumps. (Source:* Hydraulic Institute Standards, *13th Edition, Cleveland, OH, 1975).*

- Size or capacity
- Method of control
- Materials composing the liquid-handling end
- Valve styles
- Primary drive requirements
- Environmental conditions

Metering pumps fall into four basic types, defined by the method used to seal the liquid end of the pump from the power end, thus preventing leakage and pumping inaccuracies:

FIGURE 8–5   *Three metering pumps driven from a common input shaft. (Source: LEWA, Leonberg, West Germany.)*

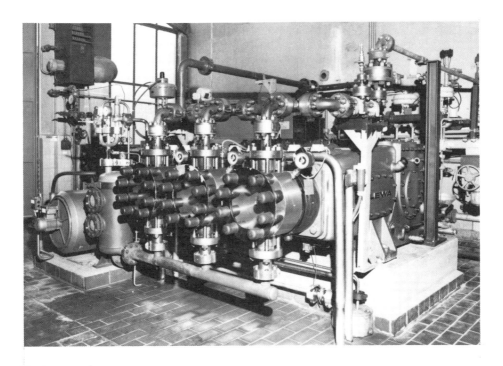

FIGURE 8–6   *Large diaphragm-type pumps in a chemical plant. (Source: LEWA, Leonberg, West Germany.)*

- Piston, packed seal
- Plunger, gland packed seal
- Mechanical diaphragm seal
- Hydraulic diaphragm seal

The power end of the metering pump is common to all four types, with various designs used to generate the reciprocating movement required to power the liquid end.

Most metering pump designs employ an electric motor as a power source. The motor speed can be reduced to pump design speed by the use of internal motor gears or through gearing built into the pump power end. This rotary power is converted to a linear motion through a crank mechanism producing power in a straight-line reciprocating motion. Depending on the type of adjustable output flow mechanism used, the power can be utilized on both the forward thrust of the crank and the back thrust of the crank. Other capacity controls, however, only take advantage of the power in one direction.

There are also metering pumps with electromagnetic and pneumatic power ends creating linear, reciprocating motion using electromagnets and pneumatic pistons, respectively. Although they use conventional liquid end designs, they use highly specialized components to produce this linear reciprocating action. However, the overwhelming majority of metering pumps use some form of crank motion, which is then linked to a crosshead device for positive alignment of the piston or plunger to its cylinder (Figures 8-7 and 8-8).

Packed Piston Pump. The packed piston metering pump uses a power end as just described. The piston is driven by either a crank, a connecting rod, or a crosshead driven by the crank. The piston is the measuring component of the metering pump, designed to displace a measured volume of liquid with a high degree of accuracy as it reciprocates within the pump (Figure 8-9). The packing used moves back and forth with the piston to effect a dynamic seal with the inside diameter of the cylinder and a static seal with the outside diameter of the piston.

The forward travel of the piston reduces the internal volume of the liquid chamber, displacing the metered liquid out the discharge check valve. The pressure required to

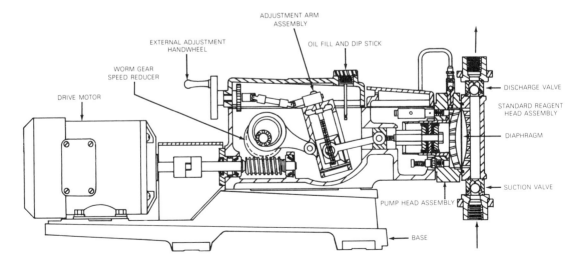

FIGURE 8-7   *Adjustable stroke reciprocating power mechanism. (Source: Pulsafeeder, Rochester, NY.)*

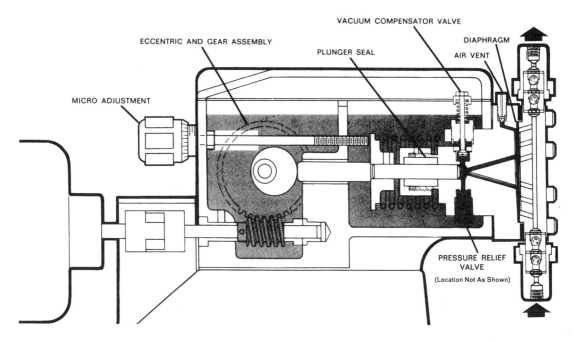

VACUUM COMPENSATOR VALVE

ECCENTRIC AND GEAR ASSEMBLY

DIAPHRAGM

PLUNGER SEAL

AIR VENT

MICRO ADJUSTMENT

PRESSURE RELIEF VALVE
(Location Not As Shown)

FIGURE 8–8   *Cam-driven reciprocating power mechanism. (Source: Pulsafeeder, Rochester, NY.)*

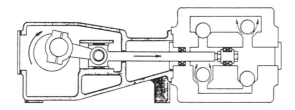

FIGURE 8–9   *Packed piston pump. (Source:* Hydraulic Institute Standards, *13th Edition, Cleveland, OH, 1975.)*

move the liquid through the discharge check valve is also applied to the suction check valve, forcing it into a closed position, ensuring correct flow direction (Figure 8–10).

The reverse travel of the piston decreases the pressure within the liquid chamber by enlarging the internal volume of the chamber. This change of pressure results in a rapid closing of the discharge check valve caused by the external pressure acting on the valve and allows the suction check valve to open because of an external pressure under the check valve that can be either above or below atmospheric pressure (Figure 8–11).

An added feature of the packed piston pump is its ability to be double acting, i.e., if so designed, to provide a discharge of fluid into a system on both the forward thrust of the piston and the back thrust of the piston (Figure 8–9). This feature provides up to twice the output capacity for the same horsepower input and minimizes the typical pulsing output flow common to reciprocating pumps.

The accuracy of the reciprocating metering pump is achieved by the previously described predetermined controlled piston travel of the pump, the control of the strokes per minute, and the precise opening and closing of the check valves. The inaccuracy, on the other hand, is caused by leakage past the piston packing and the check valves.

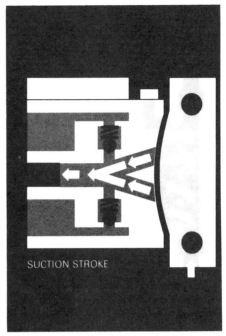

FIGURE 8–10   *Valve action during discharge stroke of diaphragm-type metering pump. (Source: Pulsafeeder, Rochester, NY.)*

FIGURE 8–11   *Valve action during suction stroke of diaphragm-type metering pump. (Source: Pulsafeeder, Rochester, NY.)*

Packed Plunger Pump.   The packed plunger pump is very similar to the packed piston pump except for the packing design and location. The packed plunger, unlike the packed piston, has the packing installed in a stationary gland in the inside diameter of the cylinder. As the plunger reciprocates within the pump, a dynamic seal is made between the outside diameter of the plunger and the inside diameter of the packing, and a static seal is made between the outside diameter of the packing and the inside diameter of the stuffing box (Figure 8–12).

The choice between a packed piston pump or a packed plunger pump is dependent on many variables including fluid compatibility with the packing, speed of the piston or plunger, allowable leakage, and pressure requirements.

Mechanical Diaphragm Pumps.   Both the packed piston pump and the packed plunger pump allow some degree of leakage past their dynamic seals. In some cases, this is not an objectionable shortcoming; in other cases, it can be very objectionable and, in most instances, costly as well. For example, a leakage rate of only 25 drops per minute, metering a fluid costing $1.00 a gallon, can cost $200 per year in lost fluid (Figure 8–13). To be considered also is the cost of pressure flushing, drainage, contamination, maintenance, and loss of accuracy.

To overcome the leakage problem, a mechanically actuated diaphragm pump can be used. The power side of the pump and the capacity control are the same as were previously described for other types of reciprocating pumps. However, in place of a piston rod or plunger, the mechanical diaphragm pump uses a connecting rod fastened to the center of a diaphragm. The configuration of the diaphragm itself can take on many forms, but the most popular designs are those illustrated in Figure 8–14—the flat disc, the convoluted disc, and the bellows.

In the mechanical diaphragm pump, the principle of positive displacement output is similar to that of the piston plunger pump except that the diaphragm becomes the

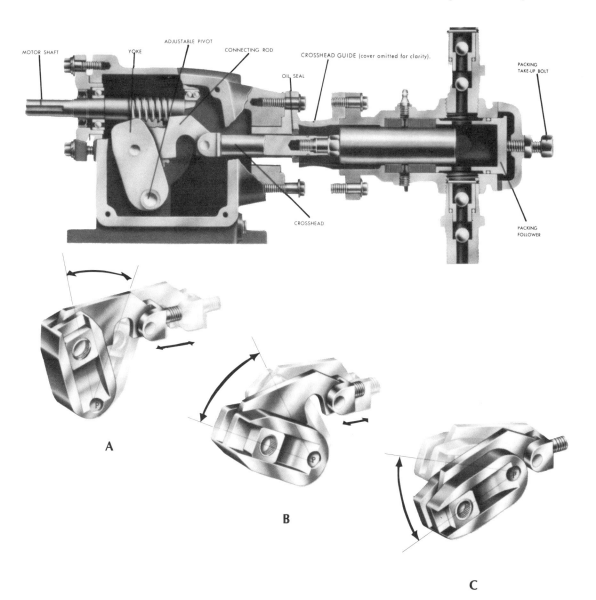

MOTOR SHAFT · YOKE · ADJUSTABLE PIVOT · CONNECTING ROD · CROSSHEAD GUIDE (cover omitted for clarity). · OIL SEAL · CROSSHEAD · PACKING TAKE-UP BOLT · PACKING FOLLOWER

A   B   C

FIGURE 8–12 *Packed plunger metering pump. Stroke length is adjusted by changing the position of the pivot P. (A) When the yoke is positioned at right angles to crosshead motion, the eccentricity is all directed to the crosshead and full stroke results. (B) For any intermediate position of the yoke, any interediate stroke length results. (C) When the yoke is positioned parallel to crosshead motion, the action of the crank is no longer directed to the crosshead, and minimum stroke results. (Source: Milton Roy Company, St. Petersburg, FL.)*

displacement measuring element, as it moves back and forth in the fluid chamber (Figure 8–15).

Hydraulic Diaphragm Pump. The hydraulically balanced diaphragm pump is a hybrid design that provides the principal advantages of the other three pump types. Like the other pumps, its power end and capacity are common. This, however, is where the similarity ends, since the piston or plunger does not come into contact with the pumped fluid, and the actuation of the diaphragm is by hydraulic power instead of mechanical power (Figure 8–7).

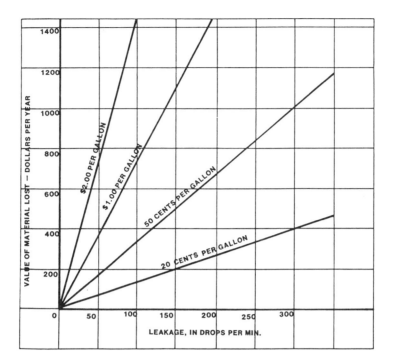

FIGURE 8–13 *Cost of leakage. (Source:* Metering Pump Handbook, *Industrial Press, Inc., New York, NY, 1984.)*

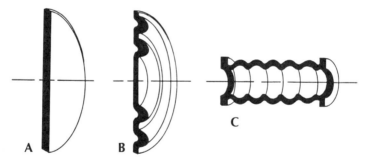

FIGURE 8–14 *Mechanical diaphragm styles. (A) Flat disk; (B) convoluted disk; (C) bellows. (Source: Pulsafeeder, Rochester, NY.)*

The measuring piston or plunger reciprocates within a precisely sized cylinder at an established stroke length, displacing a volume of hydraulic liquid, not the product liquid. The hydraulic liquid is stable and has excellent lubricating qualities. The piston uses the hydraulic oil to move the diaphragm forward and backward (Figures 8-10 and 8-11), causing a displacement that expels the product liquid through the discharge check valve and, on the suction stroke, takes in an equal amount through the suction check valve. The diaphragm isolates the liquid product being contained within the liquid chamber and check valves. These are the only parts that must be made of chemically compatible materials.

The only function of a diaphragm is to separate two liquids. The diaphragm normally does no work, carries no load, and pumps no liquid; rather, it serves as a moving barrier between liquids during periods of pressure imbalance. It is simply a moving partition with pressure hydraulically balanced on both sides; on one side is the liquid

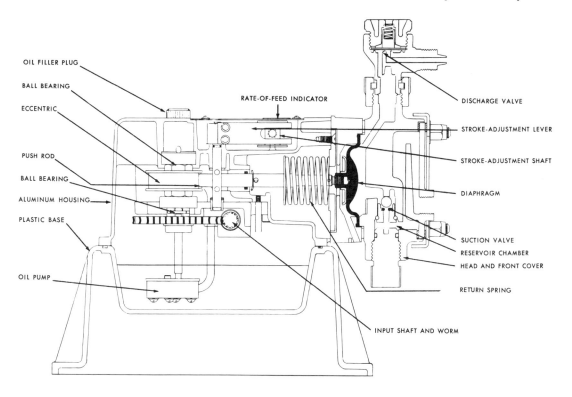

FIGURE 8–15 *Mechanical diaphragm pump. (Source: Wallace and Tiernan, Belleville, NJ.)*

product and on the other side is the hydraulic oil. At full deflection, the diaphragm undergoes total combined stresses well within the endurance limit of the diaphragm material. Contoured support plates are provided on either side of the diaphragm to ensure that stresses are kept within limits. When properly installed and working within the recommended temperature range and not affected by corrosion or abrasion, the diaphragm has an unlimited life.

As previously stated, the piston or plunger handles only hydraulic oil. Conventional seals are used on the piston or plunger, which does not require power flushing and complicated drain systems, as are found on conventional piston or plunger pumps handling corrosive or hazardous liquids.

Even the slightest leakage past the piston is replaced on the suction stroke through the automatic functioning of a compensation system, which draws in replacement oil from the oil reservoir (Figure 8–16).

Any excess pressure within the hydraulic system or the liquid product chamber is relieved through the automatic action of a pressure relief valve. This valve blows off oil, under excess pressure ahead of the piston, back into the oil reservoir (Figure 8–17).

The vacuum and pressure compensator systems actually perform three important functions that the other described types of metering pumps cannot do unless auxiliary equipment is added to their piping systems. As described previously, they compensate for any leakage occurring within the hydraulic system of the pump, ensuring a balanced diaphragm movement. In addition, they serve to protect the process system from over-pressure conditions produced by the pump. For instance, the positive displacement pump, because of its design, causes excess pressure within the system to the point of damaging the pump, bursting pipes, or damaging other downstream equipment should an operator mistakenly close a shut-off valve downstream from the pump. The hydraulic diaphragm pump will, however, relieve any pump-produced pressure beyond

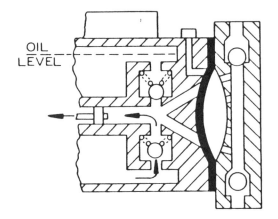

OIL LEVEL

FIGURE 8-16   *Function of oil makeup valve. (Source:* Metering Pump Handbook, *Industrial Press, Inc., New York, NY, 1984.)*

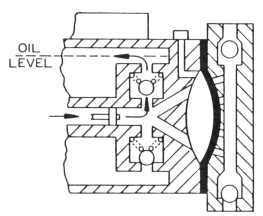

OIL LEVEL

FIGURE 8-17   *Function of pressure relief valve. (Source:* Metering Pump Handbook, *Industrial Press, Inc., New York, NY, 1984.)*

the set pressure of the pressure relief valve, thus avoiding the dangerous buildup of pressure. The compensation system also serves to protect the pump from a closed suction line or a partially clogged strainer in the suction line. Should this occur, the backward movement of the diaphragm is prevented and the vacuum relief system would automatically open to relieve the starved suction condition within the pump. In doing so, however, a surplus of hydraulic oil enters into the system between the diaphragm and piston. As the piston starts forward on its discharge stroke, the diaphragm is displaced forward and will come into contact with the contoured dish support plate in the process liquid chamber, because of the surplus oil drawn into the hydraulic chamber. At the moment of diaphragm contact with its support plate, an over-pressure condition starts to develop within the hydraulic system. The pressure relief valve now opens to relieve the surplus oil back into the hydraulic reservoir, preventing a dangerous buildup of pressure. The interaction of the two compensation systems continues stroke after stroke to activate a fluid clutch-type action to prevent overloading of the pump power end until the condition plugging the suction or discharge lines is found and corrected.

As in all properly designed hydraulic systems, an air-bleed system is required to purge air from the hydraulic system either automatically or manually.

Because the diaphragm is actuated hydraulically, this type of pump is highly adaptable. The product liquid chamber can be separated from the power end of the pump, using a pipe or tube to transfer the hydraulic power required. This feature is highly desirable when metering toxic, high- or low-temperature extremes. Additional diaphragms may also be added to accomplish various system requirements.

Summary.   In summary, the principal features and limitations of each type of metering pump are listed.

FEATURES	LIMITATIONS

Piston Pump

1. Low cost	1. Packing leakage is unavoidable, a consideration with corrosive or dangerous chemicals
2. Can be double acting	
3. Capacities from a few cubic centimeters per hour to 1200 gallons per hour (gph) (4.5×10^3 liters/hr)	2. Unsuitable for abrasive slurries
	3. Maintenance needed for piston packing
4. Packing adjustment not required	4. No built-in relief features
5. Accuracies to 1 percent over 10:1 range	5. Dynamic packing subject to differential pressure
6. Can pump high-viscosity fluids	

Packed Plunger Pump

1. Relatively low cost	1. Packing wear requires periodic adjustment
2. Capacities from a few cubic centimeters per hour to over 20,000 gph (7.6×10^4 liters/hr)	2. Packing leakage is unavoidable, a consideration with corrosive or dangerous chemicals
3. Accuracies to 1 percent over 15:1 range	3. Unsuitable for abrasive slurries
4. Pressure capability to 50,000 psi (3.5×10^5 kPa)	4. Maintenance needed for packing and plunger wear
5. Least effect from changes in discharge pressure	5. Pressure flushing or draining of packing gland required
6. Can pump high-viscosity liquids	6. No built-in relief

Mechanical Diaphragm Pump

1. Relatively low cost	1. High maintenance due to high-stress loading of the diaphragm
2. Zero chemical leakage	
3. Can pump abrasive slurries and chemicals	2. Discharge pressure limitation of 200 psi (1.4×10^3 kPa)
4. Packing adjustment not required	3. Accuracy in the 5 percent range and as much as 10 percent zero shift with pressure change from minimum to maximum
5. Can pump high-viscosity liquids	
	4. Limited capacity range
	5. Dynamic packing subject to differential pressures
	6. No built-in safety features

Hydraulic Diaphragm Pump

1. Zero chemical leakage	1. Higher initial cost
2. High adaptability	2. Subject to a predictable zero shift of 5 percent to 10 percent per 1000 psig (6.9×10^3 kPa)
3. Can pump wide range of liquids	
4. Packing adjustment not required	3. Limited to moderate-viscosity fluids
5. Low maintenance	
6. Pressures to 5000 psi (3.5×10^4 kPa)	
7. Accuracies 1 percent and capacities from a few cubic centimeters per hour to 1500 gph (5.7×10^3 liters/hr)	
8. Built-in safety features	

ROTATING POSITIVE DISPLACEMENT PUMPS*

External and internal gear pumps are prevalent not only in the process industries but in general machinery and industrial equipment as well. The *external gear pump* (Figure 8–18) is probably the most widely used rotary pump. It consists of two meshing gears in a close-fitting housing. Gear rotors are cut externally and, in this type of pump, fluid is carried between the gear teeth and displaced when they mesh.

The gears may be one of three designs: spur, helical, or herringbone. Spur gears are used on low-capacity, high-pressure applications, since they offer line contact between the teeth. This reduces the slip through the meshing line and gives better performance at higher pressure. Spur gears are noisy at high speeds, and unless some relief is provided with grooves or ports, there is a tendency to trap liquid at the meshing point and shaft deflection could result.

Single helical gears eliminate trapping by gradual engagement and disengagement of the teeth. Unfortunately, a component of the load produces an end thrust in one direction that could cause sideplate wear unless compensated for by thrust bearings. The major advantage of single helical gears over double helical or herringbone gears is cost. But herringbone gears eliminate end thrust while still retaining the advantages over spur gears of higher speed operation, pulsation-free flow, and no liquid trapping.

Some controversy exists over the direction of rotation of gear pumps. The preferred direction is with the gear apexes leading so the liquid is squeezed out from the center toward both ends of the gears. However, except for extremely viscous liquids at high pressures, there is no measurable difference in capacity, power, or noise due to direction of rotation.

Internal gear pumps (Figure 8–19) have one rotor with internally cut gear teeth meshing with an externally cut gear. On the outer sideplate is a stationary crescent. As the internal gear rotates, the idler (external) gear follows and liquid is displaced between the internal gear in the crescent and between the idler and the crescent. This type of pump is generally used for lower pressure applications at low speeds and it is generally a cantilever design.

The *vane type* category of rotating positive displacement pumps consists of external vane types or sliding vanes, as shown in Figure 8–20. The vane (or vanes) may be in the form of blades, buckets, or slippers, with a cam to draw fluid into and force it out of the pump chamber.

In the sliding vane pump, the rotor is slotted and a series of vanes follow the bore of the casing. Liquid is displaced between the vanes. This is a slow speed design and it does have lower viscosity limits than gear pumps. The vanes are, however, self-compensating for wear, and this pump, unlike a close-clearance gear design, can be used on mildly erosive liquids.

In *lobe type* pumps, Figure 8–21, liquid is carried between the rotor lobe surfaces from the inlet to the outlet. Since the lobes cannot drive each other, external timing gears are necessary. This type is a low-speed, high-displacement, low-pressure pump popular for marine or highly viscous, low-pressure process applications. There is little difference between lobe-type pumps and the lobe type blowers discussed in Chapter 15 of this text.

In *single-screw* pumps (Figure 8–22), the liquid is carried between the rotor screw threads and axially displaced as the rotor threads mesh with internal threads on the stator. This design is versatile in that the rotor and the stator can be made of many different materials so the pump can be used on certain abrasive and corrosive services. Single-screw pumps are sometimes called progressive cavity pumps and are most successfully used in the food processing industries.

---

*Sources: As acknowledged in captions to illustrations.

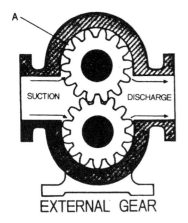

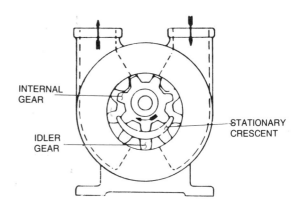

FIGURE 8–18 *External gear pump. (Source: Dresser-Worthington, Harrison, NJ.)*

FIGURE 8–19 *Internal gear pump. (Source: Dresser-Worthington, Harrison, NJ.)*

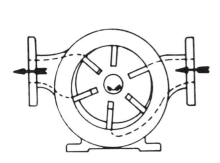

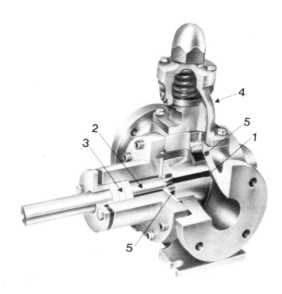

FIGURE 8–20 *Sliding vane pump. 1 = nonmetallic or metallic vanes; 2 = bearing area, internal or external to fluid pumped; 3 = soft packing or mechanical face seal area; 4 = safety valve; 5 = casing (could be jacketed); 6 = rotor. (Source: Foster Pump Works, Inc., Westerley, RI.)*

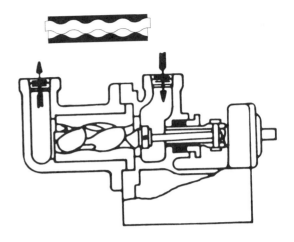

FIGURE 8–21   *Lobe-type pump. (Source: Dresser-Worthington, Harrison, NJ.)*

FIGURE 8–22   *Single-screw pump and progressive cavity detail. (Sources: Dresser-Worthington, Harrison, NJ, and Netzsch Inc., Exton, PA.)*

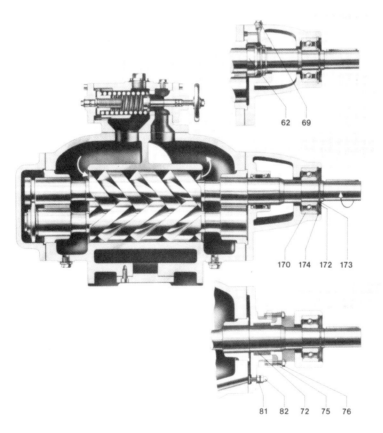

FIGURE 8–23   *Twin-screw pump. 62* = mechanical seal; 69+ = locking screw; 72+* = gland; 75 = gland; 76+ = screw; 81 = adjusting screw; 82+ = cap nut; 170+* = ball bearing; 172 = supporting disk; 173+* = circlip-driving spindle-drive side; 174+ = circlip-cover-drive-side; + = spare parts; * = DIN parts. (Source: Leistritz Pump Corporation, Allendale, N.J.)*

Multiscrew pumps are self-priming rotary displacement pumps. Theoretically speaking, they operate like a piston pump with strokes of infinite length. Numerous variations are available. In Figure 8–23, a two-flight driven spindle closely engages and rotates with a three-flight driven spindle. Both are located in a close-fitting casing. In combination, screws and casing form perfectly sealed chambers. The material confined in these chambers is continuously advanced without undue shear force and turbulence.

Peristaltic Pumps

Peristaltic pumps are positive displacement pumps incorporating a flexible hose. The model shown in Figure 8–24, essentially a development of the conventional peristaltic pumping concept, is glandless and valveless. The liquid passes straight through the pump without restrictions. Three main components produce the pumping action—a rotor with three rollers and a smooth-bored flexible hose tube element with the addition of a flexible rubber separator element. The separator is attached to the top of the pump

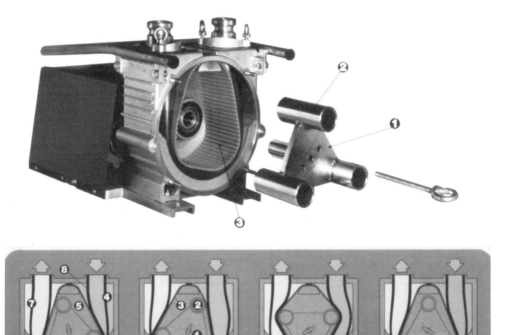

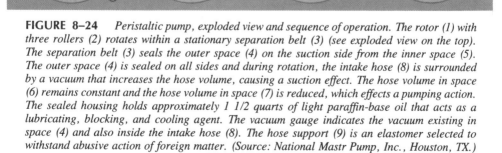

FIGURE 8–24 *Peristaltic pump, exploded view and sequence of operation. The rotor (1) with three rollers (2) rotates within a stationary separation belt (3) (see exploded view on the top). The separation belt (3) seals the outer space (4) on the suction side from the inner space (5). The outer space (4) is sealed on all sides and during rotation, the intake hose (8) is surrounded by a vacuum that increases the hose volume, causing a suction effect. The hose volume in space (6) remains constant and the hose volume in space (7) is reduced, which effects a pumping action. The sealed housing holds approximately 1 1/2 quarts of light paraffin-base oil that acts as a lubricating, blocking, and cooling agent. The vacuum gauge indicates the vacuum existing in space (4) and also inside the intake hose (8). The hose support (9) is an elastomer selected to withstand abusive action of foreign matter. (Source: National Mastr Pump, Inc., Houston, TX.)*

casing, creating a vacuum on the suction side. The hose tube passes directly through the pump between inlet and outlet interposed between the separator and the bottom of the casing. Each roll/squeeze movement of the roller displaces the pumped liquid while the vacuum surrounding the tube on the suction side ensures that it regains shape instantly when the pressure is released. This rapid tube recovery feature enhances and accelerates the normal suction effect of the roll/squeeze peristaltic movement and confers on the pump its exceptional suction characteristics. It allows the use of thin-walled tubes that create a high RPM/volume range. Perstaltic pumps are suitable for liquids, slurries, and gases. Different hose materials accommodate the various feed streams.

APPENDIX 8A

Principles of Operation of Reciprocating Pumps

How a Reciprocating Pump Works

A reciprocating pump is a positive displacement mechanism with liquid discharge pressure being limited only by the strength of the structural parts. Liquid volume or capacity delivered is constant regardless of pressure, and is varied only by speed changes.

Characteristics of a reciprocating pump are 1) positive displacement of liquid, 2) high pulsations caused by the sinusoidal motion of the piston, and 3) high volumetric efficiency.

Plunger or Piston Rod Load

Plunger or piston "rod load" is an important power end design consideration for reciprocating pumps. Rod load is the force caused by the liquid pressure acting on the face of the piston or plunger. This load is transmitted directly to the power frame assembly and is normally the limiting factor in determining maximum discharge pressure ratings. This load is directly proportional to the pump gauge discharge pressure and proportional to the square of the plunger or piston diameter.

Occasionally, allowable liquid end pressures limit the allowable rod load to a value below the design rod load. IT IS IMPORTANT THAT LIQUID END PRESSURES DO NOT EXCEED PUBLISHED LIMITS.

Calculations of Volumetric Efficiency

Volumetric efficiency (E_v) is defined as the ratio of plunger or piston displacement to liquid displacement. The volumetric efficiency calculation depends upon the internal configuration of each individual liquid body, the piston size, and the compressibility of the liquid being pumped.

Tools for Liquid Pulsation Control, Inlet and Discharge

Pulsation Control Tools ("PCT", often referred to as "dampeners" or "stabilizers") are used on the inlet and discharge piping to protect the pumping mechanism and associated piping by reducing the high pulsations within the liquid caused by the motions of the slider-crank mechanism. A properly located and charged pulsation control tool may reduce the length of pipe used in the acceleration head equation to a value of 5 to 15 nominal pipe diameters. The pulsation control tools are specially required to compensate for inadequately designed or old/adapted supply and discharge systems.

Acceleration Head

Whenever a column of liquid is accelerated or decelerated, pressure surges exist. This condition is found on the inlet side of the pump as well as the discharge side. Not only can the surges cause vibration in the inlet line, but they can restrict and impede the flow of liquid and cause incomplete filling of the inlet valve chamber. The magnitude of the surges and how they will react in the system is impossible to predict without an extremely complex and costly analysis of the system. Since the behavior of the natural frequencies in the system is not easily predictable, as much of the surge as possible must be eliminated at the source. Proper installation of an inlet pulsation control PCT will absorb a large percentage of the surge before it can travel into the system. The function of the PCT is to absorb the "peak" of the surge and feed it back at the low part of the cycle. The best installation for the PCT is in the liquid supply line as close to the pump as possible, or attached to the blind flange side of the pump inlet. In either location, the surges will be dampened and harmful vibrations reduced.

RECIPROCATING PUMPS FLOW CHARACTERISTICS

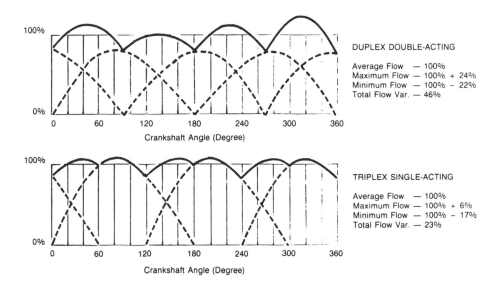

DUPLEX DOUBLE-ACTING

Average Flow — 100%
Maximum Flow — 100% + 24%
Minimum Flow — 100% – 22%
Total Flow Var. — 46%

TRIPLEX SINGLE-ACTING

Average Flow — 100%
Maximum Flow — 100% + 6%
Minimum Flow — 100% – 17%
Total Flow Var. — 23%

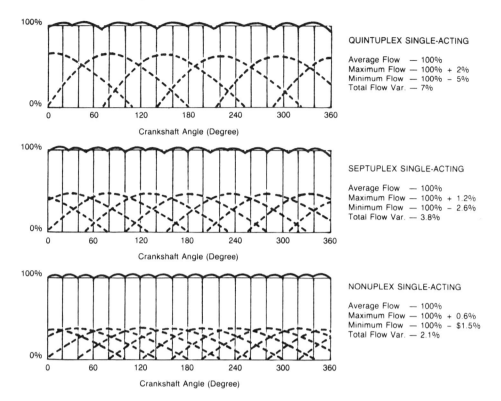

QUINTUPLEX SINGLE-ACTING

Average Flow — 100%
Maximum Flow — 100% + 2%
Minimum Flow — 100% – 5%
Total Flow Var. — 7%

SEPTUPLEX SINGLE-ACTING

Average Flow — 100%
Maximum Flow — 100% + 1.2%
Minimum Flow — 100% – 2.6%
Total Flow Var. — 3.8%

NONUPLEX SINGLE-ACTING

Average Flow — 100%
Maximum Flow — 100% + 0.6%
Minimum Flow — 100% – $1.5%
Total Flow Var. — 2.1%

REQUIRED FORMULAE AND DEFINITIONS

Acceleration Head

$$h_a = \frac{LVNC}{Kg} \qquad V = \frac{GPM}{(2.45)(D)}$$

Where

h_a = Acceleration head (in feet)
L = Length of liquid supply line (in line)
V = Average velocity in liquid supply line (in fps)
N = Pump speed (revolutions per minute)
C = Constant depending on the type of pump
   C = 0.200 for simplex double-acting
     = 0.200 for duplex single-acting
     = 0.115 for duplex double-acting
     = 0.066 for triplex single or double-acting
     = 0.040 for quintuplex single or double-acting
     = 0.028 for septuplex, single or double-acting
     = 0.022 for nonuplex, single or double-acting
K = Liquid compressibility factor
   K = 2.5 For relatively compressible liquids
       (ethane, hot oil)
   K = 2.0 For most other hydrocarbons
   K = 1.5 For amine, glycol and water
   K = 1.4 For liquids with almost no compressibility
       (hot water)
g = Gravitational constant = 32.2 ft/sec²
d = Inside diameter of pipe (inches)

Stroke

One complete uni-directional motion of piston or plunger.
Stroke length is expressed in inches.

Pump Capacity (Q)

The capacity of a reciprocating pump is the total volume
through-put per unit of time at suction conditions. It includes
both liquid and any dissolved or entrained gases at the stated
operating conditions. The standard unit of pump capacity is
the U.S. gallon per minute.

Pump Displacement (D)

The displacement of a reciprocating pump is the volume
swept by all pistons or plungers per unit time. Deduction for
piston rod volume is made on double acting piston type
pumps when calculating displacement. The standard unit of
pump displacement is the U.S. gallon per minute.

For single-acting pumps: $D = \dfrac{Asnm}{231}$

For double-acting piston pumps:

$$D = \frac{(2A - a)\,snm}{231}$$

Where

A = Plunger or piston area, square inch
a = Piston rod cross-sectional area, square inch
     (double-acting pumps)
s = Stroke length, inch
n = RPM of crankshaft
m = Number of pistons or plungers

Plunger or Piston Speed (v)

The plunger or piston speed is the average speed of the plunger of piston. It is expressed in feet per minute.

$$v = \frac{ns}{6}$$

Pressures

The standard unit of pressure is the pound force per square inch.

Discharge Pressure (Pd)—The liquid pressure at the centerline of the pump discharge port.

Suction Pressure (Ps)—The liquid pressure at the centerline of the suction port.

Differential pressure (Ptd)—The difference between the liquid discharge pressure and suction pressure.

Net Positive Suction Head Required (NPSHR)—The amount of suction pressure, over vapor pressure, required by the pump to obtain satisfactory volumetric efficiency and prevent excessive cavitation.

The pump manufacturer determines (by test) the net positive suction head required by the pump at the specified operating conditions.

NPSHR is related to losses in the suction valves of the pump and frictional losses in the pump suction manifold and pumping chambers. Required NPSHR does not include system acceleration head, which is a system-related factor.

Slip (S)

Slip of a reciprocating pump is the loss of capacity, expressed as a fraction or percent of displacement, due to leaks past the valves (including the back-flow through the valves caused by delayed closing) and past double-acting pistons. Slip does not include fluid compressibility or leaks from the liquid end.

Power (P)

Pump Power Input (Pi)—The mechanical power delivered to a pump input shaft, at the specified operating conditions. Input horsepower may be calculated as follows:

$$Pi = \frac{Q \times Ptd}{1714 \times \eta p}$$

Pump Power Output (Po)—The hydraulic power imparted to the liquid by the pump, at the specified operating conditions. Output horsepower may be calculated as follows:

$$Po = \frac{Q \times Ptd}{1714}$$

The standard unit for power is the horsepower.

Efficiencies (n)

Pump Efficiency (ηp) (also called pump mechanical efficiency)—The ratio of the pump power output to the pump power input.

$$\eta p = \frac{Po}{Pi}$$

Volumetric Efficiency (ηv)—The ratio of the pump capacity to displacement.

$$\eta v = \frac{Q}{D}$$

Plunger Load (Single-Acting Pump)

The computed axial hydraulic load, acting upon one plunger during the discharge portion of the stroke is the plunger load. It is the product of plunger area and the guage discharge presssure. It is expressed in pounds force.

Piston Rod Load (Double-Acting Pump)

The computed axial hydraulic load, acting upon one piston rod during the forward stroke (toward head end) is the piston rod load.

It is the product of piston area and discharge pressure, less the product of net piston area (rod area deducted) and suction pressure. It is expressed in pounds force.

Liquid pressure $\left(\frac{pounds}{square\ inch} \right)$ or(psi) =

Cylinder area (square inches) =
(3.1416) x (Radius(inches))² =
$$\frac{(3.1416) \times (diameter(inches))^2}{4}$$

Cylinder force (pounds)—pressure (psi) x area (square inches)

Cylinder speed or average liquid velocity through piping (feet/second) = $\frac{flow\ rate\ (gpm)}{(inside\ diameter\ (inch))^2}$ 2.448 x

Reciprocating pump displacement (gpm) =
$$\frac{rpm \times displacement\ (cubic\ in/revolution)}{231}$$

Pump input horsepower—see horsepower calculations

Shaft torque (foot-pounds) = $\frac{horsepower \times 5252}{shaft\ speed\ (rpm)}$

Electric motor speed (rpm) = $\frac{120 \times frequency\ (Hz)}{number\ of\ poles}$

Three phase motor horsepower (output) =
$$\frac{1.73 \times amps \times volts \times efficiency \times power\ factor}{746}$$

Static head of liquid (feet) =
$$\frac{2.31 \times static\ pressure\ (psig)}{specific\ gravity}$$

Velocity head of liquid (feet) =
$$\frac{liquid\ velocity^2}{g = 32.2\ ft/sec^2}$$

Absolute viscosity (centipoise) =
specific gravity x Kinematic viscosity (centistrokes)

Kinematic viscosity (centistrokes) =
$$0.22 \times saybolt\ viscosity\ (ssu) = \frac{180}{Saybolt\ viscosity\ (ssu)}$$

Absolute pressure (psia) =
Local atmospheric pressure + gauge pressure (psig)

Gallon per revolution =
$$\frac{Area\ of\ plunger\ (sq\ in) \times length\ of\ stroke(in) \times number\ of\ plungers}{231}$$

Barrels per day = gal/rev x pump speed (rpm) x 34.3

Specific gravity (at 60°F) =

$$\frac{141.5}{131.5 + \text{API gravity (degree)}}$$

Bolt clamp load (lb) =
0.75 x proof strength (psi) x tensile stress area (in²)

Bolt torque (ft-lb) =
$\frac{0.2 \text{ (or 0.15)} \times \text{nominal diameter in inches} \times \text{bolt clamp}}{12} \text{ load (lb)}$
0.2 for dry
0.15 for lubricated, plating, and hardened washers

Calculating Volumetric Efficiency for Water

The volumetric efficiency of a reciprocating pump, based on

capacity at suction conditions, using table of water compressibility, shall be calculated as follows:

Vol. Eff. =

$$1 - \frac{P_{td}\,\beta t\ \ 1 + \dfrac{c}{d}}{1 - P_{td}\,\beta t} - S$$

Where

βt = Compressibility factor at temperature t (degrees Fahrenheit or centigrade). (See Tables 1 and 2).
c = Liquid chamber volume in the passages of chamber between valves when plunger is at the end of discharge stroke in cubic inches
d = Volume displacement per plunger in cubic inches
P_{td} = Discharge pressure minus suction pressure in psi
S = Slip, expressed in decimal value

TABLE 1 Water Compressibility

Compressibility Factor βt x 10^{-6} = Contraction in Unit Volume Per Psi Pressure
Compressibility from 14.7 Psia, 32 F to 212 F and from Saturation Pressure Above 212 F

TEMPERATURE

Pressure Psia	0 C 32 F	20 C 63 F	40 C 104 F	60 C 140 F	80 C 176 F	100 C 212 F	120 C 248 F	140 C 284 F	160 C 320 F	180 C 356 F	200 C 392 F	220 C 428 F	240 C 464 F	260 C 500 F	280 C 536 F	300 C 572 F	320 C 608 F	340 C 644 F	360 C 680 F
200	3.12	3.06	3.06	3.12	3.23	3.40	3.66	4.00	4.47	5.11	6.00	7.27							
400	3.11	3.05	3.05	3.11	3.22	3.39	3.64	3.99	4.45	5.09	5.97	7.21							
600	3.10	3.05	3.05	3.10	3.21	3.39	3.63	3.97	4.44	5.07	5.93	7.15	8.95						
800	3.10	3.04	3.04	3.09	3.21	3.38	3.62	3.96	4.42	5.04	5.90	7.10	8.85	11.6					
1000	3.09	3.03	3.03	3.09	3.20	3.37	3.61	3.95	4.40	5.02	5.87	7.05	8.76	11.4	16.0				
1200	3.08	3.02	3.02	3.08	3.19	3.36	3.60	3.94	4.39	5.00	5.84	7.00	8.68	11.2	15.4				
1400	3.07	3.01	3.01	3.07	3.18	3.35	3.59	3.92	4.37	4.98	5.81	6.95	8.61	11.1	15.1	23.0			
1600	3.06	3.00	3.00	3.06	3.17	3.34	3.58	3.91	4.35	4.96	5.78	6.91	8.53	10.9	14.8	21.9			
1800	3.05	2.99	3.00	3.05	3.16	3.33	3.57	3.90	4.34	4.94	5.75	6.87	8.47	10.8	14.6	21.2	36.9		
2000	3.04	2.99	2.99	3.04	3.15	3.32	3.56	3.88	4.32	4.91	5.72	6.83	8.40	10.7	14.3	20.7	34.7		
2200	3.03	2.98	2.98	3.04	3.14	3.31	3.55	3.87	4.31	4.89	5.69	6.78	8.33	10.6	14.1	20.2	32.9	86.4	
2400	3.02	2.97	2.97	3.03	3.14	3.30	3.54	3.85	4.29	4.87	5.66	6.74	8.26	10.5	13.9	19.8	31.6	69.1	
2600	3.01	2.96	2.96	3.02	3.13	3.29	3.53	3.85	4.28	4.85	5.63	6.70	8.20	10.4	13.7	19.4	30.5	61.7	
2800	3.00	2.95	2.96	3.01	3.12	3.28	3.52	3.83	4.26	4.83	5.61	6.66	8.14	10.3	13.5	19.0	29.6	57.2	238.2
3000	3.00	2.94	2.95	3.00	3.11	3.28	3.51	3.82	4.25	4.81	5.58	6.62	8.08	10.2	13.4	18.6	28.7	53.8	193.4
3200	2.99	2.94	2.94	3.00	3.10	3.27	3.50	3.81	4.23	4.79	5.55	6.58	8.02	10.1	13.2	18.3	27.9	51.0	161.0
3400	2.98	2.93	2.93	2.99	3.09	3.26	3.49	3.80	4.22	4.78	5.53	6.54	7.96	9.98	13.0	17.9	27.1	48.6	138.1
3600	2.97	2.92	2.93	2.98	3.09	3.25	3.48	3.79	4.20	4.76	5.50	6.51	7.90	9.89	12.9	17.6	26.4	45.4	122.4
3800	2.96	2.91	2.92	2.97	3.08	3.24	3.47	3.78	4.19	4.74	5.47	6.47	7.84	9.79	12.7	17.3	25.8	44.5	110.8
4000	2.95	2.90	2.91	2.97	3.07	3.23	3.46	3.76	4.17	4.72	5.45	6.43	7.78	9.70	12.5	17.1	25.2	42.8	101.5
4200	2.95	2.90	2.90	2.96	3.06	3.22	3.45	3.75	4.16	4.70	5.42	6.40	7.73	9.62	12.4	16.8	24.6	41.3	93.9
4400	2.94	2.89	2.90	2.95	3.05	3.21	3.44	3.74	4.14	4.68	5.40	6.36	7.68	9.53	12.2	16.5	24.1	40.0	87.6
4600	2.93	2.83	2.89	2.94	3.05	3.20	3.43	3.73	4.13	4.66	5.37	6.32	7.62	9.44	12.1	16.3	23.6	38.8	82.3
4800	2.92	2.87	2.88	2.94	3.04	3.20	3.42	3.72	4.12	4.64	5.35	6.29	7.57	9.36	12.0	16.0	23.2	37.6	77.7
5000	2.91	2.87	2.87	2.93	3.03	3.10	3.41	3.71	4.10	4.63	5.32	6.25	7.52	9.28	11.8	15.8	22.7	36.6	73.9
5200	2.90	2.85	2.87	2.92	3.02	3.18	3.40	3.69	4.09	4.61	5.30	6.22	7.47	9.19	11.7	15.6	22.3	35.6	70.3
5400	2.90	2.85	2.86	2.91	3.01	3.17	3.39	3.68	4.07	4.59	5.27	6.19	7.41	9.12	11.6	15.3	21.9	34.6	66.9

EXAMPLE: Find the volumetric efficiency of a reciprocating pump with the following conditions:

Type of pump	3 in diam plunger x 5 in stroke triplex
Liquid pumped	Water
Suction pressure	Zero psig
Discharge pressure	1785 psig
Pumping temperature	140 F
c	127.42 cu in
d	35.343 cu in
S	.02

Find βt from Table of Water Compressibility (Table 1).

βt = .00000305 at 140 F and 1800 psia Calculate volumetric efficiency:

Vol. Eff. =

$$\frac{1 - \left[P_{td}\,\beta t \left(1 + \dfrac{c}{d}\right)\right]}{1 - P_{td}\,\beta t} - S =$$

$$\frac{1 - \left[(1785 - 0)(.00000305)\right]\ 1 + \dfrac{127.42}{35.343}}{1 - (1785 - 0)(.00000305)} - .02$$

= .96026

= 96 per cent

TABLE 2 Water Compressibility

Compressibility Factor $\beta t \times 10^{-6}$ = Contraction in Unit Volume Per Psi Pressure
Compressibility from 14.7 Psia at 68 F and 212 F and from Saturation Pressure at 392 F

Pressure Psia	Temperature			Pressure Psia	Temperature		
	20 C 68 F	100 C 212 F	200 C 392 F		20 C 68 F	100 C 212 F	200 C 392 F
6000	2.84	3.14	5.20	22000	2.61	2.42	3.75
7000	2.82	3.10	5.09	23000	2.59	2.38	3.68
8000	2.80	3.05	4.97	24000	2.58	2.33	3.61
9000	2.78	3.01	4.87	25000	2.57	2.29	3.55
10000	2.76	2.96	4.76	26000	2.56	2.24	3.49
11000	2.75	2.92	4.66	27000	2.55	2.20	3.43
12000	2.73	2.87	4.57	28000	2.55	2.15	3.37
13000	2.71	2.83	4.47	29000	2.54	2.11	3.31
14000	2.70	2.78	4.38	30000	2.53	2.06	3.26
15000	2.69	2.74	4.29	31000	2.52	2.02	3.21
16000	2.67	2.69	4.21	32000	2.51	1.97	3.16
17000	2.66	2.65	4.13	33000	2.50	1.93	3.11
18000	2.65	2.60	4.05	34000	2.49	1.88	3.07
19000	2.64	2.56	3.97	35000	2.49	1.84	3.03
20000	2.63	2.51	3.89	36000	2.48	1.79	2.99
21000	2.62	2.47	3.82				

Calculating Volumetric Efficiency For Hydrocarbons

The volumetric efficiency of a reciprocating pump based on capacity at suction conditions, using compressibility factors for hydrocarbons, shall be calculated as follows:

$$\text{Vol. Eff.} = 1 - \left[S - \frac{c}{d} \left(1 - \frac{\rho_d}{\rho_s} \right) \right]$$

Where

c = Fluid chamber volume in the passages of chamber between valves, when plunger is at the end of discharge strike, in cubic inches

d = Volume displacement per plunger, in cubic inches

P = Pressure in psia (P_s = suction pressure in psia; P_d = discharge pressure in psia)

P_c = Critical pressure of liquid in psia

P_r = Reduced pressure

$$\frac{\text{Actual pressure in psia}}{\text{Critical pressure in psia}} = \frac{P}{P_c}$$

P_{rs} = Reduced suction pressure = $\dfrac{P_s}{P_c}$

P_{rd} = Reduced discharge pressure = $\dfrac{P_d}{P_c}$

S = Slip expressed in decimal value

t = Temperature, in degrees Rankine
= Degrees F + 460 (t_s = suction temperature in degrees Rankine; t_d = discharge temperature in degrees Rankine)

T_c = Critical temperature of liquid, in degrees Rankine (See Table 3)

T_r = Reduced temperature

$$= \frac{\text{actual temp. in degrees Rankine}}{\text{critical temp. in degrees Rankine}}$$

$$= \frac{t}{T_c} \quad \text{(See Fig. 1)}$$

T_{rs} = Reduced suction temperature

$$= \frac{t_s}{T_c}$$

T_{rd} = Reduced discharge temperature

$$= \frac{t_d}{T_c}$$

Vol. Eff. = Volumetric efficiency expressed in decimal value.

$$= \frac{P_1}{\omega_1} \times \quad \times 62.4 = \text{density of liquid in lb per cu ft}$$

s = Density in lb per cu ft at suction pressure

d = Density in lb per cu ft at discharge pressure

= Expansion factor of liquid

$\dfrac{P_1}{\omega_1}$ = Characteristic constant in grams per cubic centimeter for any one liquid which is established by density measurements and the corresponding values of (See Table 3).

TABLE 3

Carbon Atoms	Name	T_c Degrees Rankine	P_c Lb Per Sq In	pl/wl Grams Per cc
1	Methane	343	673	3.679
2	Ethane	550	717	4.429
3	Propane	666	642	4.803
4	Butane	766	544	5.002
5	Pentane	847	482	5.128
6	Hexane	915	433	5.216
7	Heptane	972	394	5.285
8	Octane	1025	362	5.340
9	Nonane	1073	332	5.382
10	Decane	1114	308	5.414
12	Dodecane	1185	272	5.459
14	Tetradecane	1248	244	5.483

Example: Find volumetric efficiency of the previous reciprocating pump example with the following new conditions:

Type of Pump	3 inch dia plunger x 5 inch stroke triplex
Liquid pumped	Propane
Suction temperature	70 F
Discharge temperature	80 F
Suction pressure	242 psig
Discharge pressure	1911 psig

Find density at suction pressure:

$$T_{rs} = \frac{t_s}{T_c} = \frac{460 + 70}{666} = .795$$

$$P_{rs} = \frac{P_s}{P_c} = \frac{257}{642} = .4$$

$$\frac{\rho_1}{\omega_1} = 4.803 \text{ (From Table 3, propane)}$$

$$\omega = .1048$$

$$\rho_s = \frac{\rho_1}{\omega_1} \times \omega \times 62.4$$

$$= 4.803 \times .1048 \times 62.4$$
$$= 31.4 \text{ lb per cu ft}$$

Find density at discharge pressure:

$$T_{rd} = \frac{t_d}{T_c} = \frac{460 + 80}{666} = .81$$

$$P_{rd} = \frac{P_d}{P_c} = \frac{1926}{642} = 3.0$$

$$\omega = .1089$$

$$\rho_d = \frac{\rho_1}{\omega_1} \times \omega \times 62.4$$

$$= 4.803 \times .1089 \times 62.4$$
$$= 32.64 \text{ lb per cu ft}$$

Therefore

$$\text{Vol. Eff.} = 1 - \left[S - \frac{c}{d} \left(1 - \frac{\rho_d}{\rho_s} \right) \right]$$

$$= 1 - \left[.02 - \frac{127.42}{35.343} \left(1 - \frac{32.64}{31.4} \right) \right]$$

$$= .8376$$

$$= 83.76 \text{ per cent}$$

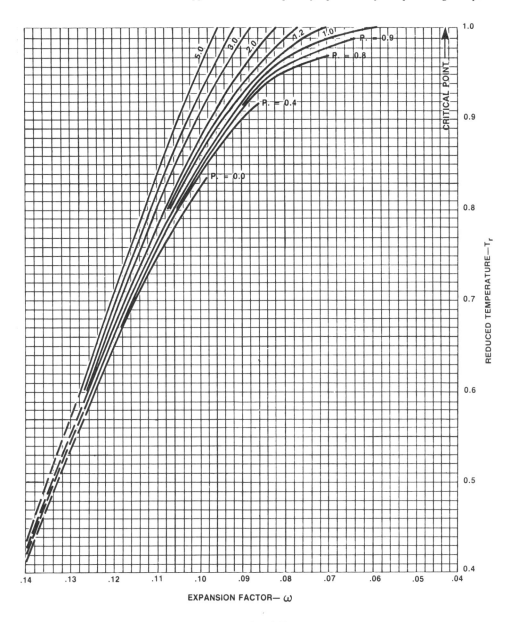

Expansion Factor vs. Reduced Temperature

Chapter 9

Vacuum Pumps*

An amazing number of process plant applications require vacuum systems for continuous or intermittent services. Central vacuum systems are often found in power stations and in industrial and marine installations for the purpose of priming large centrifugal pumps; similar systems provide vacuum for surgical and clinical purposes in hospitals, dentist's offices, and laboratories.

In sterilization processes, vacuum pumps remove air to permit rapid penetration of steam and/or gas into articles requiring sterilization. Food processing plants utilize vacuum systems for anything from poultry evisceration to final packaging in bags, windows, or blister packs for enhanced appearance and marketability. Vacuum filtration is used for sewage, chemicals, foods, and many mined or fibrous products.

Many condensing processes utilize vacuum pumps to evacuate condensers and to remove air leakage and noncondensibles. Smaller installations use liquid ring vacuum pumps in a dual role, removing condensates as well as noncondensibles, sometimes utilizing the cooled condensate as a service liquid. Vacuum pumps are also used in impregnation and metallurgical treatment processes where air and gas have to be removed prior to impregnation with appropriate chemical fluids; or prior to the application of diffusion coatings, or in advanced ion and plasma-type surfacing techniques.

Vacuum pumps are used for drying, distillation, and evaporation. Lower boiling temperatures attained under vacuum preserve nutrients and improve taste, quality, and shelf life of products such as candies, jams, pharmaceuticals, and many mild products. Deaeration is needed for products such as meat pastes, sauces, soups, cellulose, latex, bricks, tiles, sewer pipes, and pottery clay. Also, vacuum conveyance of dangerous, viscous, contaminated, powdery, flaky, bulky, or simply hard-to-handle materials or products is used. The author remembers the ease and utter simplicity with which laminated plastic toothpaste tubes are transferred in partially evacuated, transparent plastic pipes from the forming machine at one end of the plant to the filling equipment at the other end of the building.

With a profusion of processes and applications thus benefiting from vacuum pumps, it is not surprising that many different types and styles, sizes and models, configurations and variations of vacuum producing machinery are available to the user. The familiar steam, gas, and fluid jet injectors/eductors must be acknowledged as prime vacuum producers; however, we will only mention them in passing because they lack moving parts and thus do not fit our definition of "machinery."

Vacuum pumps are often classified in two broad categories: dry type and liquid type. Dry types include lobe, rotary piston, sliding vane, and even diaphragm pumps. Liquid vacuum pumps include liquid jet and liquid ring pumps. Figures 9–1 and 9–2 show the operating ranges for many of these pumps. It should be noted that there is considerable overlap among ranges.

---

*Sources: As acknowledged in captions to illustrations.

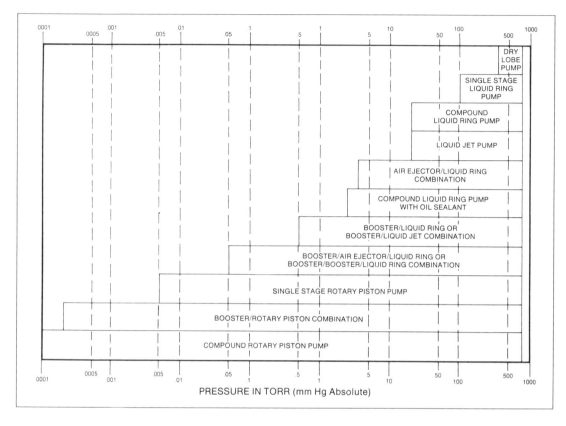

FIGURE 9–1 *Typical pressure ranges for various vacuum pumping devices. (Source: Stokes Division of Pennwalt Corporation, Philadelphia, PA.)*

The most important vacuum producers and their respective operating modes and features are of interest to us in the order listed in Figure 9–1.

SINGLE-STAGE LIQUID RING PUMPS

Figure 9–3 depicts the operating principle of a liquid ring pump. Its circular pump body (A) contains a rotor that consists of a shaft and impeller (B). Shaft and impeller center-lines are positioned parallel, but eccentrically offset relative to the centerline of the pump body. The amount of eccentricity is related to the depth of the liquid ring (C). The liquid ring is formed by introducing service liquid, normally water, via the pump suction casing (L) and through the channel (D) positioned in the suction port plate (E). The centrifugal action of the rotating impeller forces the liquid toward the periphery of the pump body. By controlling the amount of service liquid within the pump body where the impeller blades are completely immersed to their root at one extreme (F) and all but their tips exposed at the other extreme (G), optimum pumping performance will be attained.

When this pumping action is achieved, the vapor to be handled is induced through the suction port (H) when the depth of impeller blade immersion is being decreased. Then as the immersion increases, the vapor is compressed and discharged through the discharge port (J) in the intermediate port plate (K). As there is no metal-to-metal con-

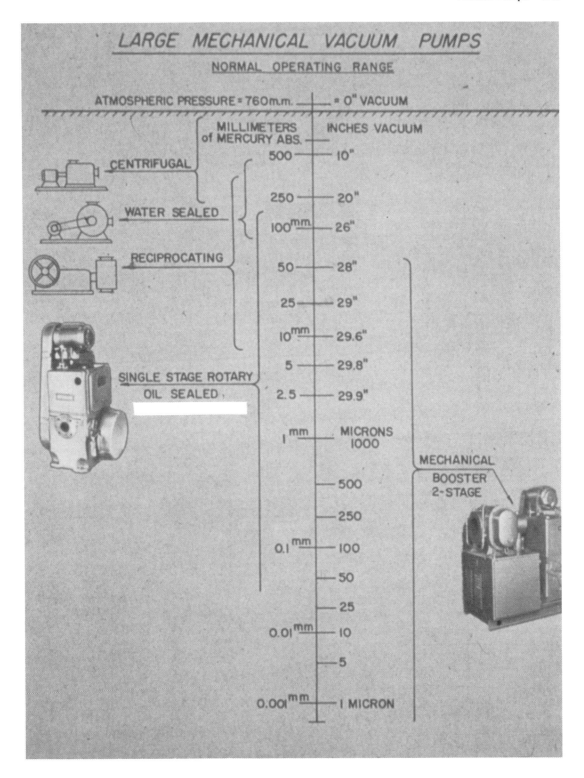

FIGURE 9–2 *Operating range of large mechanical vacuum pumps. (Source: Stokes Division of Pennwalt Corporation, Philadelphia, PA.)*

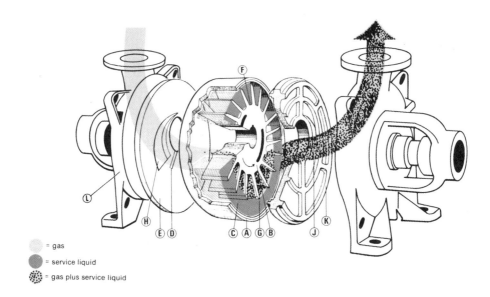

= gas

= service liquid

= gas plus service liquid

FIGURE 9–3   *Operating principle of liquid vacuum pumps. (Source: SIHI Pumps, Inc., Grand Island, NY.)*

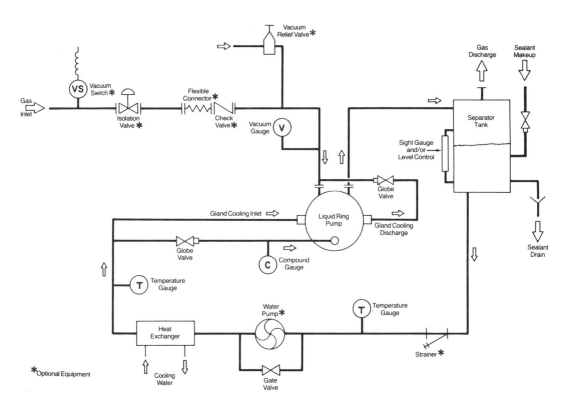

FIGURE 9–4   *Liquid ring vacuum pumping system with full sealant recovery. (Source: Kinney Vacuum Company, Boston, MA.)*

tact between the impeller and the pump body and intermediate plates, the need for lubrication is eliminated and wear is reduced to a minimum.

During the compression cycle heat is being imparted to the liquid ring. In order to maintain a temperature below the vapor point, cooling must be applied. This cooling is achieved by continuously adding a cool supply of service liquid to the liquid ring. The amount of coolant added is equal to that discharged through the discharge port (J) together with the compressed vapor. The mixture of vapor and liquid is then passed to subsequent stages and eventually through the pump discharge for separation.

An entire vacuum pumping system is shown in Figure 9-4. This so-called full sealant recovery system is used to conserve sealant and/or where suitable or compatible sealant is not available from an outside source. Periodic sealant makeup and/or purge may be required. Full recirculation of sealant is provided from the discharge separator tank. Cooling is provided by running recirculated sealant through a heat exchanger. Separate cooling liquid or gas is required.

LIQUID JET VACUUM PUMPS

A typical liquid jet pump is illustrated in Figure 9-5. A centrifugal pump circulates water (the usual hurling liquid) through the multijet nozzle and venturi and returns it to the separation chamber. The water, forced at high velocity across the gap between the nozzle and venturi, entrains the air and gases in multiple jet streams, creating a smooth, steady vacuum in the air suction line and vacuum system. This mixture is discharged through the venturi tangentially into the separation chamber, causing the water in the separation chamber to rotate, which results in a centrifugal action that forces the water to the periphery of the chamber, while the air is separated and discharged. When the hurling liquid is water, it is cooled by a continuous flow of cooling water into the separation chamber. Where process requirements allow and economy is an important factor, automatic controls and other cooling methods are often utilized.

Typical Liquid Jet Vacuum Pump

A

Cutaway View of Liquid Jet Vacuum Pump

B

FIGURE 9-5   *(A) Typical liquid jet vacuum pump. (B) Cutaway view of liquid vacuum pump. (Source: Kinney Vacuum Company, Boston, MA.)*

AIR EJECTOR AND/OR BOOSTER LIQUID RING PUMPS

Air ejectors or mechanical booster pumps are often used upstream of liquid ring pumps for applications requiring higher pumping speeds and lower pressures. These systems are particularly suited for processes where freedom from oil in the pumping system is required, such as oxygen handling, or where pump oil contamination makes the use of oil-sealed pumps impractical or expensive.

Figure 9–6 shows how the addition of air ejectors or mechanical booster pumps can extend the range of liquid ring vacuum pumps.

Ejector/Liquid Ring   Booster/Liquid Ring

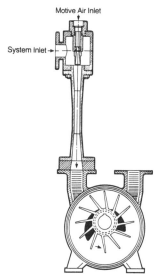

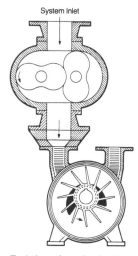

End view of air ejector backed by a liquid ring vacuum pump.

End view of mechanical booster pump backed by liquid ring vacuum pump.

Typical Combinations of Liquid Ring Pumps with Mechanical Boosters and Air Ejectors.

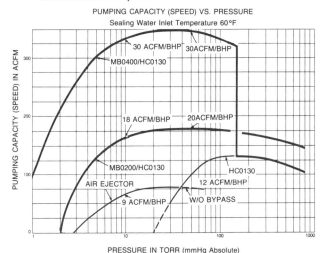

FIGURE 9–6   *Liquid ring pumps with air ejector and mechanical booster. (Source: Kinney Vacuum Company, Boston, MA.)*

FIGURE 9–7   *Multistage vacuum unit. (Source: Sulzer-Burckhardt, Zurich and Basel, Switzerland.)*

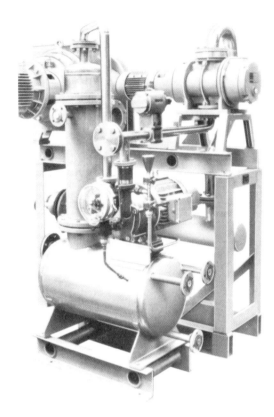

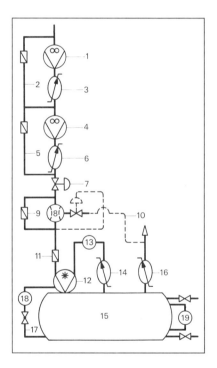

FIGURE 9–8   *Example of a four-stage vacuum unit with closed liquid ring vacuum pump circuit. 1 = Rotary blower, first stage; 2 = bypass line, first stage; 3 = intermediate cooler, first stage; 4 = rotary blower, second stage; 5 = bypass line, second stage; 6 = intermediate cooler, second stage; 7 = quick-acting valve; 8 = gas ejector; 9 = bypass line for gas ejector; 10 = drive gas line; 11 = nonreturn valve; 12 = liquid ring vacuum pump; 13 = separator; 14 = ring liquid cooler; 15 = ring liquid tank; 16 = condenser; 17 = ring liquid feed line; 18 = flowmeter; 19 = level indicator. (Source: Sulzer-Burckhard, Zurich and Basel, Switzerland.)*

MULTISTAGE COMBINATION UNITS

Multistage vacuum units consist of one or two rotary blowers, the necessary intermediate gas coolers, possibly a gas ejector, and a liquid ring vacuum pump. All this equipment is typically mounted on a single-base frame as shown in Figure 9–7.

The entire system is schematically illustrated in Figure 9–8; the rotary blower(s), gas ejector, and liquid ring vacuum pump are connected in series, whereby the extracted gases flow first through the blower, then the gas ejector, and finally the liquid ring vacuum pump. The latter can be operated with fresh water or with a closed ring-liquid circuit. This is always indicated when the evacuated gases and vapors are not

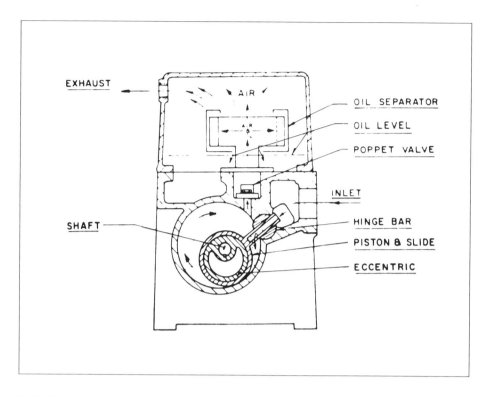

FIGURE 9-9   *Rotary piston oil-sealed vacuum pump. (Source: Stokes Division of Pennwalt Corporation, Philadelphia, PA.)*

allowed to be routed to the atmosphere or discharged into the sewage system for environmental protection reasons, or if the vapors contain valuable raw materials that have to be recovered by means of condensation. In such cases, the self-priming liquid ring vacuum pump draws the liquid out of the tank. The gas and liquid are then separated and cooled in the heat exchanger. The liquid returns to the tank and the gas passes through the exhaust gas cooler, where the condensibles are recovered before leaving the system.

ROTARY OIL-SEALED VACUUM PUMPS

Of the variety of mechanical vacuum pumps described earlier and illustrated in Figure 9-2, the unit most commonly used for high-vacuum work is the rotary piston, oil-sealed, single-stage mechanical vacuum pump (Figure 9-9). This pump type, sometimes called a cam and piston pump, is typically capable of producing ultimate pressures below 10 microns. Its normal operating range is between 0.05 Torr and 100 Torr, and typical evacuation rates range from 10 to 1000 cubic centimeters per minute.

The rotary oil-sealed pump employs an eccentrically mounted piston-slide assembly, rotating at speeds of 400 RPM and higher, depending on pump size. The vacuum "seal" is maintained by a "wedge" of oil ahead of the piston. The oil wedge prevents blow-by of gases and provides lubrication.

In operation, the gas to be evacuated enters the rotary oil-sealed vacuum pump freely through the intake port of the piston-slide. As the oil-sealed piston revolves, it closes the inlet port from the vacuum system and traps ahead of it all air or gas that has entered. With each revolution of the pump, the gas is compressed and discharged through the valve ports to the atmosphere or to a collecting system.

Chapter 10

Cooling Water Supply Systems*

The machinery most closely associated with many cooling water supply systems consists of pumps and fans. In a modern process plant, machines can take on a variety of configurations and range from small ''office-size'' and inexpensive to extremely large and expensive. We have covered pumps in Chapter 7 and will, therefore, concentrate on cooling tower fans and their drive systems.

Mechanical draft towers use either single or multiple fans to provide flow of a known volume of air through the tower. Thus their thermal performance tends toward greater stability and is affected by fewer psychrometric variables than that of the atmospheric towers. The presence of fans also provides a means of regulating air flow, to compensate for changing atmospheric and load conditions, by fan capacity manipulation and/or cycling.

Mechanical draft towers are categorized as either *forced draft* (Figure 10-1), on which the fan is located in the ambient air stream entering the tower and the air is blown through, or *induced draft* (Figure 10-2), wherein a fan located in the exiting air stream draws air through the tower.

Forced draft towers are characterized by high air entrance velocities and low exit velocities. Accordingly, they are extremely susceptible to recirculation and are therefore considered to have less performance stability than the induced draft. Furthermore, located in the cold entering ambient air stream, forced draft fans can become subject to severe icing (with resultant imbalance) when moving air laden with either natural or recirculated moisture.

Usually, forced draft towers are equipped with centrifugal blower type fans, which, although requiring considerably more horsepower than propeller type fans, have the advantage of being able to operate against the high static pressures associated with ductwork. Therefore, they can either be installed indoors (space permitting) or within a specially designed enclosure that provides significant separation between intake and discharge locations to miminize recirculation.

Induced draft towers have an air discharge velocity of from three to four times higher than their air entrance velocity, with the entrance velocity approximating that of a 5-mph wind. Therefore, there is little or no tendency for a reduced pressure zone to be created at the air inlets by the action of the fan alone. The potential for recirculation on an induced draft tower is not self-initiating and, therefore, can be more easily quantified purely on the basis of ambient wind conditions. Location of the fan within the warm air stream provides excellent protection against the formation of ice on the mechanical components. Widespread acceptance of induced draft towers is

*Source: The Marley Cooling Tower Company, Mission, KS, and Kansas City, MO. Adapted by permission.

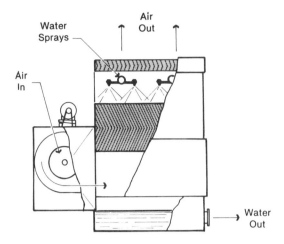

FIGURE 10–1   *Forced draft, counterflow, blower fan tower. (Source: The Marley Cooling Tower Company, Mission, KS, and Kansas City, MO.)*

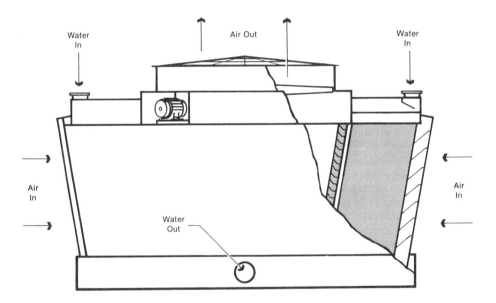

FIGURE 10–2   *Induced draft, crossflow, propeller fan tower. (Source: The Marley Cooling Tower Company, Mission, KS, and Kansas City, MO.)*

evidenced by their existence on installations as small as 15 gallons per minute (gpm) and as large as 700,000 gpm.

Hybrid draft towers (Figure 10–3) can give the outward appearance of being natural draft towers with relatively short stacks. Internal inspection (Figure 10–4), however, reveals that they are also equipped with mechanical draft fans to augment air flow. Consequently, they are also referred to as *fan-assisted natural draft* towers. The intent of their design is to minimize the horsepower required for air movement, but to do so with the least possible stack cost impact. Properly designed, the fans may need to be operated only during periods of high ambient and peak loads. In localities where

FIGURE 10–3 *Fan-assisted natural draft tower. (Source: The Marley Cooling Tower Company, Mission, KS, and Kansas City, MO.)*

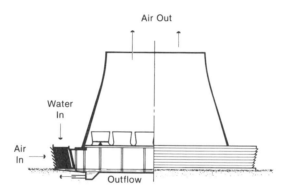

FIGURE 10–4 *Cutaway of fan-assisted draft tower. (Source: The Marley Cooling Tower Company, Mission, KS, and Kansas City, MO.)*

a low-level discharge of the tower plume may prove to be unacceptable, the elevated discharge of a fan-assisted natural draft tower can become sufficient justification for its use.

CHARACTERIZATION BY AIR FLOW

Cooling towers are also classified by the relative flow relationship of air and water within the tower, as follows:

In *counterflow* towers (Figure 10–5), air moves vertically upward through the fill, counter to the downward fall of water. Because of the need for extended intake and discharge plenums, the use of high pressure spray systems, and the typically higher air pressure losses, some of the smaller counterflow towers are physically higher, require more pump head, and utilize more fan power than their crossflow counterparts. In larger counterflow towers, however, the use of low-pressure, gravity-related distribu-

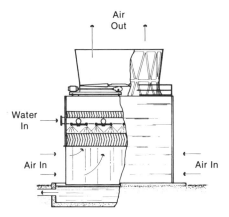

FIGURE 10–5   *Induced draft, counterflow tower. (Source: The Marley Cooling Tower Company, Mission, KS, and Kansas City, MO.)*

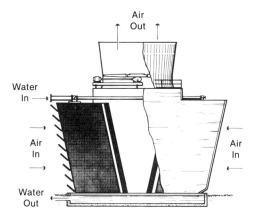

FIGURE 10–6   *Double-flow, crossflow tower. (Source: The Marley Cooling Tower Company, Mission, KS, and Kansas City, MO.)*

tion systems, plus the availability of generous intake areas and plenum spaces for air management, is tending to equalize, or even reverse, this situation. The enclosed nature of a counterflow tower also restricts exposure of the water to direct sunlight, thereby retarding the growth of algae.

Crossflow towers (Figure 10-6) have a fill configuration through which the air flows horizontally, across the downward fall of water. Water to be cooled is delivered to hot water inlet basins located atop the fill areas and is distributed to the fill by gravity through metering orifices in the floor of those basins. This obviates the need for a pressure-spray distribution system and places the resultant gravity system in full view for ease of maintenance. By the proper utilization of flow-control valves, routine cleaning and maintenance of crossflow tower distribution systems can be accomplished sectionally, while the tower continues to operate.

Crossflow towers are also subclassified by the number of fill "banks" and air inlets that are served by each fan. The tower indicated in Figure 10-6 is a *double-flow* tower because the fan is inducing air through two inlets and across two banks of fill. Figure 10-7 depicts a *single-flow* tower having only one air inlet and one fill bank, the

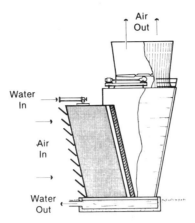

FIGURE 10-7   *Single-flow tower. (Source: The Marley Cooling Tower Company, Mission, KS, and Kansas City, MO.)*

remaining three sides of the tower being cased. Single-flow towers are customarily used in locations where an unrestricted air path to the tower is available from only one direction. They are also useful in areas having a dependable prevailing wind direction, where consistent process temperatures are critical. The tower can be sited with the air inlet facing the prevailing wind, and any potential for recirculation is negated by the downwind side of the tower being a cased face. *Spray-filled* towers have no heat transfer (fill) surface, depending only upon the water break-up afforded by the distribution system to promote maximum water-to-air contact. Removing the fill from the tower in Figure 10-5 would also make it "spray-filled." The use of such towers is normally limited to those processes where higher water temperatures are permissible. They are also utilized in those situations where excessive contaminants or solids in the circulating water would jeopardize a normal heat transfer surface.

CHARACTERIZATION BY CONSTRUCTION

Field-erected towers are those on which the primary construction activity takes place at the site of ultimate use. All large towers, and many of the smaller towers, are prefabricated, piece-marked, and shipped to the site for final assembly. Labor and/or supervision for final assembly is usually provided by the cooling tower manufacturer.

Factory-assembled towers undergo virtually complete assembly at their point of manufacture, whereupon they are shipped to the site in as few sections as the mode of transportation will permit. A relatively small tower would ship essentially intact. Larger, multicell towers are assembled as "cells" or "modules" at the factor and are shipped with appropriate hardware for ultimate joining by the user. Factory-assembled towers are also known as "packaged" or "unitary" towers.

CHARACTERIZATION BY SHAPE

Rectilinear towers (Figure 10-8) are constructed in cellular fashion, increasing linearly to the length and number of cells necessary to accomplish a specified thermal performance.

Round Mechanical Draft (RMD) towers, as the name implies, are essentially round in plan configuration, with fans clustered as close as practicable around the centerpoint

FIGURE 10–8   *Multicelled, field-erected, crossflow cooling tower with enclosed stairway and extended fan deck to enclose piping and hot water basins. (Source: The Marley Cooling Tower Company, Mission, KS, and Kansas City, MO.)*

FIGURE 10–9   *Octagonal mechanical draft (OMD) counterflow cooling tower. (Source: The Marley Cooling Tower Company, Mission, KS, and Kansas City, MO.)*

of the tower. Multifaceted towers, such as the octagonal mechanical draft (OMD) depicted in Figure 10–9, also fall into the general classification of "round" towers. Such towers can handle enormous heat loads with considerably less site area impact than that required by multiple rectilinear towers. Additionally, they are significantly less affected by recirculation.

CHARACTERIZATION BY METHOD OF HEAT TRANSFER

All of the cooling towers heretofore described are *evaporative*-type towers, in that they derive their primary cooling effect from the evaporation that takes place when air and

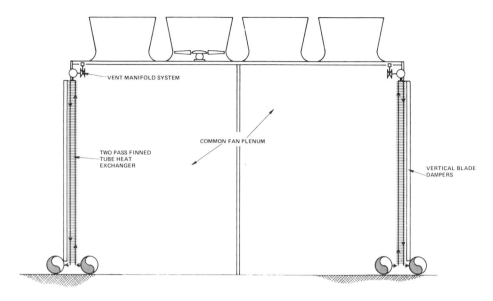

FIGURE 10–10   *Dry-type cooling tower, cross-sectional elevation. (Source: The Marley Cooling Tower Company, Mission, KS, and Kansas City, MO.)*

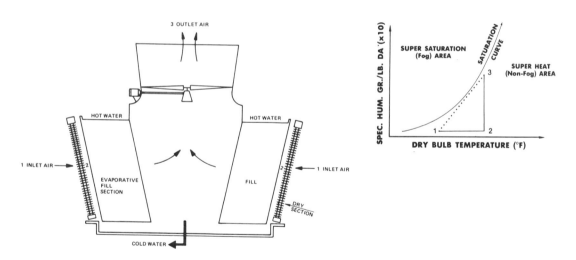

FIGURE 10–11   *Plume abatement tower and psychrometrics (coil before fill). (Source: The Marley Cooling Tower Company, Mission, KS, and Kansas City, MO.)*

water are brought into direct contact. At the other end of the spectrum is the *dry tower* (Figure 10–10), where, by full utilization of dry surface coil sections, no direct contact (and no evaporation) occurs between air and water. Hence the water is cooled totally by sensible heat transfer.

In between these extremes are the *plume abatement* (Figure 10–11) and *water conservation* (Figure 10–12) towers, wherein progressively greater portions of dry surface coil sections are introduced into the overall heat transfer system to alleviate specific problems, or to accomplish specific requirements.

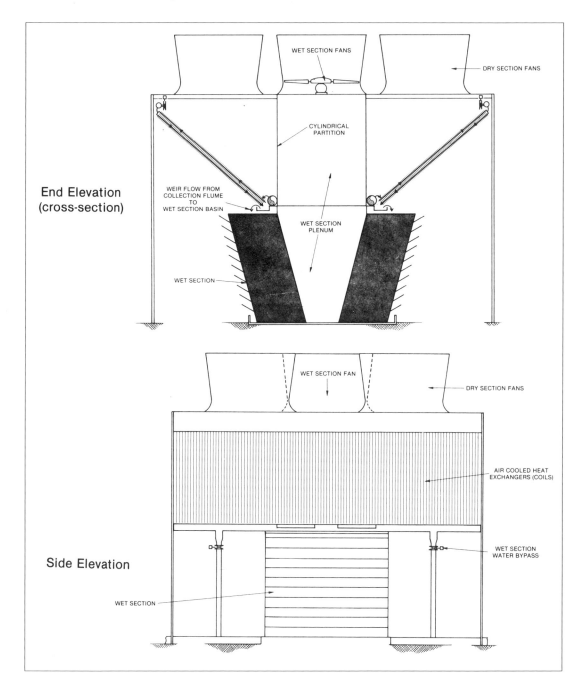

FIGURE 10–12   *Water conservation cooling tower. (Source: The Marley Cooling Tower Company, Mission, KS, and Kansas City, MO.)*

MECHANICAL COMPONENT REVIEW

Cooling tower mechanical equipment is required to operate within a highly corrosive moisture-laden atmosphere that is unique to the cooling tower industry, and the historical failure rate of commercially available components caused reputable tower manufacturers to undertake their own production of specific components some years ago. Currently, the low failure rate of manufacturer-produced components reinforces that decision. Purchasers also benefit from the advantage of single-source responsibility for warranty and replacement parts.

Exclusive of motors, the mechanical components basic to the operation of the cooling tower are fans, speed reducers, drive shafts, and water flow control valves.

Fans

Cooling tower fans must move large volumes of air efficiently and with minimum vibration. The materials of manufacture must not only be compatible with their design, but must also be capable of withstanding the corrosive effects of the environment in which the fans are required to operate. Their importance to the ability of mechanical draft cooling towers to perform is reflected in the fact that fans of improved efficiency and reliability are the object of continuous development.

Propeller Fans

Propeller-type fans predominate in the cooling tower industry because of their ability to move vast quantities of air at the relatively low static pressures encountered. They are comparatively inexpensive, may be used on any size tower, and can develop high overall efficiencies when "system designed" to complement a specific tower structure—fill—fan cylinder configuration. Most-utilized diameters range from 24 inches to 10 meters (Figure 10–13), operating at horsepowers from 1/4 to 250+. Although the use of larger fans, at higher power input, is not without precedence, their application naturally tends to be limited by the number of projects of sufficient size to warrant their consideration. Fans 48 inches and larger in diameter are equipped with adjustable pitch blades, enabling the fans to be applied over a wide range of operating horsepowers. Thus the fan can be adjusted to deliver the precise required amount of air at the least power consumption.

The rotational speed at which a propeller fan is applied typically varies in inverse proportion to its diameter. The smaller fans turn at relatively high speeds, whereas the larger ones turn somewhat slower. This speed-diameter relationship, however, is by no means a constant one. If it were, the blade tip speeds of all cooling tower fans would be equal. The applied rotational speed of propeller fans usually depends on best ultimate efficiency, and some diameters operate routinely at tip speeds approaching 14,000 feet per minute. However, since higher tip speeds are associated with higher sound levels, it is sometimes necessary to select fans turning at slower speeds to satisfy a critical requirement.

The increased emphasis on reducing cooling tower operating costs has resulted in the use of larger fans to move greater volumes of air more efficiently. Much research has also gone into the development of more efficient blade, hub, and fan cylinder designs. The new generations of fans are light in weight to reduce parasitic energy losses, and have fewer, but wider, blades to reduce aerodynamic drag. Moreover, the characteristics of air flow through the tower, from inlet to discharge, are analyzed and appropriate adjustments to the structure are made to minimize obstructions; fill and distribution systems are designed and arranged to promote maximum uniformity of air and water flow; and drift eliminators are arranged to direct the final pass of air toward

FIGURE 10–13   *Typical large-diameter fan utilized on cooling towers. (Source: The Marley Cooling Tower Company, Mission, KS, and Kansas City, MO.)*

the fan. This is recognized as the ''systems'' approach to fan design, without which the best possible efficiency *cannot* be obtained.

The intent of good propeller fan design is to achieve air velocities across the effective area of the fan, from hub to blade tips, that are as uniform as possible. The most effective way to accomplish this is with tapered and twisted blades having an airfoil cross section. Historically, cast aluminum alloys have been the classic materials used for production of this blade type. Cast aluminum blades continue to be utilized because of their relatively low cost, good internal vibration damping characteristics, and resistance to corrosion in most cooling tower environments.

Currently, lighter blades of exceptional corrosion resistance are made of fiberglass-reinforced plastic, cast in precision molds. These blades may be solid, formed around a permanent core, or formed hollow by the use of a temporary core. In all cases, they have proved to be both efficient and durable as long as the design avoided aerodynamically induced vibration resonance.

Fan hubs must be of a material that is structurally compatible with blade weight and loading, and must have good corrosion resistance. Galvanized steel weldments, gray and ductile iron castings, and wrought or cast aluminum are in general use as hub materials. Where hub and blades are of dissimilar metals, they must be insulated from each other to prevent electrolytic corrosion.

Smaller diameter fans are customarily of galvanized sheet metal construction with fixed-pitch nonadjustable blades. These fans are matched to differing air flow requirements by changing the design speed.

Automatic Variable-Pitch Fans

These are propeller fans on which a pneumatically actuated hub controls the pitch of the blades in unison (Figure 10–14). Their ability to vary airflow through the tower in response to a changing load or ambient condition—coupled with the resultant energy savings and ice control—make them an optional feature much in demand.

Centrifugal Fans

These are usually of the double inlet type, used predominantly on cooling towers designed for indoor installations (Figure 10–15). Their capability to operate against relatively high static pressures makes them particularly suitable for that type of application. However, their inability to handle large volumes of air, and their characteristically high input horsepower requirement (approximately twice that of a propeller fan), limit their use to relatively small applications.

FIGURE 10–14 *Automatic, variable-pitch fan used to adjust air flow and fan horsepower. (Source: The Marley Cooling Tower Company, Mission, KS, and Kansas City, MO.)*

Three types of centrifugal fans are available: (1) forward curved blade fans, (2) radial blade fans, and (3) backward curved blade fans. The characteristics of the forward curved blade fan make it the most appropriate type of cooling tower service. By virtue of the direction and velocity of the air leaving the fan wheel, the fan can be equipped with a comparatively small size housing, which is desirable from a structural standpoint. Also, because the required velocity is generated at a comparatively low speed, forward curved blade fans tend to operate quieter than other centrifugal types.

Centrifugal fans are usually of sheet metal construction, with the most popular protective coating being hot-dip galvanization. Damper mechanisms are also available to facilitate capacity control of the cooling tower.

FIGURE 10–15 *Blower-type cooling tower fan. (Source: The Marley Cooling Tower Company, Mission, KS, and Kansas City, MO.)*

Fan Laws

All propeller-type fans operate in accordance with common laws. For a given fan and cooling tower system, the following is true:

1. The capacity (cfm) varies directly as the speed (RPM) ratio, and directly as the pitch angle of the blades relative to the plane of rotation.
2. The static pressure (h_s) varies as the square of the capacity ratio.
3. The fan horsepower varies as the cube of the capacity ratio.
4. At constant cfm, the fan horsepower and static pressure vary directly with the air density.

If, for example, the capacity (cfm) of a given fan were decreased by 50 percent (either by a reduction to half of design rpm, or by a reduction in blade pitch angle at constant speed), the capacity ratio would be 0.5. Concurrently, the static pressure would become 25 percent of before, and the fan horsepower would become 12.5 percent of before. These characteristics afford unique opportunities to combine cold water temperature control with significant energy savings.

Selected formulas, derived from these basic laws, may be utilized to determine the efficacy of any particular fan application:

Symbols

Q = Volume of air handled (cfm). Unit: cu ft per min.
A = Net flow area. Unit: sq ft.
V = Average air velocity at plane of measurement. Unit: ft per sec.
g = Acceleration due to gravity. Unit: 32.17 ft per sec per sec.
D = Density of water at gauge fluid temperature. Unit: lb per cu ft.
d = Air density at point of flow. Unit: lb per cu ft.
h_s = Static pressure drop through system. Unit: inches of water.
h_v = Velocity pressure at point of measurement. Unit: inches of water.
h_t = Total pressure differential (= h_s + h_v). Unit: inches of water.
v_r = Fan cylinder velocity recovery capability. Unit: percent.

Thermal performance of a cooling tower depends on a specific mass flow rate of air through the fill (pounds of dry air per minute), whereas the fan does its job purely in terms of volume (cubic feet per minute). Since the specific volume of air (cubic feet per pound) increases with temperature, it can be seen that a larger volume of air leaves the tower than enters it. The actual cfm handled by the fan is the product of mass flow rate times the specific volume of dry air corresponding to the temperature at which the air leaves the tower. This volumetric flow rate is the "Q" used in the following formulas, and it *must* be sufficient to produce the correct mass flow rate or the tower will be short of thermal capacity.

Utilizing appropriate cross-sectional flow areas, velocity through the fan and fan cylinder can be calculated as follows:

$$V = \frac{Q}{A \times 60}$$

It must be understood that "A" will change with the plane at which velocity is being calculated. Downstream of the fan, "A" is the gross cross-sectional area of the fan cylinder. *At* the fan, "A" is the area of the fan *less* the area of the hub or hub cover.

Velocity pressure is calculated as follows:

$$h_v = \frac{V^2 \times 12 \times d}{2 \times g \times D}$$

If V represents the velocity through the fan, then h_v represents the velocity pressure for the fan itself (h_{vf}). Moreover, if the fan is operating within a nonflared-discharge fan cylinder, this effectively represents the total velocity pressure because of no recovery having taken place.

However, if the fan is operating within a flared, velocity-recovery-type fan cylinder (Figure 10–16), h_v must be recalculated for the fan cylinder exit (h_{ve}), at the appropriate velocity, and applied in the following formula to determine total velocity pressure:

$$h_v = h_{vf} - [(h_{vf} - h_{ve}) \times v_r]$$

Although the value of v_r will vary with design expertise and is empirically established, a value of 0.75 (75 percent recovery) is normally assigned for purposes of anticipating fan performance within a reasonably well-designed velocity-recovery cylinder.

The power output of a fan is expressed in terms of air horsepower (ahp) and represents work done by the fan:

$$\text{ahp} = \frac{Q \times h_t \times D}{33,000 \times 12}$$

Static air horsepower is obtained by substituting static pressure (h_s) for total pressure (h_t) in the formula.

A great deal of research and development goes into the improvement of fan efficiencies, and those manufacturers that have taken a systems approach to this research and development effort have achieved results that, although incrementally small, are highly significant in the light of current energy costs. Static efficiencies and overall mechanical (total) efficiencies are considered in the selection of a particular fan in a specific situation, with the choice usually going to the fan that delivers the required volume of air at the least input horsepower:

$$\text{Static Efficiency} = \frac{\text{static ahp}}{\text{input hp}}$$

$$\text{Total Efficiency} = \frac{\text{ahp}}{\text{input hp}}$$

It must be understood that input hp is measured at the fan shaft and does not include the drive-train losses reflected in actual motor brake horsepower (bhp). Input hp will normally average approximately 95 percent of motor bhp on larger fan applications.

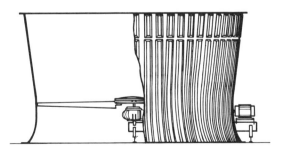

FIGURE 10–16 *Cutaway view of a velocity-recovery-type fan cylinder. (Source: The Marley Cooling Tower Company, Mission, KS, and Kansas City, MO.)*

Speed Reducers

The optimum speed of a cooling tower fan seldom coincides with the most efficient speed of the driver (motor). This dictates that a speed reduction, power transmission unit of some sort be situated between the motor and the fan. In addition to reducing the speed of the motor to the proper fan speed (at the least possible loss of available power), the power transmission unit must also provide primary support for the fan, exhibit long-term resistance to wear and corrosion, and contribute as little as possible to the overall noise level.

Speed reduction in cooling towers is accomplished either by differential gears of positive engagement or by differential pulleys (sheaves) connected through V-belts. Typically, gear reduction units are applied through a wide range of horsepower ratings, from the very large down to as little as 5 hp. V-belt drives, on the other hand, are usually applied at ratings of 50 hp or less.

Gear Reduction Units

Gear speed reducers are available in a variety of designs and reduction ratios to accommodate the fan speeds and horsepowers encountered in cooling towers (Figure 10–17). Because of their ability to transmit power at minimal loss, spiral bevel and helical gear sets are most widely utilized, although worm gears are also used in some designs. Depending on the reduction ratio required and the input hp, a gear speed reducer may use a single type gear or a combination of types to achieve "staged" reduction. Generally, two-stage reduction units are utilized for the large, slower-turning fans requiring input horsepowers exceeding 75 bhp.

The service life of a speed reducer or speed increaser is directly related to the surface durability of the gears, as well as the type of service imposed (i.e., intermittent versus continuous duty). The American Gear Manufacturers Association (AGMA) has established criteria for the rating of geared speed reducers, which are subscribed to by most reliable designers. AGMA Standard 420 defines these criteria and offers a list of suggested service factors. The gear speed reducer manufacturer should have established service factors for an array of ratios, horsepowers, and types of service, com-

FIGURE 10–17   *Gear speed reducer used for applied horsepowers above 75 HP. (Source: The Marley Cooling Tower Company, Mission, KS, and Kansas City, MO.)*

mensurate with good engineering practice. The Cooling Tower Institute (CTI) Standard 111 offers suggested service factors specifically for cooling tower applications.

Long, trouble-free life is also dependent on the quality of bearings used. Bearings are normally selected for a calculated life compatible with the expected type of service. Bearings for industrial cooling tower gear speed reducers (considered as continuous duty) should be selected on the basis of a 100,000-hour L-10 life. L-10 life is defined as the life expectancy in hours during which 90 percent or more of a given group of bearings under specific loading condition will still be in service. Intermittent duty applications provide satisfactory life with a lower L-10 rating. An L-10 life of 35,000 hours is satisfactory for an 8- to 10-hour per day application. It is equivalent, in terms of years of service, to a 100,000-hour L-10 life for continuous duty.

Lubrication aspects of a gear speed reducer, of course, are as important to longevity and reliability as are the components that compose the gear speed reducer. The lubrication system should be of a simple, noncomplex design, capable of lubricating equally well in both forward and reserve operation. Remote oil level indicators and convenient location of fill and drain lines simplify and encourage preventive maintenance. Lubricants and lubricating procedures recommended by the manufacturer should be adhered to closely. Synthetic lubricants may greatly increase both gear life and lubricant drainage intervals.

V-Belt Drives

These are an accepted standard for the smaller factory-assembled cooling towers, although most of the larger unitary towers are equipped with gear speed reducers. Correctly designed and installed, and well maintained, V-belt drives can provide very dependable service. The drive consists of the motor and fan sheaves, the bearing housing assembly supporting the fan, and the V-belts.

V-belts (as opposed to cog belts) are used most commonly for cooling tower service. A variety of V-belt designs is available, offering a wide assortment of features. Most of these designs are suitable for cooling tower use. In many cases, more than one belt is required to transmit power from the motor to the fan. Multiple belts must be supplied either as matched sets, measured and packaged together at the factory, or as a banded belt having more than one V-section on a common backing.

Various types of bearings and bearing housing assemblies are utilized in conjunction with V-belt drives. Generally, sleeve bearings are used on smaller units and ball or roller bearings on the larger units, with oil being the most common lubricant. In all cases, water slinger seals are recommended to prevent moisture from entering the bearing, and oil mist is often used as a purge medium to prevent moisture condensation in bearing housings.

Belts wear and stretch, and belt tension must be periodically adjusted. Means for such adjustment should be incorporated as part of the motor mount assembly. Stability and strength of the mounting assembly is of prime importance in order to maintain proper alignment between driver and driven sheaves (Figure 10-18). Misalignment is one of the most common causes of excessive belt and sheave wear.

Manually adjustable pitch sheaves are occasionally provided to allow a change in fan speed. These are of advantage on indoor towers, where the ability to adjust fan speed can sometimes compensate for unforeseen static pressure.

Drive Shafts

The drive shaft transmits power from the output shaft of the motor to the input shaft of the gear speed reducer. Because the drive shaft operates within the tower, it must be highly corrosion resistant. Turning at full motor speed, it must be well balanced—

FIGURE 10–18   *Adjustable motor mount for a V-belt driven fan (belt guard removed). (Source: The Marley Cooling Tower Company, Mission, KS, and Kansas City, MO.)*

and capable of being rebalanced. Transmitting full motor power over significant distances, it must accept tremendous torque without deformation. Subjected to long-term cyclical operation and occasional human error, it must be capable of accepting some degree of misalignment.

Drive shafts are described as "floating" shafts, equipped with flexible couplings at both ends. Where only normal corrosion is anticipated and cost is of primary consideration, shafts are fabricated of carbon steel, hot-dip galvanized after fabrication (Figure 10–19). Shafts for larger industrial towers, and those that will be operating in atmospheres more conducive to corrosion, are usually fabricated of tubular stainless

FIGURE 10–19   *Driveshaft in a relatively small fan drive application. (Source: The Marley Cooling Tower Company, Mission, KS, and Kansas City, MO.)*

FIGURE 10–20 *Close-up of a larger driveshaft showing guard. (Source: The Marley Cooling Tower Company, Mission, KS, and Kansas City, MO.)*

steel (Figure 10–20) or epoxy-coated wound carbon filament. The yokes and flanges that connect to the motor and gear speed reducer shafts are of cast or welded construction, in a variety of materials compatible with that utilized for the shaft.

Flexible couplings transmit the load between the driveshaft and the motor or gear speed reducer, and compensate for minor misalignment. A suitable material for use in the saturated effluent air stream of a cooling tower is neoprene, either in solid grommet form (Figure 10–21) or as neoprene-impregnated fabric (Figure 10–22), designed to require no lubrication and relatively little maintenance. Excellent service records have been established by the neoprene flexible couplings, both as bonded bushings and as impregnated fabric disc assemblies. These couplings are virtually impervious to corrosion and provide excellent flexing characteristics.

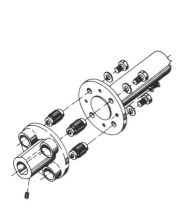

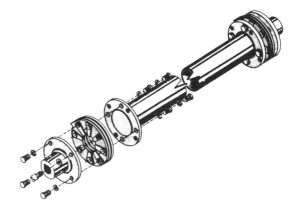

FIGURE 10–21   **FIGURE 10–22**

It is very important that drive shafts be properly balanced. Imbalance not only causes tower vibration, but it also induces higher loads and excessive wear on the mechanical equipment coupled to the shaft. Most cooling tower drive shafts operate at speeds approaching 1800 RPM. At these speeds, it is necessary that the shafts be dynamically balanced to reduce vibrational forces to a minimum.

Safety Considerations

Fan cylinders less than six feet high must be equipped with suitable fan guards for the protection of operating personnel. Drive shafts must operate within retaining guards (Figure 10-20) to prevent the drive shaft from encountering the fan if the coupling should fail. The motor shaft and outboard drive shaft coupling should either be within the confines of the fan cylinder or enclosed within a suitable guard.

Chapter 11

Centrifugal Compressors

OVERVIEW OF GAS COMPRESSION MACHINERY

The purpose of compression is simply to increase the pressure of a gas from one level to another. Depending on a host of circumstances and situations, the pressure increase imparted to a gas will be from a fraction of a pound per square inch (psi) (or a few pascals) in laboratory equipment to literally tens of thousands of psi in hypercompressors used for the manufacture of polyethylene.

Before we embark on our more thorough consideration of centrifugal compressors, we should examine gas compression machinery in general. In a typical process plant, compression services include instrument and plant air, combustion air for burners and furnaces, gas circulation, or simple elevation to pressure conditions that will allow chemical reactions to take place. Gas volumes will vary from laboratory quantities to flows well in excess of a million cubic feet per minute ($\sim$ 2 million m^3/hr).

Two principal methods are used to compress gases. The first method is to trap a volume of gas and displace it by the positive action of a piston or rotating member; we call these machines positive-displacement compressors. The second method uses dynamic compression; it is accomplished either by the mechanical action of contoured blades, which impart velocity and hence pressure to the flowing gas, or by the entrainment of gases in a high velocity jet of the same or another gas (e.g., steam). The entrainment principle is generally found in ejectors that operate with inlet pressures below atmospheric pressure, but since our definition of process machinery implies the existence of moving parts, we have elected not to cover ejector devices in this text.

The two major groupings of compression machinery, positive displacement and dynamic, can be further subdivided as shown in Figure 11–1. *Positive-displacement* units are those in which successive volumes of gas are confined within a closed space and elevated to a higher pressure. *Reciprocating compressors* are positive-displacement machines in which the compressing and displacing element is a piston that moves back and forth within a cylinder. *Rotary positive-displacement compressors* are machines in which compression and displacement is effected by the close meshing of rotating elements. *Sliding-vane compressors* are rotary positive-displacement machines in which axial vanes slide radially in a rotor. This rotor is eccentrically mounted in a cylindrical casing. Gas trapped between the vanes is compressed and displaced. *Liquid-piston compressors* are rotary positive-displacement machines in which water or other liquid is used as the piston to compress and displace the gas handled. *Roots®-type straight-lobe compressors* are rotary positive-displacement machines in which two mating lobed impellers trap gas and displace it from intake to discharge. There is no internal compression. *Helical- or spiral-lobe compressors* are rotary positive-displacement machines in which two intermeshing rotors, each with a helical form, compress and displace the gas.

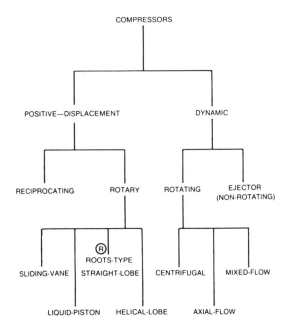

FIGURE 11–1   *Major compressor types.*

Dynamic compressors are continuous-flow machines in which either the rapidly rotating element or ejector nozzle accelerates the gas as it passes through the element. The velocity head is converted into pressure, partially in the rotating element and partially in stationary diffusers or blades. *Centrifugal compressors* are dynamic machines in which one or more rotating impellers, usually shrouded on the sides, accelerate the gas. Main gas flow is radial. *Axial compressors* are dynamic machines in which gas acceleration is obtained by the action of the bladed rotor shrouded on the blade ends. Main gas flow is axial. *Mixed-flow compressors* are dynamic machines with an impeller form combining some characteristics of both the centrifugal and axial types.

Here we will treat centrifugal compressors. Other types of compressors are covered in succeeding chapters.

CENTRIFUGAL COMPRESSORS*

Centrifugal compressors are employed in numerous fields: chemical and petrochemical industries, refineries and fertilizer plants, nuclear reactors and air separation plants, iron and steelworks, production of liquefied natural gas (LNG) and substitute natural gas (SNG), cryogenic and refrigeration plants, mining, transportation and storage of gas, onshore and offshore installations. The range of applications can be expanded still further by combining these centrifugal compressors with other compressor types such as axial-flow or reciprocating compressors.

The wide range of processes in which centrifugal compressors are employed makes varying demands on these machines. Compressor design is dependent on such factors as the fluid handled, the pressure ratio, the volume flow, the number of interstage coolers, injection and extraction of the medium, and the type of shaft sealing.

---

*Source: Mannesmann-Demag, Duisburg, Federal Republic of Germany. Reprinted and adapted from copyrighted material, by permission.

FIGURE 11–2 *Horizontally split compressors in a modern process plant. (Source: Mannesmann Demag, Duisburg, West Germany.)*

Taking all these factors into consideration, the major compressor manufacturers have developed series of centrifugal compressors offering an optimum engineering solution implemented by the use of standard components. These series include the two basic types, distinguished by horizontally or vertically split casing (see Figures 11–2 and 11–3), compressors with two or three pairs of main nozzles, and compressors with additional sidestream nozzles. A few of the many possible nozzle arrangements are depicted in Figure 11–4.

Horizontally split casings with nozzles in the lower half permit simple removal of the rotor and facilitate the checking of labyrinth clearances and O-rings. As pressure levels rise and gas molecules become smaller, vertically split casings are employed.

Horizontally Split Centrifugals

Centrifugal compressors with horizontally split casings typically permit internal pressures of 70 bar with small volume flow rates and volume flow rates of up to 300,000 m³/hr at low pressures. Drive ratings of 30 MW for single-casing machines have already been implemented.

Figure 11–5 shows a cross section of a six-stage horizontally split centrifugal compressor. Standardized components ensure high availability and easy fitting. The two halves of the casing are sealed and bolted together.

The rigid structure is supported at the centerline, thus preventing vertical shifting of the compressor shaft as a result of thermal expansion. For erection and dismantling purposes, the top half of the casing, complete with the associated stationary components, can be handled as a single unit. All types of drivers can be employed, for example, gas turbines, steam turbines, and electric motors.

Many processes require compression of a fluid in one process stage only, i.e., continuous compression from the first to the final stage with constant mass flow rate. Most

FIGURE 11-3   *Vertically split (barrel-type) centrifugal compressor. (Source: Mannesmann Demag, Duisburg, West Germany.)*

major centrifugal compressor manufacturers build machines for this field of application with up to nine or as many as twelve compression stages. Aerodynamic matching of the individual compression stages is by means of diaphragms with diffuser channels and vaned return passages. Following the final stage, the compressed gas enters a collecting chamber in the form of a volute before it reaches the discharge nozzle. The shaft is supported in bearings outside the compression space. Shaft sealing is by means of tried and proven systems such as labyrinth, floating ring, mechanical contact or non-contact seals.

The wide range of possible variations in the materials used and in the selection of the sealing system render compressors of this series suitable for virtually all fields of application in industry, chemical and petrochemical processes, and for almost all gases and mixtures of gases.

Sidestream Compressors

In multistage refrigeration processes, different mass flows pass through the various refrigeration stages. Sidestreams therefore have to be injected into or extracted from the main flow in the compressor at process-dependent intermediate pressures and

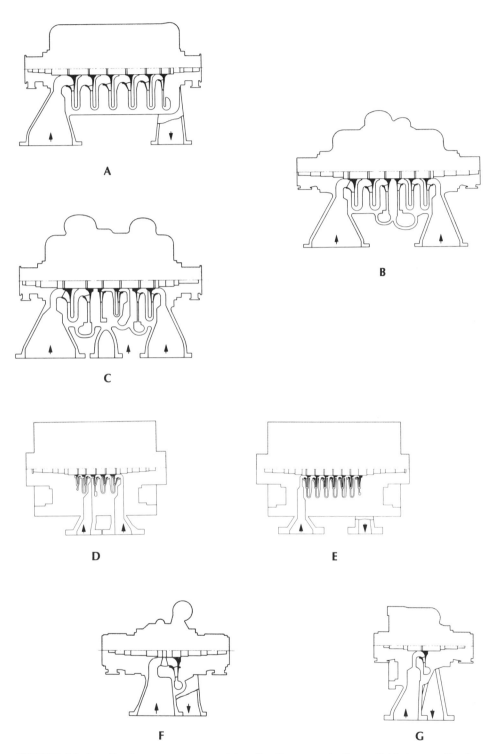

FIGURE 11–4 *Nozzle arrangements and impeller lineups typically available in centrifugal compressors. (A) Straight-through, (B) back-to-back, (C) back-to-back with sidestream entry, (D) straight-through (barrel-type), (E) back-to-back (barrel-type), (F) future addition to stage possible, (G) single stage. (Source: Mannesmann Demag, Duisburg, West Germany.)*

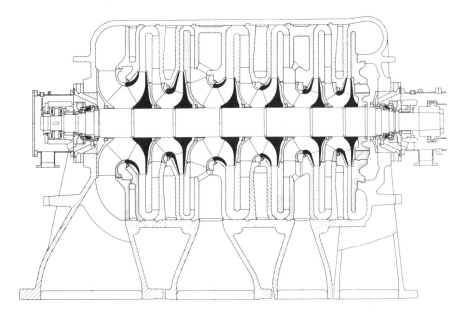

FIGURE 11–5   *Horizontally split compressor with two sidestream entry nozzles. (Source: Mannesmann Demag, Duisburg, West Germany.)*

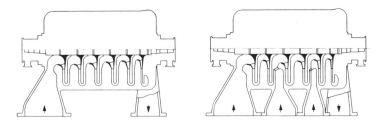

FIGURE 11–6   *Flow path through horizontally split straight-through compressor (left) and sidestream compressor (right). (Source: Mannesmann Demag, Duisburg, West Germany.)*

temperatures. Injection or extraction is by means of additional nozzles. (The flow path through a sidestream compressor is illustrated in Figure 11-6.)

In the case of gas injection, the sidestream is mixed with the main stream in the return channel. Mixing takes place over the entire periphery. When a sidestream is extracted, a separating volute removes part of the main stream. The compressor stages are designed to correspond to the stages of the refrigeration process. The working media for refrigeration processes are primarily ethylene, propylene, ethane, and propane. Refrigeration processes usually form a closed cycle, rather less frequently a semi-open cycle.

In centrifugal compressors with two main nozzle pairs, the two process stages can be arranged back to back, i.e., flow in the two process stages is in opposite directions, or they can be arranged in series. In the double-flow version, the compression process of both stages terminates in a common discharge nozzle.

Back-to-back arrangement of the first and third stages or series arrangement of the stages is also possible. Typical back-to-back arrangements are shown in Figure 11-7.

Tried and proven labyrinth seals separate the individual process stages. The choice

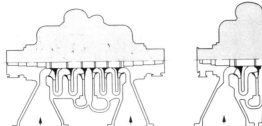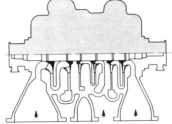

FIGURE 11–7 *Compressors with back-to-back oriented impellers. (Source: Mannesmann Demag, Duisburg, West Germany.)*

of shaft seal system is dictated by the service. Whereas casings with two main nozzle pairs are widely employed for a variety of media and processes, casings with three process stages are mainly employed for air, oxygen, and nitrogen. The medium is normally cooled outside the compressor.

Interstage cooling produces an almost isothermal compression process. This requires the least compression work. Intercooling also becomes necessary when the temperatures produced by compression have to be limited.

In most compression systems, the coolers are mounted separately, permitting a high degree of freedom in design and layout. However, compressors with internally arranged coolers are available from some manufacturers and may merit consideration when the ultimate in compactness must be achieved.

Vertically Split Compressors

Vertically split (barrel-type) centrifugal compressors are the preferred, and sometimes mandatory, design for high pressures or for compressing gases rich in hydrogen. The cylindrical casing ensures good stress distribution and extremely good gas-tightness. Unlike the casing, the stationary internal components of the compressor, with the exception of seal components, are horizontally split. During assembly of the compressor they are mounted together with the rotor and inserted axially into the casing. The end covers are retained by shear ring segments. A cross-section view is shown in Figure 11–8.

The inlet and discharge nozzles are welded to the cylindrical casing or, where heavy wall thicknesses are involved, are integral with the casing; the pipework is bolted to these nozzles. These compressors are also built for two process stages; in this case they feature two main nozzle pairs.

The main fields of application for barrel-type compressors are in handling gases rich in hydrogen; hydrogenation cracking; synthesis of ammonia, urea, and methanol; gas lift and reinjection; and transportation of gas in pipelines.

A compressor with one stage is often adequate for compression applications involving a low head. For such applications, the user may choose from compressor types that may be vertically or horizontally split. As mentioned earlier, the vertically split version is particularly suitable for high pressures and for compressing gases of low molecular weight.

Depending on the operating conditions involved, the tried and proven systems employing labyrinth, mechanical contact, or floating ring seals are used for shaft sealing. Other seal systems may go under the names of trapped bushing seal, gas phase mechanical seal, cone seal, etc., but each type represents merely a variation of the three principal configurations.

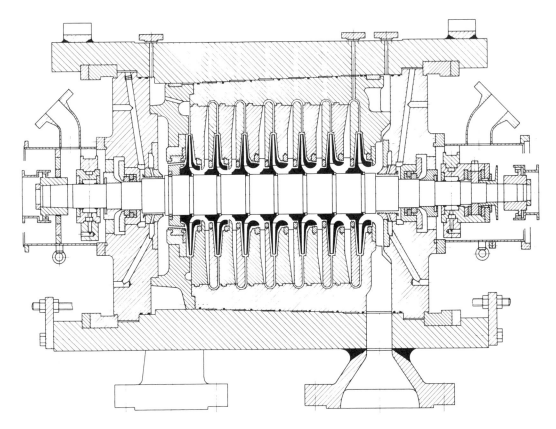

FIGURE 11–8   *Cross section of vertically split (barrel-type) compressor. (Source: Mannesmann Demag, Duisburg, West Germany.)*

Fields of application for single-stage process compressors include phthalic acid anhydride plants and cupolas, oxosynthesis, and water treatment. In addition, they are suitable for handling all gases and gas mixtures in the main and auxiliary processes in the chemical and petrochemical industries. They also find application in the transportation of natural gas and mineral oil gas as well as in closed-loop systems in nuclear technology.

Compressor Trains

Large pressure ratios cannot be handled by one casing alone. Similarly, it is not possible to split the compression cycle into more than two or three process stages within one casing. The major compressor manufacturers therefore build compressor trains that may consist of up to four separate casings. A train with three separate casings is shown in Figure 11–9. These separate compressors, which need not be of the same type, are interconnected by couplings; they can be powered by a common driver. When additional transmission gearing is used, the compressor casings may also be run at different speeds. The train is designed so that a minimum of dismantling is necessary for maintenance, i.e., when a vertically split casing is used, it is located at the opposite end to the driver.

The compressor train can be arranged on a common baseframe. In special applications, for instance, offshore installations, this baseframe can be of torsionally stiff design.

FIGURE 11-9 *Centrifugal compressor train. (Source: Mannesmann Demag, Duisburg, West Germany.)*

Construction and Mode of Operation

Each application requires its own casing configuration. In spite of this, the internal design and construction of a given manufacturer's centrifugal compressor is often essentially the same. This allows the use of standard components.

The components that are important for the compression function are the rotor and the energy-converting parts, as illustrated in Figures 11-10 and 11-11.

The rotor consists of the shaft and the impellers. The number of impellers is determined by the thermodynamic operating conditions, but it is limited by the mechanical and dynamic behavior of the rotor. The shaft is carried in pressure-lubricated tilting-pad bearings; one of these is purely a radial bearing, while the other is either a separate or a combined radial and thrust bearing. The shaft is generally provided with a balance piston to reduce axial thrust.

Shaft seals separate the gas spaces from the oil-lubricated bearings and the atmosphere. Simple labyrinth seals, multiported labyrinths with buffer gas, mechanical contact, or floating ring seals are employed, the choice being dictated by the process involved and the fluid handled.

The materials for the rotor, internals, and casing are selected on the basis of their mechanical properties and compatibility with the fluid to be compressed.

A lube oil system and a seal liquid system supply the bearings and the liquid seals with the required volumes of oil and seal liquid.

The fluid to be compressed passes through the inlet nozzle and aerodynamically

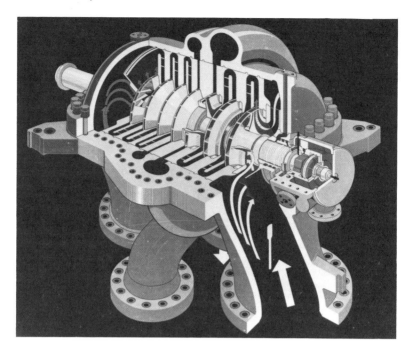

FIGURE 11–10   *Cutaway view of horizontally split centrifugal compressor. (Source: Mannesmann Demag, Duisburg, West Germany.)*

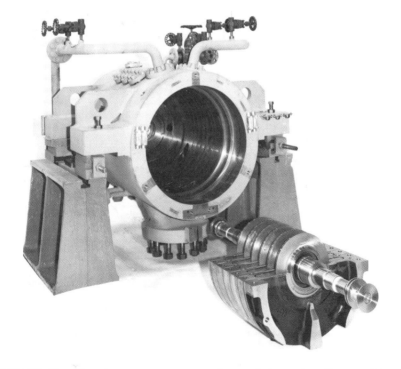

FIGURE 11–11   *Barrel-type compressor casing and internals. (Source: Mannesmann Demag, Duisburg, West Germany.)*

designed inlet channel into the first impeller. The first impeller may be preceded by an adjustable inlet guide vane unit. Impellers and the diffusers following them are designed so as to provide an optimum low-loss compression cycle. After the diffuser channel, the gas enters the return bend, which guides it to the vaned return channel, and it then reaches the impeller of the next stage with the correct angle of incidence.

Immediately following the final stage or an intermediate stage, after which the compressed gas leaves the compressor, the diffuser opens out into a volute that widens gradually in the direction of flow to match the increase in volume.

Impellers

The head to be produced and the volume flow to be handled provide the design criteria for impellers. The total energy y that can be transferred by the impeller is a function of the peripheral velocity u_2 and the pressure coefficient ψ:

$$y = \psi \cdot \frac{u_2^2}{2}$$

An upper limit is set to the permissible peripheral velocity by the type of impeller and the mechanical properties of the materials from which it is made. The pressure coefficient increases with increasing blade outlet angle β_2. However, the impeller geometry is not exclusively determined by the maximum energy that can be transferred.

Further criteria include good overall efficiency of the stage and a broad envelope of curves with as steep a rise as possible toward the surge limit, thus ensuring good partial load controllability and stable operation within the plant (immunity to pressure fluctuations).

The standard impeller in many centrifugal compressors is the type with backward-leaning blades and cover (Figure 11–12). It represents a good compromise to meet the aforementioned requirements. This type of impeller may feature blades curved in two or three dimensions. Impellers with three-dimensional blading (Figure 11–13) have high capacity limits, due to the large eye diameter relative to the outside diameter and a large outlet width. Three-dimensional impellers are among the standard features available for modern centrifugal compressors.

Manufacture of Impellers. Five principal manufacturing methods are typically available for the efficient production of impellers: milling and riveting, milling and brazing, milling and welding, welding and welding, and casting.

Impellers fabricated by milling and riveting have the gas passages milled from the solid impeller disc. The cover is riveted in place. This long-established method combines great mechanical strength and reliability with high aerodynamic quality, i.e., dimensional accuracy (particularly important for narrow gas passages) and good sur-

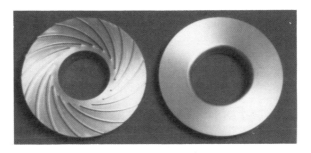

FIGURE 11–12 *Impeller with backward-leaning blades, shown before cover assembly. (Source: Mannesmann Demag, Duisburg, West Germany.)*

FIGURE 11–13 *Impeller with three-dimensional blades. (Source: Mannesmann Demag, Duisburg, West Germany.)*

face finish. More recent mechanical analyses have especially demonstrated the system-damping properties of the riveted fixing, properties that afford enhanced protection against fracture due to alternating stress cycles.

Milled and brazed impellers are used when thin blades and narrow gas passages with good aerodynamic properties are necessary. As in the case of riveted impellers, the gas passages are milled from the solid impeller disc; the cover is then brazed into place with a gold-nickel brazing alloy, using a high-temperature, high-vacuum process. The strength of the brazed joint is equal to that of the parent metal. It is totally immune to H_2S stress corrosion cracking.

When access to the gas passages is good, i.e., when the passages are wide, the impellers are of milled and welded or welded and welded construction. The blades are either milled from the solid impeller disc or welded to it. In both instances, the cover is welded to the blades by a continuous weld.

If a number of identical impellers are required, casting is the most economical method of manufacturing them. Large impellers are cast in sand molds. Precision casting processes that ensure unusually good quality of the product are being used to an increasing extent for small impellers.

Impeller Testing. Tests are carried out on material specimens prior to and after manufacture to prove the required properties. On completion of manufacture, the dimensions of the impeller are checked, after which the impeller is then tested for cracks, using dye penetration or magnetic particle methods. The impeller is then initially balanced and run at overspeed. This speed may be so high that flow and cold strain hardening may take place at the stress peaks. After overspeeding, the impeller is once again checked for dimensional accuracy and tested for cracks. On successful completion of these tests, the impeller is rebalanced.

Rotor

The shaft carries the impellers and the balance piston, as illustrated in Figure 11–11. It is supported in tilting-pad, plain, or modified (contoured) sleeve bearings. The impellers and balance piston are shrunk onto the shaft. Multipart rings or similar components locate the impellers in the direction of the axial thrust. The shrink fit offers the advantage of uniform stress distribution over the whole circumference and a constant self-centering effect. This shrink fit is designed so that after the bore has expanded due to centrifugal force at maximum speed, sufficient shrinkage effect still exists to transmit the torque and the axial thrust. At the same time, the impellers can be removed whenever necessary without damage.

The balance piston balances part of the axial thrust produced by pressure differentials across the impellers. Part of the axial thrust is automatically balanced with a back-to-back layout. All component parts such as shaft, impellers, balance piston, and couplings are separately balanced, after which the complete rotor is assembled. Each time another component is added, concentricity of running is checked. The stresses produced in the shaft by the shrink-fit method of mounting the impellers and balance piston are relieved by running the complete rotor up to operating speed.

Rotor Dynamics. Detailed vibration analyses as early as the preliminary design stage ensure the operational availability of modern, well-engineered centrifugal compressors. These analyses investigate the following vibration phenomena: resonance behavior, stability, and torsional analysis.

Investigation of the resonance behavior includes computation of the first and second lateral critical natural frequencies based on the bearing stiffness and damping for each casing. Precise determination of the critical natural frequencies, which are principally lateral vibrations of the rotor caused by unbalance forces, allows multistage compressors to be operated at a speed between the first and second lateral critical speeds.

The stability behavior describes the behavior of the rotor when the vibrations are self-excited. This essentially includes vibrations excited by the specific spring and damping characteristic of the bearings (oil whip) and rotor-dynamic effects of labyrinths and sealing elements.

The torsional analysis covers torsional vibrations produced by electric motors when switching occurs in the supply network and when a synchronous motor is run up to operating speed asynchronously.

A precise knowledge of these interrelated factors ensures trouble-free operation and a high degree of operational availability of major, unspared centrifugal compressors.

Bearings

Tilting-pad bearings support and locate the compressor rotor. They employ the hydrodynamic principle and are designed in the light of the most recent scientific and engineering knowledge in this field. The running surfaces of the bearings are divided into segments and inserted into the horizontally split bearing bracket. This bearing bracket is positioned in the bottom half of the bearing housing (Figure 11–14) and is typically secured in place by a bolted bearing retainer. Properly designed compressor bearings can be inspected without the compressor casing having to be opened.

All bearings have pressure-oil lubrication. A lubricating oil system supplies them with a flow of oil sufficient to form an oil film on the running surfaces and to dissipate the heat produced by friction. Oil is piped centrally to the bearings. Retaining rings fitted at both sides control the rate of discharge of oil from the bearing via the gap set.

Radial Bearing. The radial bearing—often specially developed by capable compressor manufacturers—is a multipad bearing with four or five pads arranged so that the stationary shaft rests on one of them or, in some designs, between pads. In a multipad bearing, the reaction forces act over the entire circumference and therefore stabilize the position of the shaft. In addition, the tilting pads adjust automatically to suit the operating conditions; optimum load distribution is therefore achieved at all times. With the exception of certain special designs, these pads are symmetrically supported and therefore are unaffected by the sense of rotation. Figure 11–15 depicts a radial tilting-pad bearing.

Thrust Bearing. At the nondrive end of the compressor, the radial bearing is combined with a thrust bearing—also frequently specially developed by the compressor manufacturer. This bearing employs the principle of a double adjustable pad-thrust bearing and absorbs the residual rotor thrust resulting from the unbalanced gas forces acting on the impellers and balance piston.

FIGURE 11–14   *Compressor bearing housing with combination thrust and radial bearings. (Source: Mannesmann Demag, Duisburg, West Germany.)*

FIGURE 11–15   *Tilt-pad radial bearing. (Source: Mannesmann Demag, Duisburg, West Germany.)*

The tilting pads are sometimes asymmetrically supported. The axial thrust is determined with the most modern methods available, taking into consideration all the aerodynamic effects that arise. A certain amount of residual thrust has a stabilizing effect, since the rotor is then in contact with a specific side of the bearing. The design chosen and the load-bearing capacity of the thrust bearing must ensure the operational readiness of the compressor even if thrust reversals and sudden loads occur during the widely varying operational phases the compressor may experience. Figure 11–16 illustrates a combination radial/thrust tilting-pad bearing.

FIGURE 11-16 *Combined radial and thrust tilt-pad bearing for centrifugal compressor. (Source: Mannesmann Demag, Duisburg, West Germany.)*

Sealing Elements

The long-term, reliable operation of centrifugal compressors must be ensured by thoroughly well-proven sealing elements.

Labyrinth seals (Figure 11-17) minimize the flow around the impellers and hence also minimize leakage losses. These labyrinths are located over the rim of the impeller eye on the inlet side and close to the shaft over the hub at the back of the impeller on the discharge side. Centrifugal compressors with several nozzle pairs employ labyrinth seals to separate the individual process stages. Labyrinth seals that prevent lube oil and oil mist from escaping from the bearing chambers are also used to seal the bearing housings. Thoroughly proven sealing systems seal the shaft exits. Depending on

FIGURE 11-17 *Labyrinth seals surround each impeller of this centrifugal compressor rotor. (Source: Mannesmann Demag, Duisburg, West Germany.)*

FIGURE 11-18   *Multiported oxygen compressor labyrinth seal. (Source: Mannesmann Demag, Duisburg, West Germany.)*

specific requirements, multiported labyrinth, floating ring, or mechnical contact systems are employed for this purpose.

Multiported labyrinth seals (Figure 11-18) with buffer gas injection are used when buffer gas can be allowed to mix with the process gas or when leakage of process gas is permissible. Provision for ejection can be made in order to avoid excessive leakage of process gas. In all other instances, floating ring or mechanical non-contact or contact seals are used, the former being employed for use with high pressures, the latter being used with clean gases.

Labyrinth Seal.   The labyrinth seal is a noncontacting seal. It consists of a number of sealing strips in an insert in the stationary part of the compressor. This insert is horizontally split and easily replaced. Sealing strips are also sometimes fitted to the shaft. "Straight" or stepped labyrinths may be employed, depending on the specific need.

The labyrinth seal forms a series of throttling points, at each of which the pressure differential is decreased. The smaller the clearance—the distance between the labyrinth strip and the surface of the shaft—the more the leakage is reduced. The turbulence zones between the strips enhance the throttling effect. The pressure difference involved dictates the number of labyrinth strips. If absolutely reliable separation of the oil and process gas spaces is necessary, a multiported labyrinth seal (Figure 11-19) with injection of buffer gas under controlled pressure is used. Part of this gas flows outward and prevents atmospheric air from entering, while the remainder passes into port 2 at lower pressure. This space is connected to a leak-off line.

Ports 3 on the suction and discharge sides of the compressor are interconnected and can be controlled and monitored jointly. The pressure is adjusted so that, although it is lower than the pressure of the buffer gas in port 1, it is higher than the pressure in port 2, causing some gas to flow into that port and mix with the buffer gas there. This arrangement ensures that, in the event of a fault in the discharge line, neutral buffer gas may enter the compressor via port 2, although the gas being compressed inside the centrifugal compressor cannot escape to the atmosphere. Ingress of buffer gas into the compression spaces is normally prevented by the pressure drop across port 2.

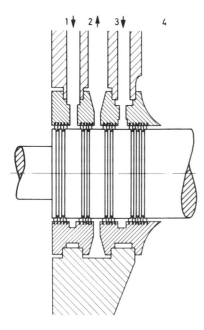

FIGURE 11-19   *Multiport labyrinth seal, cross- section view. (Source: Mannesmann Demag, Duisburg, West Germany.)*

The multiported labyrinth system is very adaptable. For instance, a seal employing buffer gas only and without port 2, with a smaller flow of buffer gas into the compressor, is just as feasible as a seal without buffer gas from which a mixture of atmospheric air and the compressed medium are evacuated via port 2.

Floating Ring Seal.   In conjuction with a seal liquid introduced under pressure, the floating ring seal illustrated in Figure 11-20 prevents process gas from escaping at even the highest of operating pressures. It operates without mechanical contact and therefore without wear. This type of seal consists of the inner ring between the process gas space and the seal liquid space, and the outer ring that permits enough seal liquid to escape outward to ensure adequate cooling of the seal. The form of the intermediate ring leads to intensive heat dissipation away from the inner ring. Sealing is maintained by the controlled seal liquid pressure being above the process gas pressure at all times. Clearance between the inner ring and the shaft is such that only very little seal liquid passes through to the gas side. Since this clearance is smaller than the bearing clearance and also because of rotor dynamics considerations, the sealing rings are designed to float, i.e., they can follow any radial shaft deflections without acting like bearing supports.

The seal liquid enters the seal via supply pipe 1. Most of it is discharged outward through the radial gap formed by the outer ring. A small volume flows into port 5 via the radial gap formed by the inner ring. Seal liquid is prevented from entering the compressor by a constant flow of buffer gas from the supply pipe 7 to port 5 via labyrinths. The mixture of seal oil and buffer gas is led via connection 6 to an automatic separator.

Mechanical Contact Seal.   The mechanical contact seal shown in Figure 11-21 employs a stationary carbon ring in sliding contact with a rotating ring manufactured from high-quality material with a special finish. A seal liquid is employed. This type of seal is also effective when the compressor is at standstill and the oil pumps have been shut down.

The main components are the carbon ring and the rotating ring for inward sealing. In the outboard direction, a floating ring controls the flow of the seal liquid that cools the seal.

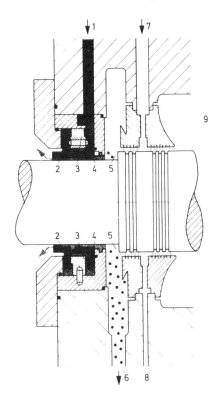

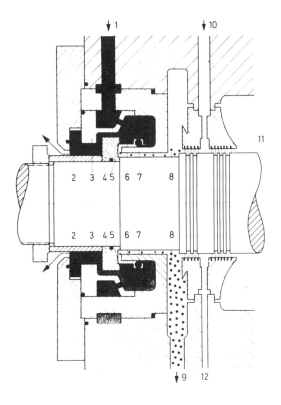

FIGURE 11–20   *Cross section of floating ring seal. (Source: Mannesmann Demag, Duisburg, West Germany.)*

FIGURE 11–21   *Mechanical contact seal, cross section view. (Source: Mannesmann Demag, Duisburg, West Germany.)*

The seal liquid enters the seal via the supply pipe 1 and flushes the seal ring components via the holes in the distributor ring. The pressure of the liquid is higher than that of the gas, so that the carbon ring, under constant spring pressure, is always kept in sliding contact with the rotating ring. Some of the liquid wets the sliding surface 5 and reduces wear. Only a very small proportion of this liquid passes to the gas side. A controlled flow of buffer gas flowing from the supply pipe 10 through a labyrinth to port 8 entrains this leakage liquid and leads it via outlet 9 to the separator.

O-rings fitted externally and within the seal reliably separate the buffer gas and seal liquid spaces.

Mechanical Non-Contact Seals ("Gas Seals").   These novel seals are very similar in function and geometry to mechanical contact seals. Instead of a sealing liquid, they use a small quantity of a clean sealing gas. The seal faces operate without actually making contact. The escaping seal gas separates the seal faces by a fraction of a thousandth of an inch.

Casings

The two halves of horizontally split compressors are joined together by hydraulically pretensioned bolts. The joint is sealed by a suitable sealant, where necessary of the string type. The vertically split casings are provided with end covers retained by shear rings. O-rings or O-ring joints between inner casing and barrel casing provide proper sealing of the gas zone from the atmosphere.

Casings are supported at the centerline. The pedestal supports at the drive end are fixed points, so that axial thermal expansion is in the direction of the free end of the shaft. Lateral alignment is provided by guide lugs in the vertical center plane of the compressor. Thermal expansion of shaft and casing is compensated by the rotor being located by the thrust bearing at the free end of the casing.

Positioning of the suction and discharge nozzles can be arranged to suit requirements. Compressor casings are hydraulically tested at 1.5 times the maximum operating pressure.

Stationary Components

The term "stationary components" refers to the inlet channel, the diffuser, and the return channel.

Impellers achieve optimum efficiency and operating characteristics only if the inlet flow is free of disturbance, i.e., unswirled, and exhibits a uniform velocity profile. Factors such as these are extremely important for impellers of high suction capacity.

Single-shaft process compressors have radial intakes, and the flow therefore has to be deflected through 90° prior to entry into the first impeller. High-quality compressors are produced with inlet channel designs that meet the above requirements. These incorporate standard, stationary blades, which, distributed over the periphery, exhibit defined angular settings. Similarly, these centrifugal compressors would have as a standard vaneless annular diffusors that have parallel walls or profiled cross sections. This feature ensures a wide range of regulation with almost constant optimum efficiency.

Leading manufacturers have developed special cross-sectional profiles that produce wider ranges of regulation and better efficiencies than parallel-wall diffusors—especially with high-capacity impellers. As the gas passages become smaller in cross-sectional area, the surface finish of the diffusor has a decisive effect on the stage efficiency. In special cases, the surfaces are therefore coated. Particular importance is attached to smooth inflow into the following impeller in the design of the return channels, the blading of which is generally profiled. Like the impellers, they are normally milled from the solid material to ensure a superior surface finish and dimensional accuracy.

Auxiliary Equipment

The driver, suitable gearing, and the couplings have to be matched, with the compressor, to the specific application requirements of the plant and to the conditions on the site. It is only then that the maximum of operational reliability is ensured.

One of the requirements for smooth and economical compressor operation is control and regulation to match operational needs plus inherent functional reliability of the coolers and lubrication and seal liquid systems.

Monitoring of the bearings via the oil temperature, for example, and measurements of shaft vibrations and shaft position ensure that potential trouble during compressor operation is recognized at an early stage. The monitoring equipment, together with the alarm systems and controls, is accommodated in control panels.

Drive Components

Couplings interconnect the various units of a compressor train. Diaphragm or curved-tooth gear couplings are employed as a rule. Couplings of these types allow angular deflection and axial deflection, but they are nevertheless torsionally stiff. Gear couplings with convex tooth flanks require precision manufacture. The lube oil system supplies them with lubricant. Coupling hubs are located and secured by keys or oil-hydraulic means to the shaft.

Torsionally elastic couplings are employed if shock torque loads occur, for example, when starting up synchronous motors or if there is a brief drop in voltage. The shock torque load is absorbed by the deformation of elastic components such as rubber buffers. Owing to the lower torque transmission capacity of these components, torsionally elastic couplings are larger.

Spur gears or epicyclic gears are employed between driver and compressor or between two casings in order to drive centrifugal compressors at the optimum speed.

Coolers

Process coolers are employed for a variety of reasons:

- to limit the maximum temperature of the process gas
- to limit the maximum temperature of the components for safety reasons
- to minimize the compressor power consumption by aiming at isothermal compression

Process coolers are employed as intercoolers and aftercoolers. The principal cooling medium used is water, including seawater. There are various criteria for deciding whether the gas is to flow through or around the tubes of the cooler. These are the quality of the gas, the quality of the cooling medium, and the pressure level involved.

If either the process gas or the cooling medium contains contaminants, the aim is to pass these through the tubes and the relatively clean gas or cooling medium around the externally finned tubes. This is because internal cleaning of the tubes is much easier.

When pressures are high, the process gas will be routed through the tubes because of mechanical strength considerations.

The individual tubes form a bundle, the tube ends of which are expanded into the tube end plates, brazed or welded to produce a gas-tight fit. Depending on the design used, the tubes are combined into bundles or single elements. The single-element design is a lightweight, easily portable unit often allowing economical maintenance of spares. Baffle plates determining the flow of the process gas are provided inside the cooler housing.

Air-cooled process gas coolers are used in special applications. The process gas flows through the tubes around which cooling air flows boosted by fans. Air-cooling requires much larger equipment, and a number of elements are therefore arranged in parallel.

Control

Centrifugal compressors always match the process requirements. They form an integral part of the process plant, the operating characteristics of which are a function of volume flow and pressure.

Centrifugal compressors can be controlled so that they maintain constant pressure at the intake or at a preceding process point, or at the discharge nozzle or following that point. When process control is staged, intermediate pressure can also be maintained. Requirements for constant volume flow can also be met. Daily and annual meteorological influences are taken into consideration. In addition, the compressors have to be protected against unstable operating conditions.

If these requirements are to be met, the operating characteristics of the compressor have to be controlled. This can be done either by varying impeller speeds, by varying the angle of incidence of the gas, or by throttling.

Speed Variation.   When speed variation is used, all velocity components are equally affected. This requires a speed-controllable driver or suitable intermediate element.

The principal advantage of speed regulation is that only as much energy is required as is needed for the process.

Variation of Angle of Incidence. The compressor characteristic for constant speed can be directly influenced by using an inlet guide vane unit to vary the angle of incidence. Apart from providing economical operation under partial load, this also allows an extension of the characteristic above normal. With constant conditions at the impeller outlet, the changed angle of incidence causes the specific work to be influenced by the relative and absolute velocities at the inlet. With a positive guide vane setting, the total energy is reduced because of the lower relative speed at the inlet. A negative setting has the opposite effect. The influence of a negative guide vane setting generally reaches a peak value between 20° and 30°. After this setting is exceeded, the angled guide vanes choke the flow cross section to such a degree that throttling with local separation phenomena occurs, and this causes losses to increase rapidly.

Toward the lower end of the volume flow range, a positive guide vane setting has an advantageous effect, particularly in the region close to the design point. The direction of the relative velocity is turned to coincide with the design angle of the leading edge of the blade, and the shock loss otherwise occurring due to an unfavorable approach angle is reduced. In the remainder of the partial load range, a throttling effect resulting from constriction of the flow cross section by the angled guide vanes is superimposed on the aerodynamic effect.

One special advantage of guide vane adjustment is that the stable operating range is extended toward smaller volume flow rates, although this applies to the particular stage only. With multistage compressors, the effect is reduced in proportion to the number of uncontrolled stages.

Throttling. With throttling, the compressor characteristic remains unchanged. When throttling is performed on the suction side, a throttle valve reduces the compressor inlet pressure when the volume flow rate is increased. Downstream of the compressor, a lower discharge pressure is produced, which corresponds to the pressure ratio associated with the point on the compressor characteristic or compressor performance curve. In relation to the useful compression work, throttling requires maximum specific energy.

Throttling on the suction side and variation of inlet guide vane angle are also used to expand the operating range of compressor plants with speed-controlled drivers.

Antisurge Control. In order to protect the machines from excessively high mechanical loading due to unstable operating conditions (surging), which primarily stress the bearings, compressors can be equipped with antisurge control. One such control is depicted in Figure 11–22.

If the delivery required is below the minimum delivery volume of the compressor, the surplus is led away as a sidestream, via a valve. Depending on the nature of the gas, the surplus is either discharged to the atmosphere or it is cooled and returned to the suction side. In the example shown, the valve is operated by a controller that uses the volume flow rate and the discharge pressure as input parameters. The blow-off or recycle limit in the compressor curve envelope is normally an approximate simulation of the surge limit. Pneumatic, hydraulic, electropneumatic, or electrohydraulic control systems may be employed. Depending on the cross-sectional area of the discharge, control valves or flaps are employed as regulating units.

Discharge Pressure Control. The need for constant compressor discharge pressure can be met, irrespective of the delivery volume, by using a guide vane unit. The signal for the actual value is taken from the discharge line and fed to a PI controller. After comparing this reading with the set value, the controller adjusts the setting of the guide vanes via a servo-cylinder.

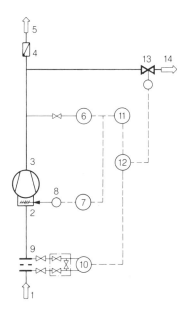

FIGURE 11-22   *Simple example of combined antisurge and discharge pressure control. 1 = inlet; 2 = inlet guide vane unit; 3 = compressor; 4 = nonreturn valve; 5 = discharge line; 6 = pressure transmitter; 7 = discharge pressure controller; 8 = servocylinder; 9 = orifice measurement; 10 = differential pressure transmitter; 11 = computer; 12 = surge limit controller; 13 = blow-off valve; 14 = blow-off line. (Source: Mannesmann Demag, Duisburg, West Germany.)*

Lube Oil System

The lube oil system supplies oil to the compressor and driver bearings and to the gears and couplings. Figure 11–23 illustrates a typical lube oil system. The lube oil starts off in the reservoir, from where it is drawn by the pumps and fed under pressure through coolers and filters to the bearings. On leaving the bearings, the oil drains back to the reservoir.

The reservoir is designed to permit circulation of its entire contents between eight and twelve times per hour. Oil level and temperature are constantly monitored. The oil can be preheated electrically or indirectly by steam for starting up at low temperatures. A thermostat with surface temperature limiter prevents overheating of the oil. The reservoir is vented.

Oil is normally circulated by the main oil pump. An auxiliary pump serves as a standby. These two pumps generally have different types of drive. When both are driven electrically, they are connected to separate supply networks. On compressors with step-up gearboxes, the main oil pump may be driven mechanically from the gearbox. The auxiliary pump then operates during the start-up and run-down phases of the compressor plant. Relief valves protect both pumps from the effects of excessively high pressures. Nonreturn valves prevent reverse flow of oil through the stationary pumps.

Heat generated by friction in the bearings is transferred to the cooling medium in the oil coolers. The return temperature is monitored by a temperature switch. Air-cooled oil coolers may be employed as an alternative to water as coolant. The former have long been used in regions where water is in short supply. Twin coolers with provision for changeover have filling and venting connections so that the standby cooler can

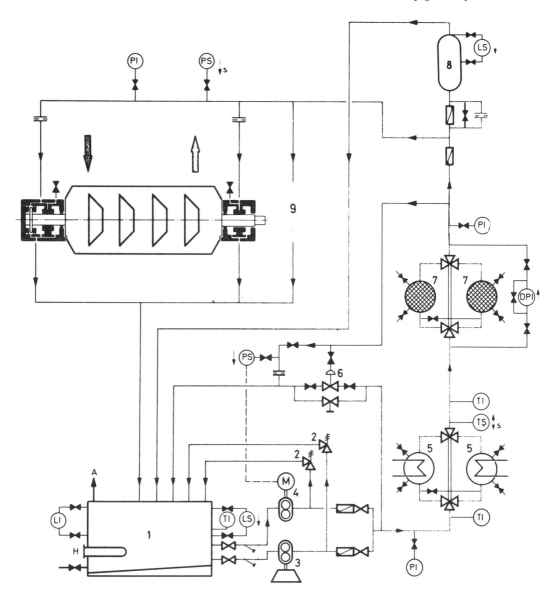

FIGURE 11-23 *Lube oil schematic for a centrifugal compressor. 1 = Reservoir; 2 = safety valve; 3 = main oil pump; 4 = auxiliary oil pump; 5 = cooler; 6 = pressure-regulating valve; 7 = filter; 8 = overhead tank; 9 = driver and other users; PI = pressure indicator; DPI = differential pressure indicator; PS = pressure switch; TI = temperature indicator; TS = temperature switch; LI = level indicator; LS = level switch; H = heater; A = reservoir vent. (Source: Mannesmann Demag, Duisburg, West Germany.)*

be filled with oil prior to changing over. This eliminates the possibility of disturbances and damage due to air bubbles in the pipework system. Twin oil filters with provision for changeover have the same facilities.

A pressure-regulating valve is controlled via the pressure downstream of the filters and maintains constant oil pressure by regulating the quantity of bypassed oil. The auxiliary oil pump is switched on by a pressure switch if the oil pressure falls. A second pressure switch shuts down the compressor plant if the pressure still continues to fall.

The filters clean the lube oil before it reaches the lubrication points. A differential pressure gauge monitors the degree of fouling of the filters.

An overhead oil tank can be provided to ensure a supply of lubricant to the bearings in the event of faults while the compressor is being run down. A continuous flow of oil through an orifice maintains the header oil constantly at operating temperature. Should the pressure in the lube oil system fall, the nonreturn valve beneath the tank opens to provide a flow of oil.

The flow of oil to each bearing is regulated individually by orifices, particularly important for lubrication points requiring different pressures. Lube oil for the driver and other users is taken from branch lines.

When a hydraulic shaft position indicator is used, this is supplied with oil from the lube oil system.

Temperatures and pressures are measured at all important locations in the system; the readings can be taken locally or transmitted to a monitoring station.

Except for a few components, the lube oil system is a conveniently installed packaged unit supplied complete and ready for installation. Oil pumps, coolers, and filters are grouped around the oil reservoir on a common baseplate. Design and construction of the lube oil system must take into account the relevant regulations and any special requirements. One such requirement might be blanketing with inert gas; another might be the on-stream purification of lube oil by modern vacuum dehydrator units.

Seal Liquid System

The seal liquid system supplies the mechanical contact and floating ring seals with an adequate flow of seal liquid at all times, thus ensuring that they function correctly (Figures 11–24A and B). An effective seal is provided at the settling-out pressure when the compressor is not running.

Starting in the main oil reservoir, the medium passes to the seals via the pumps, the twin oil coolers, and the twin filters.

Instruments for monitoring the oil level and temperature are mounted on the reservoir. If necessary, the seal oil is heated; a thermostat with surface temperature limiter protects against excessively high temperatures.

Every system has a main oil pump and an auxiliary oil pump with independent drives. They are designed for a higher delivery rate than is actually needed by the seals. To protect the pumps and downstream equipment, safety valves are fitted. Nonreturn valves after each pump prevent seal oil from flowing back to the reservoir through the stationary pumps.

The coolers dissipate the heat transferred to the seal oil. A temperature switch monitors the permissible temperature range.

The filters retain all impurities, the pressure drop across them being checked by a differential pressure indicator.

The floating ring seals are supplied with seal oil at a defined differential pressure above the reference gas pressure (pressure within the inner seal drain).

This is schematically represented in Figure 11–24A. Figure 11–24C shows a lube oil system with seal oil booster arrangement; a seal oil system with pressure reduction for the lube oil portion is illustrated in Figure 11–24D.

The flow of seal oil is regulated by a differential pressure-regulating valve, which, if there are changes in the reference gas pressure, regulates the pressure of the seal oil, or, as shown in the diagram, by a level-control valve that maintains a constant level in the overhead tank. The oil in the overhead tank is in contact with the reference gas pressure via a separate line. The static head provides the required pressure differential. In addition, the oil in the overhead tank compensates for pressure fluctuations and serves as a run-down supply if pressure is lost. If the level in the tank falls excessively, a level switch shuts down the compressor plant. There is a constant flow of oil through the overhead tank, and this heats the oil at all times.

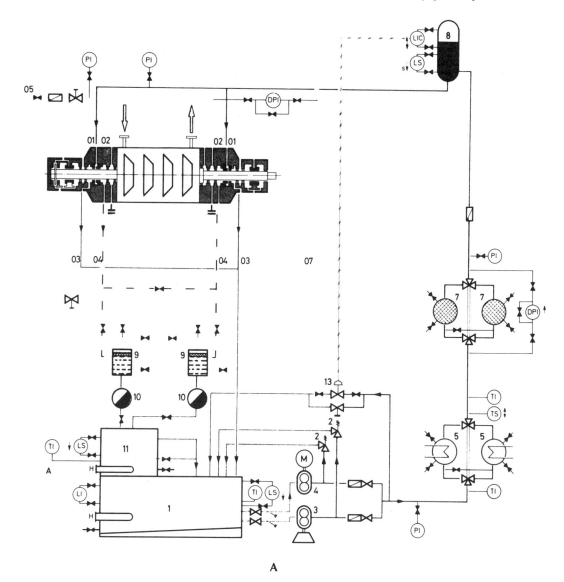

A

FIGURE 11–24 (Continued on following pages)   *Seal liquid system schematics. A = Seal oil system for floating ring seals; B = seal oil system for mechanical contact seals; C = lube oil system with seal oil booster system for floating ring seals; D = seal oil system for floating ring seals with pressure reduction for the lube oil system (combined system); 1 = oil tank; 2 = relief valve; 3 = main oil pump (high pressure); 4 = auxiliary oil pump (high pressure); 5 = oil cooler (high pressure); 6 = regulating valve; 7 = high-pressure filter; 8 = seal oil overhead tank; 9 = demister separator; 10 = condensate trap; 11 = degassing tank; 12 = sour oil tank; 13 = level control valve; 14 = pressure-reducing valve; 15 = lube oil overhead tank; 16 = main oil pump (low pressure); 17 = auxiliary oil pump (low pressure); 18 = oil cooler (low pressure); 19 = low-pressure filter; 01 = seal oil supply; 02 = buffer gas supply; 03 = outer drain; 04 = inner drain (seal oil/buffer gas); 05 = buffer gas supply; 07 = reference pressure line; 08 = lube oil supply; PI = pressure indicator; PS = pressure switch; DPI = differential pressure indicator; TI = temperature indicator; TS = temperature switch; LI = level indicator; LS = level switch; LIC = level controller; H = heater; A = reservoir vent. (Source: Mannesmann Demag, Duisburg, West Germany.)*

For the mechanical contact seal, the seal oil is kept at a constant differential pressure with respect to the reference gas by a regulating valve (Figure 11–24B). As the name indicates, the mechanical contact seal provides a mechanical seal when the compressor plant is shut down.

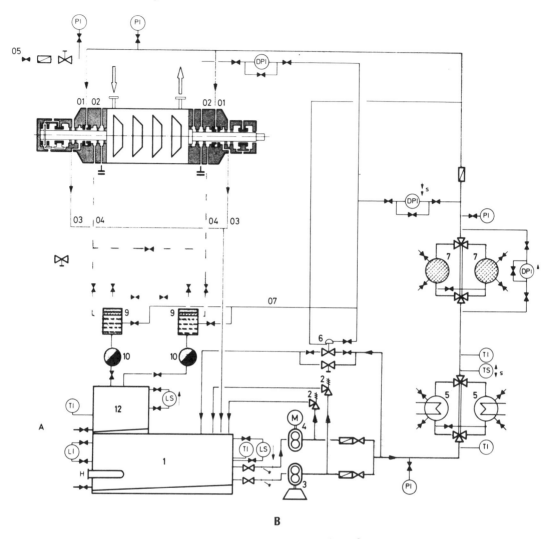

FIGURE 11-24   (Continued)

To prevent oil from gaining ingress to the compressor, the space between the oil drain and compression space is sealed by a flow of gas. The pressure of this sealing gas or buffer gas is above the pressure of the reference gas. A differential pressure indicator monitors the pressure differential.

The flow of seal oil divides in the compressor seals. Most of the flow returns under gravity to the reservoir. A small quantity passes through the inner seal ring to the inner drain, where it is exposed to the gas pressure. This oil, mixed with the buffer gas, is led to the separator system. On each side this consists of a separator and a condensate trap. The separated gas is led either to the flare stack or to the suction side of the compressor. The oil flows into a tank for degassing.

If oil is used as sealing liquid and can be used again, degassing is accelerated by exposure to a partial vacuum and by heating. The oil is then returned to the reservoir.

If the oil becomes unusable, it is led away for separate treatment.

The quantities of oil passing through the inner drain in well-designed centrifugal compressors are very small.

Temperature and pressure measuring points with local or remote reading are provided at all major points of the seal liquid system.

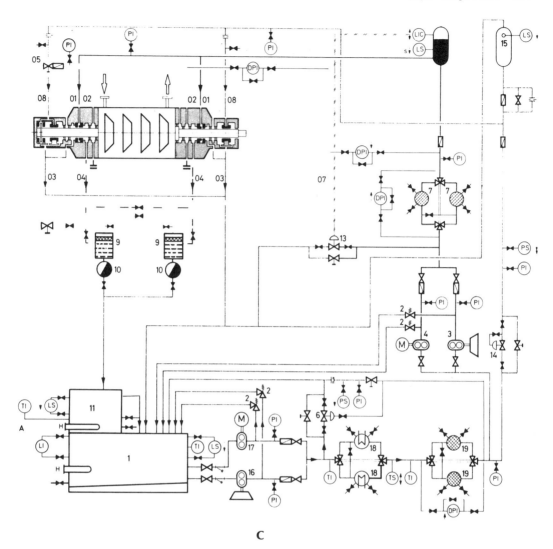

C

FIGURE 11-24 (Continued)

The seal oil system may be combined with the lube oil system if the gas does not adversely affect the lubricating qualities of the oil or provided the oil made unserviceable by the gas does not return into the oil system.

There are two methods of combining lube oil and seal oil systems. In the first of these, the oil can be raised to the pressure required for lubrication purposes and part of it then raised further to the pressure needed for sealing (booster system). Alternatively, all the oil is initially raised to the seal oil pressure and the flow of oil required for lubrication is then reduced in pressure (combined system).

Compressor Monitoring and Safety Equipment

In order to ensure the operational reliability and safety of centrifugal compressors, all major mechanical and thermodynamic parameters must be constantly scanned and evaluated. Only fast detection of changes makes possible timely intervention, thus protecting the compressor plant and the process from potentially damaging effects.

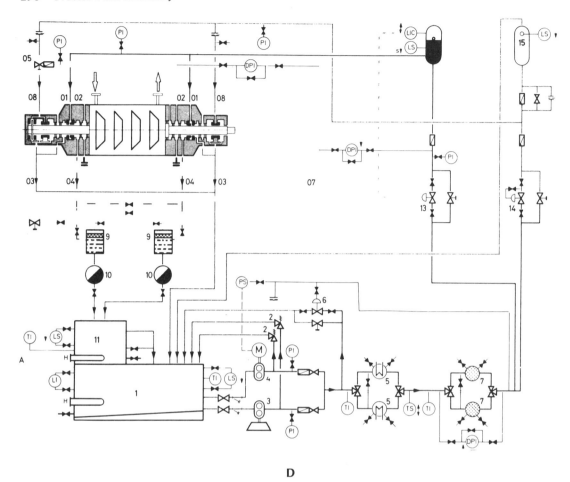

D

FIGURE 11–24   (Continued)

Prominent manufacturers therefore design and supply complete monitoring and instrument panels for compressor and driver. These contain all the instrumentation for monitoring, including control equipment, an alarm system, and interlock and shut-down systems.

Moreover, these manufacturers often furnish fully automatic compressor plants with automatic start/stop control for single compressor sets or, if required, including the higher-level station control facility. The latter switches the individual compressors on or off as required when such machines are operating in parallel. The control equipment needed for compressor and driver, fittings and auxiliaries is fully taken into consideration. Installations of this kind already supplied by major manufacturers include those with air compressors in large compressed air systems, sewage treatment plants, and offshore platforms with gas compressors.

Bearing Monitors

Smooth running of the shaft can be guaranteed only by bearings that function properly. The simplest check is monitoring of the lube oil draining from the bearings; wear of or damage to the running surfaces or insufficient bearing clearance mean greater friction and therefore an elevated oil discharge temperature. The high temperatures present another hazard, since they reduce the viscosity of the lube oil and hence its lubricating effect. Thermocouples, mercury thermometers, and resistance thermometers may be used for monitoring oil temperature. The permissible temperature range is limited by switching contacts; if the set limits are exceeded, an alarm is initiated or the compressor is tripped. It is also possible to measure the bearing temperatures directly in the bearing pads.

Measurement of Vibration and Shaft Position

Mechanical vibration of shaft and casing and axial shaft displacements are important indications of possible hazard to compressor operating reliability. Modern technology renders it possible to detect even the slightest of changes in these parameters.

Shaft vibrations may be due to a number of causes: deposits or other impeller unbalances; shaft distortion as a result of thermal stresses or shock loads; changed bearing condition; effects emanating from the driver or gearing. Measurements are performed by probes that are calibrated for a certain distance between shaft and probe head. These probes produce an electrical signal proportional to the amplitude of vibration. The permissible shaft vibration depends on the speed and the rotating mass. Low speeds allow higher vibration. Apart from measuring the total level, which is the sum of the amplitudes at different frequencies, it is also possible to analyze the vibrations on the basis of frequency. This may give a clue to the exciting frequencies or other causes of vibration.

Seismic pickups are employed for measuring casing or bearing pedestal vibrations. This method does not supply such precise results as shaft vibration measurement, but it offers the advantage of permitting measurements to be performed on running machines without having to make modifications first.

The shaft position indicator monitors the axial position of the shaft relative to the casing. The cause of axial displacement of the shaft may be wear of the thrust bearing or sudden loads that may occur when the compressor is operating in the unstable region. Detection of axial shaft shift is usually by electrical, or less frequently, by hydraulic means. Electrical measurement basically involves the same probes as are used for contactless vibration measurement.

All devices can be fitted with switches that trigger alarms or initiate shut-down of the plant to prevent damage when limit values are exceeded. Modern shut-down devices use the input from several sensors into a logic module. The machine would be automatically taken off-line if, for instance, a temperature sensor and a vibration sensor would independently confirm a violation of two setpoints.

APPENDIX 11A

Compressor Design*

The operating conditions of a plant determine the compressor design. The impellers are selected on the basis of the thermodynamic data. Selection is influenced by peripheral conditions such as materials data, aerodynamic and thermodynamic limits, or the operating characteristics.

THERMODYNAMICS

Centrifugal compressors are designed to convert velocity energy into pressure energy. They thus change work introduced through the impellers into the required gas-side operating conditions; this primarily means the required discharge pressure. The minimum driver size or rating needed is dictated by the energy required to accomplish this conversion, in which the losses should obviously be as small as possible.

The work performed to accomplish the change in volume results in an equivalent quantity of heat energy being transferred to the gas, causing an increase in temperature. Compressor design takes into account the fact that critical temperatures should not be exceeded if they might initiate polymerization or reaction of the medium with the compressor materials.

AERODYNAMICS

The compression cycle is designed to prevent the emergence of near-sonic velocities. These could initiate shock waves at the leading edges of the impeller blades, which then choke the flow cross sections and manifest themselves as irregular pressure rises. Near-sonic velocities may easily occur with compression media of high molecular mass, small isentropic exponent, and low temperature.

OPERATING CHARACTERISTICS

Optimized centrifugal compressors are designed so that their characteristics match the process requirements.

The total required energy is divided among the various stages in accordance with a number of criteria. Lightly loaded stages favor a stable working range that is as wide as possible, although the number of stages becomes larger and the work involved in building the compressor therefore more extensive. If a compressor is designed for a narrow volume flow range, highly loaded impellers are more economical. This enables the number of stages and hence the size of the compressor to be drastically reduced.

---

*Source: Mannesmann Demag, Duisburg, West Germany.

In addition to fluctuations in pressure and temperature, changes in the composition of the gas also occur. If the change is toward greater molar mass, a speed-controlled driver can be employed to reduce the drive speed and prevent surge in the final stages.

Design Diagrams

The principal characteristics of a compressor can be approximated from the diagram reproduced in Figure 11A–1 for the power, number of stages, and speed in conjunction with the auxiliary diagram for quantity relationships (Figure 11A–2). The first of these diagrams is valid for isentropic, i.e., uncooled, compression. The specific energy y_s and compressor efficiency are not shown, but they are implicit as a function of their influences.

Compressors with pressure ratios of up to 7:1 are covered by the diagram, since cooling is almost always necessary for higher pressure ratios. In order to determine the compressor data in such cases, the compression process is split at the point where intermediate cooling takes place; the two separate sections of the compression process can then be summed after each has been determined separately.

Calculation Example

Using metric units, the prescribed operating data are as follows:

Medium: a gas mixture consisting of inert gases and hydrocarbons

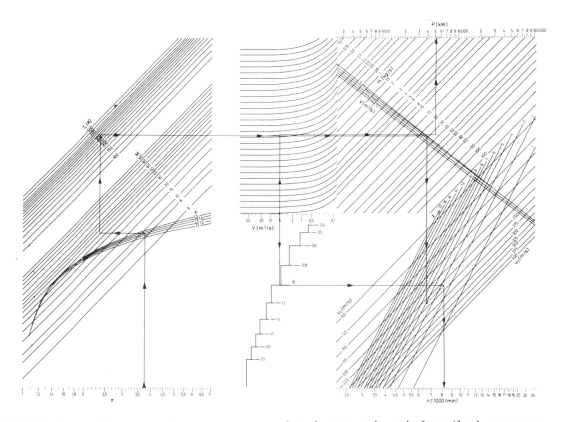

FIGURE 11A–1 *Diagram for determining power, number of stages, and speed of centrifugal compressors. (Source: Mannesmann Demag, Duisburg, West Germany.)*

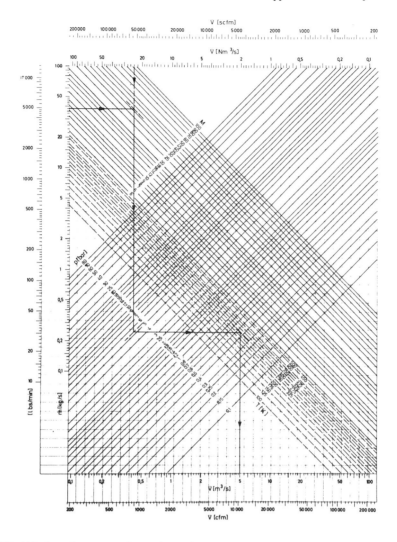

FIGURE 11A–2 *Quantity relationships for compressed gases. (Source: Mannesmann Demag, Duisburg, West Germany.)*

Molar mass M_{id}	= 36.5
Real Gas factor Z	= 1.0
Isentropic exponent $\kappa = c_p/c_v$	= 1.31
Mass flow rate $\dot{m}$	= 134,046 kg/h
Intake pressure p_1	= 5.14 bar
Intake temperature T_1	= 303 Kelvin (K)
Discharge pressure p_2	= 19.45 bar
Maximum tip speed (assumed) u_{2max}	= 240 m/s

The design procedure is as follows:

1. From the intake pressure p_1 and discharge pressure p_2, the following pressure ratio is obtained:

$$\pi = p_2/p_1 = 19.45/5.14 = 3.784$$

2. Correction of the ideal molar mass M_{id}, using the real gas factor Z:

$$M = M_{id}/Z = 36.5/1 = 36.5$$

3. The data covering the volume flow rate $\dot{V}$ and the mass flow rate $\dot{m}$ for any quantity can be taken from the adjacent auxiliary diagram for the quantity relationship for the given operating conditions. Taking M, p, and T into consideration, the following volume flow rate is obtained:

$$\dot{V} = 5 \text{ m}^3/\text{s}$$

4. Starting from pressure ratio $\pi = 3.784$, the arrow points vertically upward to the point of intersection with the curve for the isentropic $\kappa = 1.31$; from there the next arrow leads horizontally toward the point of intersection with the curve for the same molar mass M = 36.5. Vertically above that point, the temperature line T_1 = 303 K is reached, after which the next arrow points horizontally toward the range of influence of the volume rate $\dot{V}$. The heavy line then runs parallel to the family of curves until the specified volume flow rate $\dot{V} = 5$ m³/s is reached, after which the heavy line runs horizontally again until it intersects with the corresponding mass flow rate line $\dot{m} = 37.24$ kg/s. Vertically above this point of intersection the power requirement P = 5150 kW can be read off.

5. The heavy line is now extended horizontally through the range of influence of the volume flow rate $\dot{V}$ until the line reaches the volume-dependent lines of opposite inclination. From this point it is extended vertically downward as far as the stage number line z. For the assumed tip speed of $u_2 = 240$ m/s we then obtain z = 4 as the number of stages.

6. Starting from the volume flow rate scale $\dot{V} = 5$ m³/s, the heavy line in the direction of the arrow then impinges vertically on a specific size of compressor casing. From that point the heavy line is taken horizontally to the right until it intersects the line $u_2 = 240$ m/s established under step 5 (above). Vertically beneath this point of intersection, the speed n= 8200 RPM is read off.

Calculation Procedure

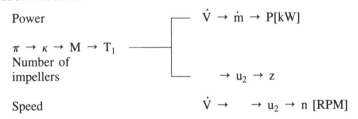

Calculation Steps

1. π = 19.45/5.14   = 3.784
2. M = 36.5/1.0   = 36.5
3. $\dot{m}$ = 134,064/3600   = 37.24 kg/s
    $\dot{V}$ = 5 m³/s
4. P = 5150 kW
5. z = 4
    u_2 = 240 m/s
6. n   = 8200 RPM

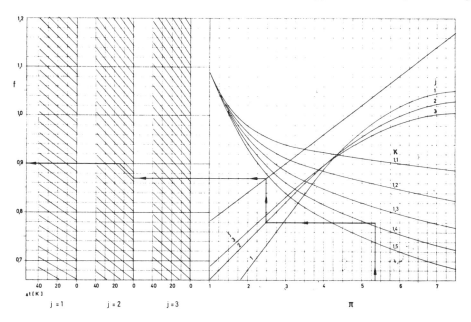

FIGURE 11A–3   *Influence of intercooling on gas compression efficiency. (Source: Mannesmann Demag, Duisburg, West Germany.)*

Influence of Intercooling

Energy saving is the main advantage and aim of cooled compression. Since the energy required is proportional to the intake temperature, the compression process is divided into a number of steps; the intake temperature of each of these steps is reduced in an intercooler.

The advantage of intercooling becomes apparent when the thermodynamically attainable energy savings are contrasted with expenditure on the cooling medium and coolers. The thermodynamic study must take into account the flow resistance of the cooler. This pressure loss is compensated by additional compression work.

The number of intercooling stages therefore depends on the overall pressure ratio π, the isentropic exponent κ, which is determined by the temperature rise during compression, and the temperature differential Δt between the intake temperature of the first stage and the recooling temperature in the subsequent stages. The recooling temperature is determined by the temperature of the cooling medium and the heat exchange surface of the cooler.

The possible advantage of intercooling can be visualized with the help of Figure 11A–3.

Example:
Pressure ratio $\pi = 5.35{:}1$
Isentropic exponent $\kappa = 1.38$
Number of intercooling stages $j = 2$
Result: Power factor $f = 0.9$, i.e., for the case in question, intercooling twice would result in a power saving at the compressor coupling of about 10%

For further information on compressor calculations, see Appendix 11B.

APPENDIX 11B

High-Speed Centrifugal Compressors*

As mentioned at the beginning of this chapter, dynamic gas compression is achieved by the mechanical action of a rotating impeller that imparts velocity energy to the gas. This velocity energy is then converted to pressure rise of the gas in a diffusion section that can be of a partial emission, vaneless, or vaned diffuser design. The rotating element or impeller of the centrifugal compressor was historically limited in rotational speed by the motor or turbine driver speed capability. With the use of separate speed-increasing gearboxes or high-speed drivers, the rotor speed was gradually increased to a maximum of 10,000 to 12,000 RPM. These are primarily multistage, medium-speed, high-flow centrifugal compressors. In the late 1950s, commercial work was compelled on centrifugal compressor designs that used an integral speed-increasing gearbox with the single impeller of the compressor rotating at speeds of 34,000 RPM and more. This provided the process industries with a design for low flows with high head capabilities.

Flow rates for high-speed centrifugal compressors can range from 10 ACFM to more than 100,000 ACFM. Power capability ranges from 15 to 2500 HP or greater. See Figure 11B–1 for the composite performance envelope of high-speed centrifugal compressor designs.

Integrally geared high-speed compressors (Figure 11B–2) are available for use with any type of gas that a process design might involve. These gases can range from air or nitrogen to such exotic gases as hydrogen bromide.

To define where a high-speed centrifugal compressor is primarily used, the concept of specific speed and specific diameter should be understood and applied. Specific speed, Ns, is a dimensionless index number for the impellers or rotors of various types of compressors and pumps. Using English/US values, the definition is the same for both pumps and compressors:

$$s = \frac{N\sqrt{Q}}{(H)^{3/4}}$$

Another dimensionless quantity for impellers or rotors is termed the specific diameter, Ds, defined by:

$$Ds = \frac{D(H)^{1/4}}{\sqrt{Q}}$$

In both formulas: H = Head in ft-lbf/lbm, or foot-pounds (force) per pound (mass)
Q = Flow capacity in ft³/second at inlet conditions
N = Rotational speed in revolutions/minute
D = Diameter of impeller or rotor in feet

*Source: Sundstrand Fluid Handling, Arvada, CO, and Larry E. Glassburn (National Business Consultants, Inc.).

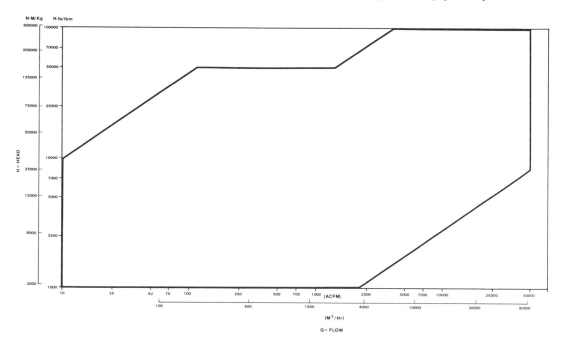

FIGURE 11B–1   *Typical head versus flow capability of high-speed centrifugal compressors. (Source: Sundstrand Fluid Handling, Arvada, CO.)*

The process or unit size determines the head (H) and capacity (Q) for a given system design. Components N and D are defined by the availability of mechanical designs from the various manufacturers. For high-speed centrifugal compressors, the normal range for these variables is N from 5,800 rev/min to 50,000 rev/min and D from 5 inches (127 mm) to 36 inches (914 mm).

After the process design requirements and the need for the compression equipment have been defined, it is necessary to determine the equipment best suited for the process by using the quantities of specific speed and specific diameter. Using Balje's chart (Figure 11B–3) with specific speed and specific diameter calculated for a particular application, it is possible to determine if the application is a good fit for a single-stage, high-speed, centrifugal compressor. High-speed centrifugal compressors can be of partial emission, radial flow, or mixed flow design. By adding additional parallel stages or series stages of compression, the performance of a single-stage machine can be increased in flow or head. This can be done integrally within the same compressor frame, or it can be done by separate compressor units.

The detailed sizing of a high-speed centrifugal compressor is no different than sizing any centrifugal compressor (for further information see previous chapter and also Suggested Reading, Lapina). Virtually every compressor manufacturer has literature available that will assist in sizing the specific unit that would be required for a given process. The available literature usually contains performance envelopes showing the manufacturer's capabilities based on a single gas for a range of flows (Q) in ACFM or cubic meters per hour and heads (H) in foot-pounds per pound mass or newton-meters per kilogram. This enables the user to compare anticipated flow rates and heads with the machinery offered by a given manufacturer. To obtain a budgetary quotation or further information from the manufacturer, the user will have to provide the following data: inlet pressure, P_1, either in gauge or absolute; discharge pressure, P_2, either in gauge or absolute; inlet temperature, T_1, °F or °C; required capacity or flow rate, Q, preferably in pounds per hour or any weight flow unit; and the molecular

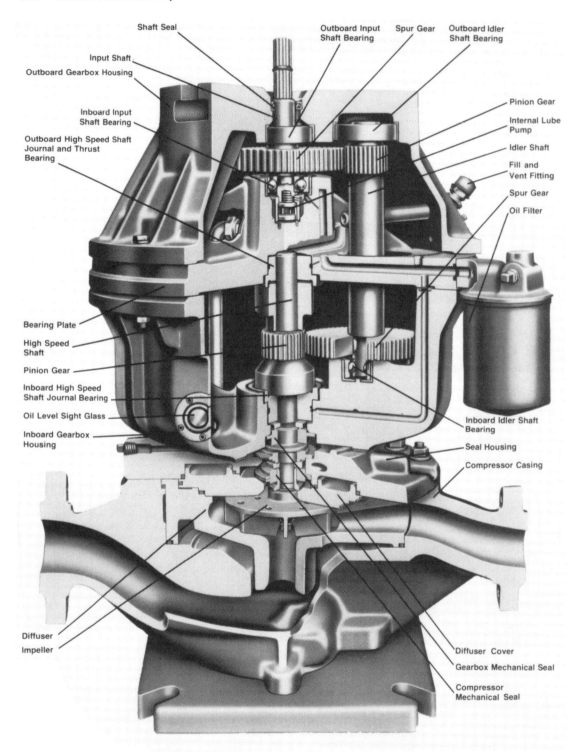

Shaft Seal

Outboard Input Shaft Bearing

Spur Gear

Outboard Idler Shaft Bearing

Input Shaft

Outboard Gearbox Housing

Pinion Gear

Internal Lube Pump

Inboard Input Shaft Bearing

Idler Shaft

Outboard High Speed Shaft Journal and Thrust Bearing

Fill and Vent Fitting

Spur Gear

Oil Filter

Bearing Plate

High Speed Shaft

Pinion Gear

Inboard High Speed Shaft Journal Bearing

Oil Level Sight Glass

Inboard Gearbox Housing

Inboard Idler Shaft Bearing

Seal Housing

Compressor Casing

Diffuser

Impeller

Diffuser Cover

Gearbox Mechanical Seal

Compressor Mechanical Seal

FIGURE 11B–2   *Vertical high-speed centrifugal compressor with integral gearbox. (Source: Sundstrand Fluid Handling, Arvada, CO.)*

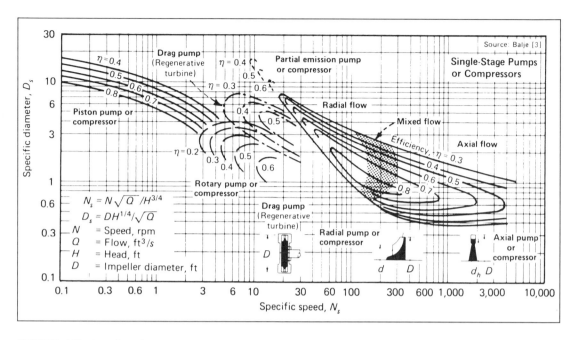

FIGURE 11B–3 *Specific speed versus specific diameter for initial selection of a type of single-stage compressor. (Source: O.E. Balje.* A Study on Design Criteria and Matching of Turbomachines—Part B.*)*

weight, MW, of the total process gas or the individual components of the total process gas mixture. The specific heat ratio, K, is also required, as is the inlet compressibility factor, Z_1. The compressor manufacturer can define compressibility factors and specific heat ratios for the majority of gases or gas mixtures. Proprietary or nonstandard gases must have the properties defined for proper selection.

Adiabatic head for single-stage compression is usually calculated from the following expression:

$$H_{AD} = Z_1 \left(\frac{1545}{MW} \right) (T_1) \left(\frac{K}{K-1} \right) \left[\left(\frac{P_2}{P_1} \right)^{\frac{K-1}{K}} - 1 \right] \text{ ft-lbf/1bm}$$

Dividing the discharge pressure (absolute) by the inlet pressure (absolute), one obtains the pressure ratio. Selection charts from the various manufacturers display either head or pressure ratio. The weight flow can be converted to volume flow rate by using the following formula:

$$Q = \frac{(\dot{W}) \, (10.729) \, (T_1) \, (Z_1)}{(P_1) \, (MW)} \text{ actual ft}^3/\text{min (ACFM)}$$

With these two values, the head and flow, it is normally possible to determine if the available compressor meets the user's needs. The design horsepower will be

$$BHP = \frac{(\dot{W}) \, (H_{AD})}{(33,000) \, (\eta_{AD})} + \text{Frictional horsepower} = \text{Total horsepower}$$

The chart in Figure 11B–4 is an approximating method for high-speed centrifugal single-stage compressor selection based on head and flow. The chart applies to two

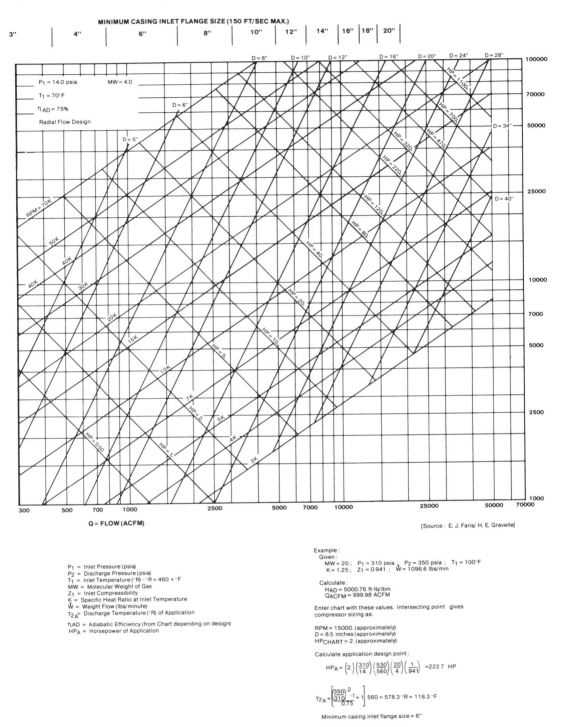

MINIMUM CASING INLET FLANGE SIZE (150 FT/SEC MAX.)

P_1 = Inlet Pressure (psia)
P_2 = Discharge Pressure (psia)
T_1 = Inlet Temperature (°R) · °R = 460 + °F
MW = Molecular Weight of Gas
Z_1 = Inlet Compressibility
K = Specific Heat Ratio at Inlet Temperature
$\dot{W}$ = Weight Flow (lbs/minute)
T_{2A} = Discharge Temperature (°R) of Application

η_{AD} = Adiabatic Efficiency (from Chart depending on design)
HP_A = Horsepower of Application

Example:
Given:
MW = 20 ; P_1 = 310 psia ; P_2 = 350 psia ; T_1 = 100°F
K = 1.25 ; Z_1 = 0.941 ; $\dot{W}$ = 1096.6 lbs/min

Calculate:
H_{AD} = 5000.76 ft-lb/lbm
Q_{ACFM} = 999.98 ACFM

Enter chart with these values. Intersecting point gives compressor sizing as:

RPM = 15000. (approximately)
D = 8.5 inches (approximately)
HP_{CHART} = 2. (approximately)

Calculate application design point:

$$HP_A = \left(2.\right)\left(\frac{310}{14.}\right)\left(\frac{530}{560}\right)\left(\frac{20}{4}\right)\left(\frac{1}{.941}\right) = 222.7 \text{ HP}$$

$$T_{2A} = \left[\frac{\left(\frac{350}{310}\right)^{.2} - 1}{0.75} + 1\right] 560 = 578.3 \text{ °R} = 118.3 \text{ °F}$$

Minimum casing inlet flange size = 6"

[Source: E. J. Faris/ H. E. Gravelle]

FIGURE 11B–4   *High-speed centrifugal compressor estimating selection chart/procedure. (Source: Sundstrand Fluid Handling, Arvada, CO.)*

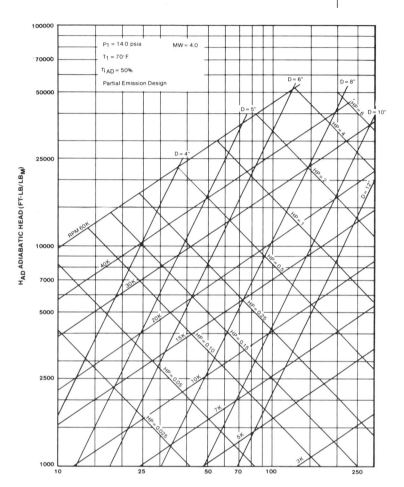

Calculate Head and Flow at Compressor Operating Point:

$$H_{AD} = Z_1 \left(\frac{1545}{MW} \right) (T_1) \left(\frac{K}{K-1} \right) \left[\left(\frac{P_2}{P_1} \right)^{\frac{K-1}{K}} - 1 \right]$$

$$Q_{ACFM} = \frac{(\dot{W})\,(10.729)\,(T_1)\,(Z_1)}{(P_1)\,(MW)} = \frac{(SCFM)\,(0.0283)\,(T_1)\,(Z_1)}{(P_1)}$$

Enter Chart and Define Intersecting Point,

Chart Ouput:

RPM = Impeller Rotational Speed (RPM) · (K is 1000)
D = Impeller Diameter (Inches)
HP = Chart HP based on MW = 4.0

Calculate Specific Application Design Point Output:

$$HP_A = \left(HP_{chart} \right) \left(\frac{P_1}{14.0} \right) \left(\sqrt{\frac{530.}{T_1}} \right) \left(\frac{MW}{4.} \right) \left(\frac{1.}{Z_1} \right)$$

$$T_{2A} = \left[\frac{\left(\frac{P_2}{P_1} \right)^{\frac{K-1}{K}} - 1}{\eta_{AD}} + 1 \right] T_1$$

different impeller/diffuser designs from Balje (Figure 11B–3), the partial emission design for low flows, and the radial/mixed flow design for higher flows.

To use Figure 11B–4, calculate the required head and flow for a given application, and enter the chart locating the intersecting point. This point will define the approximate speed, impeller diameter, and power for the compressor. The chart will also provide the discharge pressure (P_2) and temperature (T_2) by the formulas on the chart. The power curves in the chart are based on MW = 4. To calculate power for other MW gases, use the correction formula as shown. When calculating the value of T_2 for another MW gas, note the two different values based on each design type (efficiency). Power values are based on aerodynamic work only. Frictional losses are not included, but they will not greatly affect the estimate.

For a given mechanical design of a high-speed centrifugal compressor (speed, impeller diameter, etc., are all fixed), there will exist only one performance curve of head versus flow based on the inlet conditions (P_1, T_1, MW, K, and Z_1) of the process gas selected. The head/flow curve will relate discharge conditions to the given inlet condition. If inlet conditions change for the fixed mechanical design, the head/flow curve remains the same; only the discharge conditions will change. This head/flow curve is fixed by the mechanical design of the high-speed centrifugal compressor to accommodate the worst combination of operating conditions. Any other off-design conditions that do not fall on the head/flow curve must be met by control methods as described later. The head generated by a single impeller stage of a high-speed compressor is quite high for a centrifugal design. The high-speed compressor design will not match the head capability of a positive displacement design. However, unless high heads are required, the benefits of the centrifugal design, such as flow variation at constant pressure, often outweigh those of positive-displacement machines.

Typical applications for the high-speed centrifugal compressor involve those that require pressure ratios ranging from approximately 1.005 to 3.5. These are classified as recycle-, regeneration-, or booster-type applications. Pressure ratios in excess of 3.5 can be achieved by series arrangement of the high-speed centrifugal design. These arrangements can be either individual compressor units or multiple impeller stages mounted on the same speed-increasing element. Typical applications for the high-speed centrifugal compressor are molecular sieve absorption regeneration systems, vapor recovery systems, gas injection systems, gas recycle systems, booster compressors, chlorine transfer, and numerous other types of applications that require low pressure rise. High-speed centrifugal compressor designs are also used for inert gases such as air and nitrogen.

Available metallurgies for high-speed centrifugal compressors include carbon steel, 316 stainless steel, and 17–4 PH stainless steel, which are all quite compatible with normal gases. When the gases become more corrosive or erosive, as might be the case with chlorine, bromine, hydrogen sulfide, etc., special materials such as Hastelloy B or C, Inconel®, or titanium may be required and are available from most manufacturers.

Any type of process gas can be effectively handled by the high-speed centrifugal compressor design, from hydrogen to Freon®, because the components of the high-speed centrifugal compressor that come in contact with the process gas are relatively few and small. Corrosive or erosive gases can also be handled quite effectively through the use of special metallurgy. Similarly, these dynamic compressors can be built with gas containment seals suitable to meet environmental and loss prevention concerns.

A number of seal designs are available for high-speed centrifugal compressors. They can range from simple labyrinth design for nonhazardous low-pressure applications to the typical mechanical contact face-type seals in either single, double, or tandem arrangements. The seals can be either gas- or liquid-type. Any type of seal arrangement must have some leakage across the seal face for it to function properly.

This leakage can be either the process gas or a buffer fluid (gas or liquid), depending on the environmental requirements.

All high-speed centrifugal compressors consist of three major components: the compressor, the speed-increasing element, and the main driver. The main driver can be an electric motor, steam turbine, air motor, gas turbine, combustion engine, or some other prime mover. Driver selection is generally related to the utility balance of the process plant installation. The compressor or process fluid element is fairly typical for all high-speed centrifugal compressors and consists of an impeller and diffuser contained within a pressure casing. The speed-increasing element between the driver and the compressor impeller is typically a unique integral gearbox that is provided by the same manufacturer (Figure 11B-2). The gearbox has been specifically designed to provide reliable and efficient operation of the unit.

High-speed centrifugal compressors are manufactured with either vertical or horizontal in-line, or horizontal axial inlet, and with top or side discharge. Other arrangements are possible and depend on physical size restrictions of the installation.

The following control methods are available for high-speed centrifugal compressors:

- Speed variation
- Suction pressure throttling
- Inlet guide vanes
- Flow control
- Variable diffuser
- Discharge pressure throttling

Control and operation of individual series and/or parallel units is essentially managed the same as with other dynamic compressors. Surge control must also be provided to protect the high-speed centrifugal compressor from serious mechanical damage. Further details on these subjects can be found in the chapter on axial compressors.

As with any mechanical device, routine maintenance of the high-speed centrifugal compressors is required. Depending on the process application and local environment where the high-speed centrifugal compressor is installed, yearly inspection and/or replacement of lubricating oil, oil filtration cartridges, ball bearings, and mechanical seals is recommended. Some applications may require lube oil analysis and/or replacement at a six-month interval. This can be determined by operating experience or discussion with the individual compressor manufacturer. The high-speed centrifugal compressor does not require any more maintenance than any other piece of mechanical equipment within the typical process plant. Experience shows that with proper selection, installation, and operation, the high-speed centrifugal compressor will provide many years of satisfactory operation and performance for any application in a gas compression system.

If a high-speed centrifugal compressor has been defined as a potential candidate for a given process, the design advantages may be quite numerous: oil-free operation (no contamination of the process gas), compact design (requires very small floor space and has minimal foundation requirements), simple process piping arrangements are allowed, smooth continuous flow (no pulsations into the downstream process), high inlet pressure capability, simple control systems are required, a wide range of metallurgies to handle various process gases is readily available, and there are no close clearances between rotating and stationary components impeding the passage of small contaminants.

Potential concerns with the high-speed centrifugal compressor design should be reviewed when these designs are compared with the conventional designs described earlier in the chapter. An inherently narrow, stable operating range must be considered, as must the fact that like all centrifugal compressors, high-speed machines must be designed for the worst combination of operating conditions. Also, depending on the specific process conditions, high-speed centrifugal compressors may exhibit lower efficiencies than comparable types of equipment that may be available for a particular application. The user must judge if these characteristics are significant to the potential installation.

SUGGESTED READING

Ingersoll Rand Co.: Compressed Air & Gas Data.

Lapina, Ronald O.: Estimating Centrifugal Compressor Performance. Gulf Publishing Company, Houston TX, 1982.

Balje O.E.: A study on Design Criteria and Matching of Turbomachines—Part B. Trans. ASME, J Engr Power, January 1962.

Chapter 12

Axial Flow Compressors*

As stated in the introduction to the preceding chapter, dynamic compressors are machines in which air or gas is compressed by the mechanical action of rotating components imparting velocity and pressure to the air or process gas. In an axial compressor, as the name implies, flow is in the axial direction, i.e., parallel to the axis of rotation. Axial compressors are basically high-flow, low-pressure machines, in contrast to the lower flow, high-pressure centrifugal compressors.

Figure 12–1 shows the performance characteristics of a centrifugal and an axial compressor at constant speed for the same operating conditions. From this figure, a direct comparison of the characteristics is easy. The "turndown" capability of the centrifugal is much larger than that of the fixed geometry axial. The range of operation is greatly increased through the use of variable geometry.

FIELD OF APPLICATION

Axial flow compressors have found wide use in refineries, petrochemical plants, and steel mills. Particularly in refineries, applications formerly handled by centrifugal units are now handled by axial flow compressors. This is due to several trends: first, plant sizes are growing dramatically, which brings the air requirements up into a desirable range for axial compressors; second, due to rising energy costs, there exists an increasing trend toward higher efficiencies; and third, technological improvements have made axial compressors more reliable than ever before.

Axial compressors are generally more efficient than centrifugal compressors in the common flow range, depending on conditions. An axial compressor will also generally be smaller than a centrifugal compressor designed for the same flow rate. Although the axial flow compressor requires more stages due to the lower pressure rise per stage, the diametral size is much greater in a centrifugal compressor in order to pass the required air flow. The axial compressor must operate at significantly higher speeds for the same condition and is usually more costly than a comparable centrifugal compressor. In applications where speed is not a major consideration, an efficiency and size versus cost evaluation must be made.

Petroleum Refineries

These are probably among the largest current users of axial flow compressors for providing the air for catalytic cracking. This service requires 50,000 cfm to 300,000 cfm at discharge pressures from 25 to 50 pounds per square inch gauge (psig). Figure 12–2 depicts a typical installation.

---

*Source: Dresser-Rand Company, Phillipsburg, N.J.

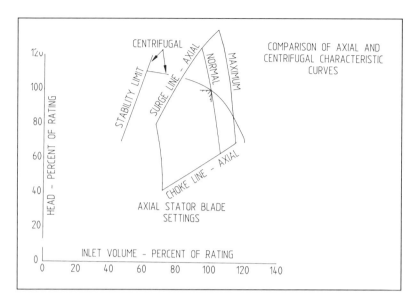

FIGURE 12-1   *Comparison of axial centrifugal characteristic curves. (Source: Dresser-Rand Co., Phillipsburg, N.J.)*

Butadiene Plants

Air capacities from 60,000 to 150,000 inlet cfm and discharge pressures from 20 to 30 psig with atmospheric intake enable the axial compressors to perform well in this application.

Nitric Acid Plants

In plants with capacities in excess of 500 tons per day, axial-centrifugal compressor combinations are frequently used to handle flows from 20,000 to greater than 80,000 cfm (approximately 100 cfm per ton-day) and discharge pressures from 110 to 130 psig for the combined compressor string. Power recovery expanders utilizing process "tail gas" usually drive the compressor string.

Air Separation Plants

Axial compressors are used almost exclusively for higher flow air services. Services range to 100,000 cfm and discharge pressure up to 100 psig.

Blast Furnaces

Axial compressors are replacing many of the older, less efficient centrifugal blowers. The compressors provide air at discharge pressures from 30 to 90 psig and flows from 125,000 to 350,000 cfm.

FIGURE 12-2   *The first power recovery train in fluid catalytic cracking service in a refinery and the equipment train. The equipment train comprises tandem motor-axial compressor-steam turbine-hot-gas expander units. The compressor is rated at 127,900 inlet cfm with a discharge pressure of 34 psig. The train power rating is approximately 15,000 HP. (Source: Dresser-Rand Co., Phillipsburg, N.J.)*

In addition to the common processes described above, axial compressors are also often used for wind tunnel service, waste treatment facilities, specialized testing facilities, and are being developed for co-generation combustion service.

BASIC AXIAL COMPRESSOR
PERFORMANCE CAPABILITIES

As described earlier, the axial flow compressor is a machine with wide ranges of capacity, i.e., flow, pressure, and horsepower required.

Axial compressor flow capabilities on the low end of the range, say 20,000 to 75,000 cfm, obviously overlap the higher range of centrifugal compressor coverage. It is within this low flow region where cost and size evaluation, along with driver considerations, must control the selection. Above this range, however, axial compressors are often the obvious choice. The physical size of the axial compressor is far smaller than the comparable centrifugal machine that would be required. In many high-flow situations, the axial is a better match for the drivers that will probably be selected.

To increase the pressure capability of the axial flow compressors, multiple casing designs have also been developed. These are known as biaxials and triaxials. As their names imply, these are two- or three-body axial compressor trains capable of pressure ratios up to approximately 12 to 1. The machines were developed for use in nitrogen injection services.

Horsepower requirements for axial flow compressors range from 3000 HP to 65,000 HP for single casing units. Horsepower inputs vary with the flow and pressure requirements of the service. A simple formula for the approximate power requirement of an axial flow compressor would be:

$$\text{Horsepower} = \frac{WW \times R \times T \times (PR^{.283} - 1)}{182.5}$$

where WW = Wet weight flow, lb/min
    R = Gas constant, ft−lbf/lbm−°R
    T = Inlet temperature, °R
    PR = Pressure ratio

FUNDAMENTALS OF AXIAL COMPRESSOR DESIGN

A multistage axial flow compressor has two or more rows of rotating components operating in series on a single rotor in a single casing. The casing includes the stationary vanes (the stators) for directing the air or gas to each succeeding row of rotating vanes. These stationary vanes, or stators, can be fixed- or variable-angle, or a combination of both.

A typical axial flow compressor cross section is shown in Figure 12–3; Figure 12–4 shows an axial compressor with the top half removed. The major components and their nomenclature are depicted in Figure 12–3 for reference use throughout this chapter.

There are two basic types of blading that are employed in an axial flow compressor; these are obviously rotating and stationary. A brief overview of these parts is presented next.

Stationary Blades

Inlet Guide Vanes

The first row of stationary blades is unique. These blades are referred to as inlet guide vanes. These vanes are designed to provide prerotation to the air or gas stream prior to entry into the rotor blades. Blade profiles have airfoil-shaped cross sections.

Stator Vanes

The majority of the stationary blades within the compressor are simply called stators. There exist two types of stator vanes, variable and fixed.

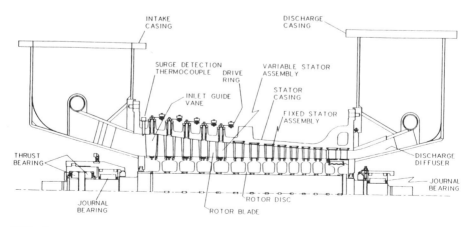

FIGURE 12–3 *Typical axial flow compressor cross section. (Source: Dresser-Rand Co., Phillipsburg, N.J.)*

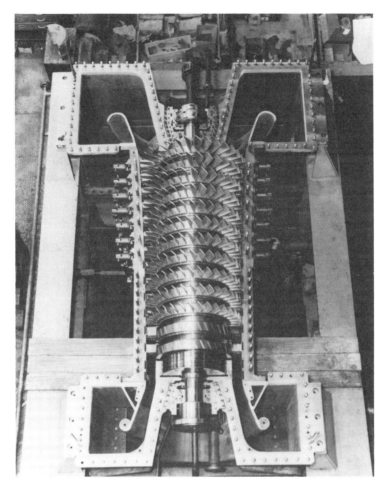

FIGURE 12–4   *Overhead view of an axial flow compressor with the top half of the casing removed. (Source: Dresser-Rand Co., Phillipsburg, N.J.)*

Variable Stator Vanes.   Variable stator vanes fit through the stator casing or a blade carrier of some kind (depending on the manufacturer's design) and are attached to a drive mechanism that moves the vanes with respect to the air flow. A more detailed description of the actuating system is provided below. The inner end of each vane can be shrouded to improve the stress condition and to reduce the interstage losses through sealing strips mounted in the inner shroud.

The actuation system used to move the variable stator section is usually a combination of linkages designed to move the vanes simultaneously. One type of linkage system is shown in Figure 12–5. Each variable stator vane is connected to a driving ring by a small link. These rings, one for each stage of blades, are individually connected to a main driving shaft so that the stages move simultaneously. The drive shaft is connected to a hydraulic (or pneumatic) power piston, which, through push-pull effect, opens and closes the stator vane.

Different designs are chosen for this system by the various manufacturers. Desirable features would include the following:

1. Solid or "tight" linkage systems to prevent slow or inefficient actuation of the vane position.

FIGURE 12–5   *A typical variable stator vane actuation system and linkage arrangement. (Source: Dresser-Rand Co., Phillipsburg, N.J.)*

2. A minimum number of ''joints'' in the system that can wear with time and become loose or seize due to the presence of dirt.
3. Dual power cylinders (one on each side of the unit) to provide even movement of all vanes.

Fixed Stator Assembly.   The fixed stators are typically welded assemblies comprising the vanes and inner and outer shrouds. These assemblies are fitted into machined grooves in the stator casing. The fixed stator assembly is also fitted with sealing strips for leakage reduction.

Rotating Blades

The rotating blades within the axial compressor are appropriately called rotor blades. These are National Aeronautics and Space Administration (NASA)-developed tapered and contoured airfoil sections. The rotor blades have an attachment on one end to allow for assembly within the rotor.

A simplified partial section of an axial flow compressor flow path is shown schematically in Figure 12–6. The basic components would typically include the following:

*   An inlet duct to collect and accelerate the gas toward the inlet guide vanes with minimum pressure losses.
*   A row of inlet guide vanes to impart prewhirl to the gas stream in the direction of rotation for smooth entry to the rotor blades and for the control of the inlet relative Mach number.
*   A multiplicity of stages, each consisting of a row of rotor blades and a row of stator vanes of airfoil shape, to increase the static and/or total pressure of the flow. The

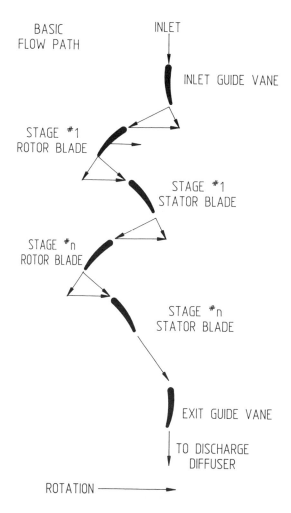

BASIC
FLOW PATH

INLET

INLET GUIDE VANE

STAGE #1
ROTOR BLADE

STAGE #1
STATOR BLADE

STAGE #n
ROTOR BLADE

STAGE #n
STATOR BLADE

EXIT GUIDE VANE

TO DISCHARGE
DIFFUSER

ROTATION ⟶

FIGURE 12–6 *Schematic presentation of an axial flow compressor flow path. (Source: Dresser-Rand Co., Phillipsburg, N.J.)*

total energy transfer to the gas stream is accomplished by the rotor blades. The hub stagger (the angle between the blade chord and the axis of rotation) is fixed, thereby fixing the amount of work done by each stage and consequently fixing the number of stages necessary to achieve the required discharge pressure. The standard frame design is adjusted to meet the required air flow by varying the rotor and stator blade heights and the unit operating speed.

- A row of exit guide vanes, oriented to remove the whirl component from the flow leaving the last stage stator vanes, and to begin deceleration of the flow.
- A discharge diffuser to further decelerate the flow and to convert the residual velocity energy into static pressure rise.

Natural Frequencies and Resulting Stresses

Due to the inner and outer shrouding of the fixed stators and the internal shrouding and casing support of the variable stators, vibration occurring at component natural frequencies and the resulting stresses in these components are not of major concern. If a specific manufacturer does not use shrouds on both inner and outer surfaces, it is important to review the frequency analysis of the stationary vanes in the same manner as discussed below for the rotor blades.

Since the rotor blades are mounted in a cantilever beam arrangement, i.e., one end unsupported, natural frequency excitation and the resulting stresses must be thoroughly analyzed by the designer. Clear definition of the natural frequencies of the blading, possible sources of excitation within the unit, and the resulting stress levels should be provided by the manufacturer. This should be presented in the form of Campbell and Goodman Diagrams for the compressor blading. (Refer to Chapter 5 for a discussion of these diagrams.) Evidence of the accuracy of the information, in the form of test data or successful long-term operating experience, should be made available by the manufacturer and should be reviewed by the purchaser. Similar scrutiny is appropriate for thrust bearing designs.

OPERATIONAL LIMITATIONS

The successful prediction of the performance of an axial compressor requires knowledge of the flow behavior likely to be encountered in the machine. In certain areas of the operating characteristic, accurate prediction of the performance is not possible. Violent overall instabilities identified as stall, surge, and choke complicate these predictions.

Stall

Stall is a commonly used term with regard to axial flow turbomachinery. It is too often incorrectly used as a cause of problems, mainly due to a misunderstanding of the true flow behavior during a stalled condition. Stall occurs when the flow separates from the surface of the blade. An aerodynamic disturbance is formed downstream of the point of separation.

Stall is a generalized term. It is often used with additional descriptors to explain the flow condition that causes the separation. For instance, the separation can occur on either side of the airfoil. High positive incidence stall, caused when the angle between the flow and the inlet to the rotor blade is too large, causes separation of the airflow from the suction, or convex, side of the airfoil. High negative incidence stall causes separation of the airflow from the pressure, or concave, side of the airfoil.

Conditions of rotating stall can be established when a group of blades becomes stalled. This is a phenomenon of stall cells being created within a stage. As the first blade stalls, it causes a disturbance to the airflow of the adjacent blade, eventually causing it to become stalled. As each subsequent blade becomes stalled, the last blade in the patch of stalled blades begins to recover. Thus, the effect is a rotating patch of stalled blades.

Whether a few blades experience stall or a rotating stall cell is formed, the mechanical damage possibly caused by this unstable aerodynamic condition can be significant. It is difficult to determine the actual loads induced on the blades during such an event. Laboratory testing has shown that loading levels can reach ten times normal levels.

Surge

As with any dynamic compressor, surge occurs when the slope of the pressure ratio versus capacity curve becomes zero. It is associated with the complete breakdown of flow through the machine, and it takes place when several adjacent stages are subjected to high positive incidence stall. At any given speed, as the inlet flow is reduced, a point of maximum discharge pressure is reached. As flow is further reduced, the pressure developed by the compressor tends to be lower than the pressure in the discharge line and a complete flow reversal of an oscillatory nature results. The reversal of flow tends to lower the pressure in the discharge line and normal compression resumes. If no change to either the system back pressure or the operation of the compressor occurs, the entire cycle is repeated. This cycling action is an unstable condition varying in intensity from an audible rattle to violent shock, depending on the energy level of the machine. Intense surges are capable of causing serious damage to the compressor blading and seals. The uncertainties surrounding this oscillating flow are cause for concern.

Surge Control

It is standard procedure at process plants to install reliable antisurge control equipment in the compressor piping to prevent operation in the surge region. A typical surge control system should incorporate or encompass recycle loops, i.e., valved bypass piping to provide sufficient flow through the compressor to keep it away from surge. Experience points to the following requirements:

- The system is to be electronic rather than pneumatic for the fastest response time, and the surge valve must be interlocked with the trip circuit such that it immediately opens on a train trip.
- The control system logic requires flow, pressure, and stator vane position input.
- The surge valve positioner-operator system must be capable of driving the valve fully closed to fully open in one second and fully open to fully closed in ten seconds.
- The surge valve should open on the loss of any input signal or operator medium.
- The surge valve should be sized to pass full flow at any point along the surge line with the valve at 60% open and with full consideration given to the downstream system pressure drop. The point on the surge line requiring the largest valve and discharge system is normally at maximum speed with stators full open, but a point at lower flow with a lower discharge pressure may in some instances dictate size.
- A check valve should be installed close to the compressor discharge just *downstream* of the surge valve connection in the discharge line.

In addition to the antisurge control system discussed above, a thermocouple detection system can also be employed to determine the presence of the surge recycling effect.

When an axial flow compressor experiences surge, it essentially undergoes a momentary internal gas flow reversal. This flow reversal will slightly elevate the stator casing inlet gas temperature. The increase results from the intermediate gas, which has been heated by compression, flowing back into the inlet area.

This detection system should be implemented in addition to a primary antisurge control system. It consists of thermocouples installed in the airstream just upstream of the inlet guide vanes. Upon reaching a predetermined setpoint, which indicates surge, a signal is sent commanding the surge control valves to go to the full open position. Upon reaching an acceptable level of temperature, i.e., the compressor has recovered from the surge, the surge valves are permitted to return to the normal position and control of the surge valve is returned to the primary surge control system.

Choked Flow

Choking occurs when the slope of the pressure ratio versus capacity curve approaches infinity. It occurs at the point where a further increase in mass flow through the cascade is not possible. Choked flow is associated with the flow velocity reaching a Mach number of 1.0 at some cross section within the machine.

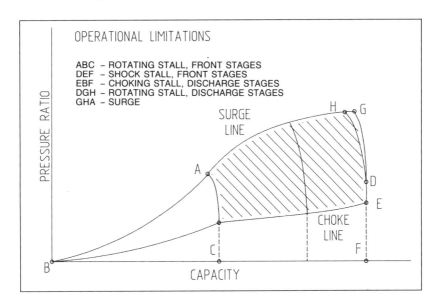

FIGURE 12–7 *Operating envelope for axial compressors. (Source: Dresser-Rand Co., Phillipsburg, N.J.)*

Unlike with surge, there is no accompanying increase in noise level or machine vibration amplitude. Choke is a "quiet" phenomenon, which, when operation continues for extended periods of time, can cause damage to the rotating blades, with eventual failure possible.

Choke Control

A choke control system is needed to avoid operation within the detrimental region. A typical system would encompass a choke valve designed for minimum pressure drop in the open position and capable of full response in ten seconds. The valve would open in the event of signal failure and would respond to a control system using differential pressure (flow) and discharge pressure as inputs.

Surge and choke conditions are affected by geometry, speed, and ambient conditions. Each of the aerodynamic instabilities is most likely to occur within a particular primary region of the performance characteristic. Figure 12–7 shows these various regions.

STANDARD MAINTENANCE CONSIDERATIONS

Like all machinery, axial flow compressors require both periodic preventive and corrective maintenance. A daily review of key operating data is quite often the best preventive maintenance strategy for axial flow compressors. These data should include various ambient and process parameters that define the operation of the unit. In addition, machinery vibration and bearing temperature data should be logged. While these daily (or weekly) records may fail to tell the entire operating history, they can nevertheless establish trends of operation, i.e., has the present problem been progressing slowly for several days or weeks, or is it a sudden change?

Quite often, these records can provide the information essential to avoiding unscheduled shutdowns simply through the establishment of normal trends. Just as

Table 12.1 Parameters for Condition Monitoring of Axial Compressors

Inlet temperature
Inlet pressure
Inlet relative humidity (dew point temperature)
Unit flow (at inlet or delivered)
Discharge pressure
Discharge temperature
Stator vane position
Unit speed
Radial vibration amplitude
Axial position
Surge valve position
Choke valve position
Radial bearing pad temperature
Thurst bearing pad temperature
Lube oil supply temperature
Lube oil supply pressure
Lube oil drain temperatures
Lube oil total flow

important, an abrupt reading might be cause for immediate investigation. Such a situation may call for an unscheduled shutdown to reduce risking catastrophic failure.

To maximize the benefits available from these records, it will be necessary to be consistent. Records must be maintained daily, since without the proper parameters being accurately recorded, the validity of the record could rightly be questioned.

A typical listing of the parameters that should be included in the daily log is outlined in Table 12.1. It is recommended that the operators discuss the list with the equipment manufacturer and add any items that both parties feel are critical to the analysis of operational fitness of the machinery. Data logging could be done manually or automatically, using either a process computer or a dedicated machinery condition computer.

Typical items to visually inspect on a regular (daily) basis are listed below.

- Check the unit for oil leaks at flanges, instrumentation outlets, vane actuator connections, etc.
- Observe the operation of the stator vane linkage during a change of its position. Check for binding of any components. Ensure smooth movement. "Jumping" or sporadic movement usually indicates mechanical binding or foreign matter in linkage joints.
- Listen to the unit for unusual sounds, such as rubbing (seals or blades) or leaking gaskets. If the sound intensity is significant, investigate the probable causes.
- Check bearing oil drain sight flow indicators to ensure good oil flow through the bearings. Note any significant change in the running level, e.g., much fuller than normal, less flow than normal.

Compressor Internal Cleaning

Modern axial flow compressors will operate for long periods between shutdowns. Well thought-out metallurgy is essential in the design and manufacture of rotor blade components. External surface coatings are applied to protect the blades from corrosive and erosive attack. Along with the addition of coatings, compressor blade life may be increased by installing inlet air filtration systems.

On-line cleaning of internals is usually considered after severe degradation of compressor performance is noted. With a properly sized and operating inlet filtration system, fouling should be minimized. However, on-line cleaning is treating a symptom rather than curing the cause. The problem is more directly addressed through investigation of the air quality entering the compressor. Proper design of the inlet filtration system has always been important to the manufacturer: proper maintenance of the system must become similarly important to the operator.

There are several on-line cleaning methods employed by operators depending on the process and the available cleaning methods. The question of whether or not axial flow compressors should be cleaned during on-line operation is a complex one.

Due to possible problems with each of the cleaning methods currently used, it is appropriate to consult the manufacturer. Below is a short description of some of the systems currently in use. The possible problems, from both the process operation and machine reliability points of view, are also highlighted.

The most effective cleaning procedure for the compressor is a low-speed water/kerosene wash. It is to be performed at approximately 30% of normal speed. The compressor is essentially soaked in the cleaning solution for approximately thirty minutes. A rinsing cycle removes residual cleaning fluid, and returning to full speed effectively dries the internal components of the compressor.

This is technically the safest and most effective cleaning method. It is, however, the least desirable from the operational view, since it requires the unit to be removed from the process for approximately one hour. This is the approximate time required to complete the procedure.

A second process employed today is a water/solvent spray wash system. It is to be used at full speed and, in most cases, is not detrimental to the process. This system requires the addition of a spray nozzle assembly into the inlet casing of the compressor unit. Commercially available cleaning fluids are used.

Possible problems include the incomplete atomizing of the fluid prior to entry into the compressor. The blading could be damaged by the impingement of large water particles. In addition, pulsations created on the rotor blades from the spray nozzles add another excitation to be considered in the natural frequency and stress analysis. Most importantly, while this system is somewhat effective on the first stage on the machine, the cleaning efficiency is drastically diminished at each successive stage. The cleaning medium is simply centrifuged to the casing wall and has little effect on either the rotating or stationary blades downstream.

The third method of on-line cleaning is one that has been employed in similar equipment for some time. This procedure entails the introduction of crushed walnut shells (or apricot pits) into the airstream. These "cleaning media" are introduced into the piping upstream of the inlet casing. They are fed at a rate of approximately 50 pounds every two minutes.

The possible problems with such a system are very similar to those encountered with the spray wash system. The solids are centrifuged to the casing wall so quickly that the effectiveness of the system is severely diminished beyond the first stage. In units employing blade coatings, nut shells or pits can cause accelerated erosion of the blade coatings. In any event, strict control of the injection rate will be required.

Corrective Maintenance

Corrective maintenance may become necessary every three to five years. At that time, inspection and replacement of wearing parts is often appropriate. Prior to the inspection shutdown, an inventory of the available spare parts should be performed. Discussions with the manufacturer should take place to determine that proper quantities of spares are available to ensure a complete and timely turnaround of the machine.

A typical inspection and replacement shutdown may require approximately two weeks. During that time, typical inspections would include the following:

- Rotor blade cleaning and magnetic particle nondestructive testing to ensure the integrity of parts if a complete spare rotor is not available.
- Nondestructive testing of rotor discs, particularly in the area of blade attachment, to check for evidence of stress-related damage. The normal method is dry powder magnetic particle inspection.
- Liquid penetrant nondestructive testing of stationary blading to ensure the integrity of the welded joints within the assemblies.
- Bearing clearance check on the previously installed bearings as well as the new bearings. Visual and dimensional inspection of the bearings for evidence of rubbing, wiping, and unusual wear.
- Rotor check balance to correct any unbalance introduced by rotor blade replacement. Rotor tip clearance check to ensure that proper running clearances are established for safe operation.
- Variable stator vane linkage check to replace any worn bushings or locking/locating pins.
- Instrumentation check to ensure the operational indicators to be recorded are accurate and available.
- Coupling inspection for tooth wear on gear couplings and diaphragm check on diaphragm couplings.
- Axial rotor-to-stator clearance check during reinstallation of the rotor assembly to ensure the proper rotor-to-thrust bearing positioning.
- Seal clearance check on all shaft and bearing assemblies. The shaft should be inspected on removal from the unit for any signs of seal contact.
- Inlet filtration system check to ensure the filter elements are clean and secure. In addition, the inlet piping should be inspected from the inside to ensure no loose pieces exist or foreign objects are inside, which could enter the unit and cause damage to the compressor.

Chapter 13

Propeller, Axial, and Centrifugal Fans*

Fan applications in modern process plants range from handling fresh air to moving large volumes of corrosive and abrasive gas streams. This chapter deals with the fundamentals of selection and operation of fan systems.

Industrial fans can be classified into three basic groups: propeller, axial and centrifugal.

PROPELLER FANS

Propeller fans utilize long slender blades twisted in such a manner as to provide some angle of attack on the gas being moved. The blades are fixed to a hub, and the entire assembly rotates in a housing. The housing has little or no effect on controlling the gas flow.

Typical applications for propeller fans are wall- and ceiling-mounted exhausters (Figure 13-1) and cooling tower and air-cooled heat exchangers. Pressures generally range from 0 to 1 inch of water column with efficiencies ranging from 10 percent to 35 percent. Tip speeds are often limited so as to minimize noise generation.

AXIAL FANS

Axial fans are basically propeller fans with shorter rigid blades assembled in a tubular housing to provide some degree of controlling and/or streamlining the gas flow. Figure 13-2 shows a small, simple, single-stage unit, while Figure 13-3 depicts a large, heavy-duty industrial version. This type of fan propels gas axially through its housing, which acts as an integral part of the ducting. Axial fans can be further subdivided into two types, namely, vane axial and tube axial. Because of the type of construction employed, these fans lend themselves easily to multistaging if higher pressures are required. Figure 13-4 illustrates a two-stage axial fan.

Vane axial fans have vanes installed before and/or after the rotor to direct and streamline the gases for better fan action. The vanes are installed on the stator housing and employ some kind of control mechanism for varying the flow pattern. Vane axial fans are used primarily in clean gas service, handling pressures generally up to 20 inches of water column (single-staging), but higher pressures are available. Multistaged vane axial fans can achieve efficiencies in excess of 85 percent and are generally considerably more efficient than tube axial fans.

---

*Source: Garden City Fan and Blower Company, Niles, MI (except as noted). Adapted by permission.

FIGURE 13–1   *Wall-mounted exhaust fan (propeller-type). (Source: ACME Engineering & Manufacturing Co., Muskogee, OK.)*

FIGURE 13–2   *Small axial fan. (Source: ACME Engineering & Manufacturing Co., Muskogee, OK.)*

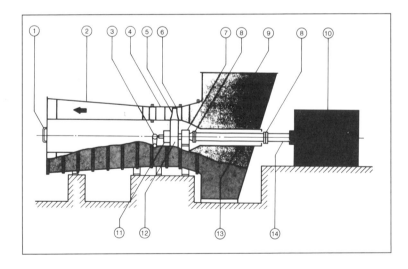

FIGURE 13–3   *Heavy-duty single-stage axial fan. 1 = Access door; 2 = diffuser; 3 = external pitch control lever; 4 = stationary blades; 5 = variable pitch rotating blades; 6 = removable upper fan housing; 7 = main bearing assembly; 8 = coupling; 9 = inlet box; 10 = motor; 11 = blade pitch control mechanism; 12 = rotor assembly; 13 = shaft tube; 14 = drive shaft. (Source: CEMAX Fans, Boston, MA.)*

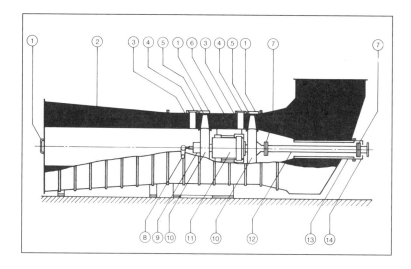

FIGURE 13–4 *Heavy-duty two-stage axial fan. 1 = Access door; 2 = diffuser; 3 = removable stationary blade; 4 = bolted blade retainers; 5 = removable variable pitch rotating blades; 6 = removable upper fan housing; 7 = coupling; 8 = external pitch control lever; 9 = blade pitch control mechanism; 10 = rotor assembly; 11 = main bearing assembly; 12 = drive shaft; 13 = shaft tube; 14 = shaft seal. (Source: CEMAX Fans, Boston, MA.)*

Tube axial fans employ the same basic construction as vane axial fans except for the guide vanes and are generally less efficient than vane axial fans. Recently, variable-pitch rotor blade construction has been offered as a standard package for both tube and vane axial fans.

Typical performance characteristics of axial flow fans are shown in Figure 13–5. Note the pronounced "dip," or stall range, in both the static pressure and brake horsepower (BHP) curves. This represents the unstable range in which surge phenomena occur and where fan operation should be avoided. It also explains why axial fans operating in parallel are difficult to start up. Although the absorbed horsepower is initially high, its subsequent decline makes it a non-overloading characteristic.

CENTRIFUGAL FANS

Centrifugal fans (Figure 13–6) are prevalent in virtually all of the industry. These fans are applied in building and equipment ventilation, hot-gas recirculation, dust-handling systems, and furnace and boiler forced draft/induced draft services. The basic difference between centrifugal fans and the types previously mentioned is the fact that gas enters the impeller axially and leaves the rotating element radially. The casing is designed and installed around the rotating element to direct the discharge gas in a manner suited for the particular application. Centrifugal fans operate with typical efficiencies in excess of 80 percent and an achievable limit of 91 percent with airfoil blade construction. Discharge pressures can be as high as 90 inches of water column or higher and cover a wide operating range. Centrifugal fans can be categorized by three basic construction styles depending on blade orientation with respect to impeller rotation. These styles are radial blade, backward-inclined blade, and forward-inclined blade. Subtle variations of each style are available.

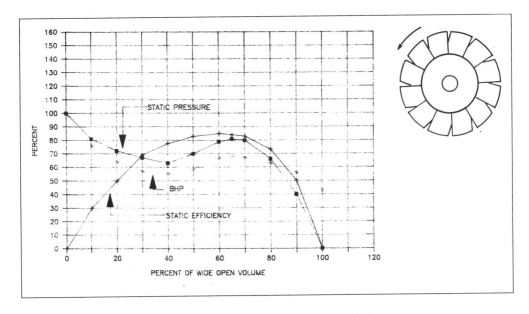

FIGURE 13–5   *Axial flow fan characteristics.*

FIGURE 13–6   *Centrifugal fan for typical process plant. (Source: ACME Engineering & Manufacturing Co., Muskogee, OK.)*

Radial Blade

As the name implies, the blades on this type of impeller are radially oriented. This type of impeller construction can be further subdivided into straight radial blade and radial-tip blade styles. The merits of each of these different blades can be observed in Figures 13–7 through 13–10. The straight radial bladed fan illustrated in Figure 13–7 has a best efficiency of approximately 72 percent. The fan has an overloading horsepower characteristic, but it features stable operation over the entire performance range. Because of low specific speed, straight radial bladed fans have relatively good resistance to abrasion from entrained particles.

The radial-tip fan of Figure 13–8 minimizes gas turbulence by curving the blades. It achieves higher efficiencies, with 78 percent to 83 percent being not uncommon. Fan horsepower is nonoverloading in the high flow range. This fan also has low specific speed and medium resistance to abrasion.

Figure 13–9 illustrates an open radial bladed fan with a geometry that readily allows the addition of simple wear plates for highly abrasive applications. It is lowest in efficiency, with 65 percent being typical. This fan has an overloading power characteristic.

In summary, radial blade fans are commonly used for induced draft service and are especially suited for contaminated gas streams where airborne particles and fouling could be a problem.

Backward-Inclined Blades

The fan blades on this style of rotor are oriented backwards with respect to the direction of rotation. Three variations of backward-inclined blades are available. Figures 13–10 through 13–12 illustrate both geometry and performance of these variations: backward-inclined flat plates, curved plates, and airfoil blades, with efficiencies increasing in that order. Typical efficiencies for flat plate blades range from 77 percent to 80 percent and 84 percent to 91 percent for airfoil blades. The flat and curved plate blades can handle

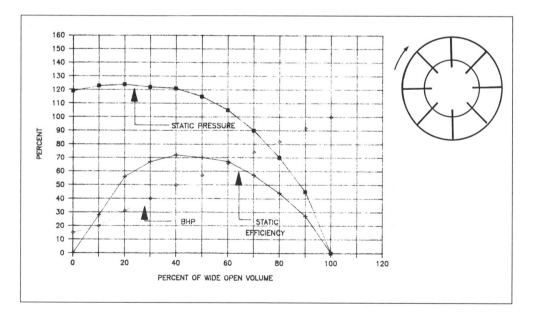

FIGURE 13–7   *Radial blade fan characteristics.*

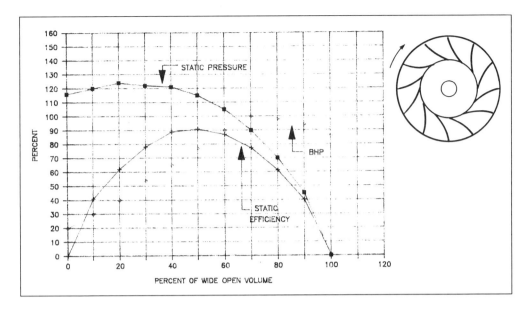

FIGURE 13-8 *Radial-tip blade fan performance characteristics.*

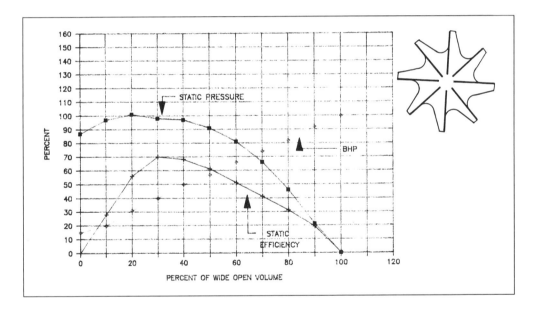

FIGURE 13-9 *Open radial blade fan characteristics.*

fine airborne particles that would normally damage the airfoil design. These fans are commonly used in forced draft services.

Modification of the backward-inclined flat plate profile of Figure 13-10 reduces the unstable range to moderate instability in the backward-curved blade (Figure 13-11) and virtually no instability in the backward-inclined airfoil blade (Figure 13-12). Efficiencies show slight increases in the order mentioned; however, the level of tolerance for handling abrasive gas streams decreases as one progresses from backward-inclined

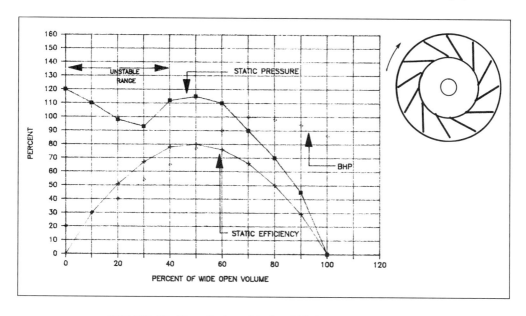

FIGURE 13–10 *Backward-inclined blade fan characteristics.*

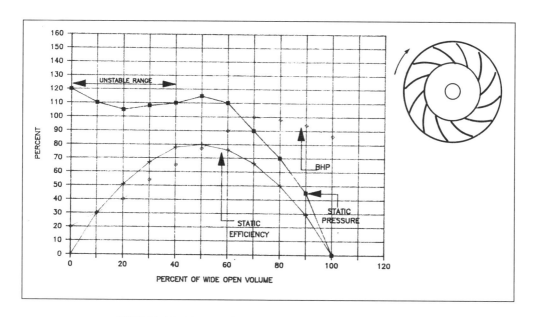

FIGURE 13–11 *Backward-curved blade fan characteristics.*

to airfoil geometries. All of the backward-inclined blade fans exhibit non-overloading horsepower characteristics.

Forward-Inclined Blades

These fans are styled such that the blades lean forward with respect to the direction of rotation. The two typical variations are the flat plate and curved plate. Figure 13–13

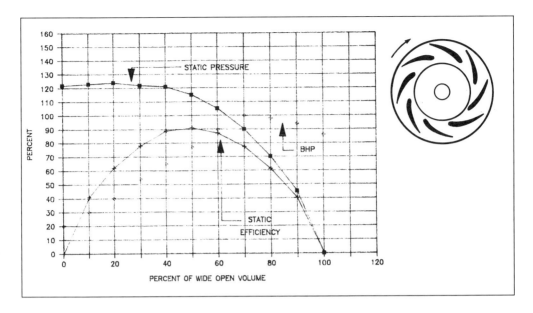

FIGURE 13–12   *Backward-inclined airfoil blade fan characteristics.*
characteristics.

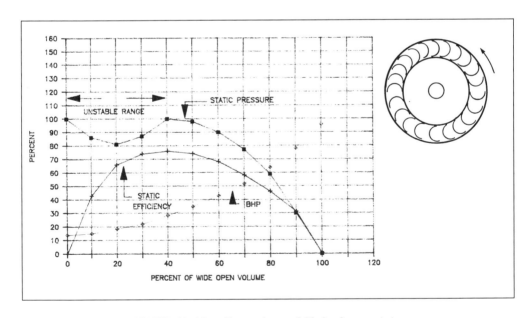

FIGURE 13–13   *Forward-curved blade characteristics.*

shows geometry and performance of the forward-inclined curved blade style. This type of fan is generally limited to high-volume applications and clean services such as residential heating and air conditioning. Efficiencies range from 72 percent to 76 percent; also, a distinct unstable region exists in the performance curve for this fan type. It has an overloading horsepower characteristic.

A summary of the relative characteristics of the principal fan types is given in Table 13.1.

Table 13.1 Relative Characteristics of Principal Fan Types

| Characteristics | Centrifugal | | | Axial | Propeller |
	Backward Vane	Radial Vane	Forward Lean		
Efficiency	High	Medium	Low	High	Low
Stability of operation	Good	Good	Fair	Good	Good
Tip Speed	High	Medium	Low	High	Medium
Noise generation	Low	High	Low	Medium	Medium
Abrasion resistance	Medium	Good	Low	Medium	Medium
Tolerance for polymers	Medium	Good	Low	Medium	Low

FAN FUNDAMENTALS

A brief review of commonly accepted definitions will assist us in highlighting the fundamental principles involved in fan selection and operation.

cfm: Volume flow rate expressed in cubic feet per minute at outlet conditions.

SCFM: Volume flow rate expressed in standard cubic feet per minute at a density of 0.075 lb/cu ft.

VP: Velocity pressure is the pressure caused by the average outlet velocity of the gas stream expressed in inches of water column. It is measured in the outlet duct by the differential reading between an impact tube facing the air flow and a static reading normal to the air flow.

SP: Static pressure is that pressure expressed in inches of water column and measured in a manner to exclude the velocity pressure of the gas stream. Static pressure of a fan is the difference in the reading of a Pitot tube placed in the inlet duct and facing in the direction of gas flow and the static reading obtained in the outlet duct. The pressure is usually referenced with respect to 70 °F and a density of 0.075 lb/cu ft.

TP: Total pressure is the algebraic sum of the static pressures and the velocity pressures and is basically the rise from inlet to outlet measured by two impact tubes. It is expressed in inches of water column, much as static pressure.

BHP: The horsepower required to move a specified volume of gas against the indicated static pressure. This is usually referred to air at standard conditions and is sometimes termed *air horsepower*. Note that for proper sizing of the driver, this should be calculated with respect to total pressure rather than static pressure and should include the ratio of actual density/air density at standard conditions.

$$\text{BHP} = \frac{62.3 \times \text{SP} \times \text{cfm}}{12 \times 33{,}000 \times \text{SE}}$$

ME: Mechanical efficiency is the ratio of the horsepower delivered to the input horsepower.

SE: Static efficiency is the mechanical efficiency multiplied by the ratio of static pressure to total pressure.

$$\text{SE} = \text{ME} \times (\text{SP})/(\text{TP})$$

SND: Static no delivery or zero flow.

WOV: Wide open volume (zero static pressure).

OV: Outlet velocity, cfm/outlet area.

$$\textbf{VP:} \quad = \frac{\text{Gas density} - \text{lb/cu ft} \times (\text{average outlet velocity} - \text{ft/min})^2}{1.203 \times 10E6}$$

Note for standard conditions, the density of air = 0.075 lb/cu ft. The unit of pressure is expressed in inches of water column where the density of water is specified as 62.3 lb/cu ft.

Specific Speed: The speed in RPM at which a fan would operate if reduced proportionally in size to deliver 1 cu ft/min against a static pressure of one inch of water.

$$\text{Specific speed} = \frac{\text{RPM} \times \text{cfm}^{0.5} \times (\text{gas density}/0.075)^{0.75}}{\text{SP}^{0.75}}$$

Fans are governed by certain basic laws known as the fan laws. These laws are reasonably accurate and can be stated as follows:

For variation in speed:

1. The volume delivered is proportional to the speed.
2. The static pressure across a fan is proportional to the square of the speed.
3. Brake horsepower is proportional to the cube of the speed.

For variation in size:

1. Volume is proportional to the cube of the wheel diameter.
2. Static pressure is proportional to the square of the wheel diameter.
3. Brake horsepower is proportional to the fifth power of the wheel diameter.

If both fan size and speed are changed, then the combination factors should be calculated.

FAN PERFORMANCE AND SYSTEM EFFECTS

The performance of the centrifugal fan is represented by either a multirating capacity table indicating various points on a family of constant speed performance curves or by a single graph developed by plotting at a fixed speed the static pressure (SP) in inches of water column versus volume flow (cfm) expressed in cubic feet per minute.

The fan develops pressure for various volumes of air flow through it and the proposed system. The characteristic fan performance curves shown earlier are constant speed (RPM) performance curves for operation from wide open volume position (zero static pressure, maximum volume) to static-no-delivery position (zero volume). Each performance curve is plotted for a specific RPM and fan size as well as a fixed air or gas density.

A simple system might consist of ductwork connected to either the fan inlet, discharge, or both, in addition to elbows, dust collectors, scrubbers, furnaces, etc., through which air may be forced to flow. The prime mover, of course, is the fan or blower in the duct system, which provides the energy to the airstream to overcome the resistance to flow of its various components. Each fan and blower system is unique, i.e., they have a combined system resistance to a specific flow that varies from system to system and is dependent on the resistance to air flow for each individual component of the system. In a fixed system, the volume flow rate (cfm) will have a corresponding pressure loss (system resistance), generally expressed as static pressure, the resistance to the flow varying directly as the square of the increased or decreased flow rate. The equation defining the relationship between flow and pressure drop in a fixed system is as follows:

$$(\text{cfm-final/cfm-original})^2 = (\text{SP-final/SP-original})$$

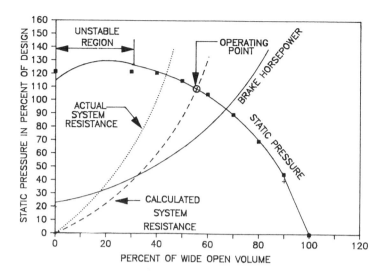

FIGURE 13–14   *Effect of system resistance on fan performance.*

Figure 13–14 represents a fixed system resistance curve to air flow and is calculated as a parabolic equation relating flow to pressure as shown in the above expression. If the actual resistance curve is substantially different from the calculated one, the performance will, of course, deviate, and in extreme cases, it will risk operation in the unstable zone.

Just as different designs of fans have different characteristic performance curves, different types of systems have different characteristic resistance curves. The more complex the system, the more involved the characteristic curve. However, the complex systems can be separated into their component parts, the individual characteristics of which are known, and a summation of the resistance of several components of a system will give the composite resistance of the overall system.

The terminology *resistance to flow* is generally accepted as equivalent to the phrase *pressure loss* for the individual system. The determination of pressure loss may be accomplished by use of handbook values for loss through ducting and branch piping as well as by published or computed values for resistance to flow through the individual pieces or components of the system. It is important to remember that the resistance for the system should be expressed in terms of standard air defined as 0.075 lb/cu ft density. The latter resistance may then be related to fan static pressure requirements that are read directly from fan manufacturers' published performance data based on standard air with a density of 0.075 lb/cu ft.

The Air Moving and Control Association (AMCA) has long been aware of the problem in applying fans in systems where inlet and outlet conditions vary from test facilities. The association is publishing a manual indicating guidelines or factors to be used in computing fan static pressure requirements when inlet and outlet connections are less than the desired design. Frequently, the system designer will add a factor of safety to his or her calculations of the system resistance in an attempt to adjust for inaccurate evaluations or to compensate for poor inlet or outlet connections. These safety factors rarely compensate for the "system effect" on the fan performance, and it is necessary either to make adjustments in the system by improving inlet and/or outlet conditions or by increasing the fan speed to obtain the desired fan capacity. However, it will be necessary to establish whether or not the increased fan speed is within the structural limits of the equipment before deciding to increase the speed of the fan. Obviously, poor inlet and outlet conditions should be corrected in order to keep operating horsepower and cost to a minimum.

A few of the most common conditions that result in deficient performance of the fan include (1) prerotation of the air into the fan inlet due to a vortex condition that occurs when the fan inlet is placed too close to an obstruction or wall, (2) nonuniform flow due to elbows prior to fan inlet, and (3) omission of an outlet duct on the fan discharge. These conditions will definitely alter the characteristic of the fan performance so that its full volume flow potential is not realized. This system effect in the performance of the fan and blower must be considered in the evaluation of the pressure loss for the total system.

If the application engineer or system designer has accurately determined the resistance to flow through a given system, the fan engineer will be in a position to properly evalute and select the fan that will develop the required flow and pressure to meet system requirements. When placed into a system, the fan will operate only where its characteristic performance curve intersects the actual system resistance curve. Figure 13–14 shows the fan characteristic performance curve with the calculated system resistance curve superimposed and the resultant point of intersection that should be the design operating point (55 percent of wide open volume and 107 percent of design pressure). If the system resistance has been accurately determined and the fan properly applied, the design performance will be at the intersection of the two curves. When the pressure losses have not been accurately determined for the system, the intersection of the fan performance curve and the actual system curve will not coincide with the design point. Occasionally, fans operating in a position other than design point may be subject to unstable or pulsating flow, which may damage the fan and the system components.

Should the estimate of a system resistance be less than actually required for the flow rate, the resultant flow will be less than required. This is also illustrated in Figure 13–14. When the actual system resistance is less than the calculated value, the flow will be higher than the design flow rate and will require additional horsepower to drive that particular fan.

Frequently, in the case of direct-connected fans, the design flow and calculated resistance of the system will not fall exactly on the constant speed fan performance curve. In this particular situation, the fan will deliver more or less air, depending on where the system curve intersects the fan curve, and adjustments will be necessary to obtain the required performance. The adjustments may involve the use of a damper to throttle the flow to design requirements or to reduce the speed of the fan until the required flow is achieved. Should the system resistance curve intersect the fan curve at a lower value than design flow, it will be necessary either to reduce the resistance by the use of turning vanes or splitters in the duct elbows, etc., or to speed the fan up to achieve the design flow requirements. This topic will be addressed later.

The foregoing comments and figures are based on fan inlet and outlet connections in accordance with good design practice as applied in actual laboratory tests. When fan inlet or outlet conditions vary from good design practice, design performance may not be obtained even though the system resistance has been properly calculated.

The system effect on fan performance is dependent on the type of inlet or outlet condition as well as gas properties and velocity. The system effect, therefore, will not be uniform throughout the complete range of the fan performance curve.

Performance Corrections

When the fan will be handling a gas at a density other than standard (0.075 lb/cu ft), it may be necessary to use correction factors so that the published tables may be used accurately. Frequently, fan operating specifications are given at temperature, altitude, and/or air density other than standard conditions. The AMCA publishes standards and rating tables for fan equipment. Since fans will operate only where the system resistance curve intersects the fan performance curve, it is important to make certain the fan performance and system resistance are compared on the same basis, i.e., values shown in the same basic units at identical operating conditions.

Example 1: Temperature and Altitude

Required: A fan to exhaust 10,000 cfm of 650° F air at a static pressure (system resistance) of 4.0 inches at 650° F and at an elevation of 2000 feet.

Step 1. The volume of air will be 10,000 cfm at 650° F, and since no reference is made to weight, this value is correct for selection directly from fan rating tables.

Step 2. Static pressure is specified at 650° F and 2000 ft elevation. Therefore, it will be necessary to correct this value in order to make use of the performance tables.

Step 3. The density ratio obtained from Table 13.2 for gas density at 650° F and 2000 ft elevation show 0.444. To obtain 4.0 inches SP at 650° F, we must select a fan at the adjusted static pressure from the standard tables determined as follows:

$$\frac{SP_c}{SP_s} = \frac{P_c}{P_s}$$

$$SP_s = SP_c \times \frac{1}{P'}$$

$$= 4.0 \times \frac{1}{.444} = 9.0 \text{ in}$$

where SP_c = static pressure at site conditions
SP_s = static pressure at standard conditions (70° F and 0.075 lb/cu ft density)
P_c = density of gas at site conditions
P_s = density of gas at standard conditions
P' = density ratio P_c/P_s (see Table 13.2)

Select a fan from the manufacturer's published rating tables for 10,000 cfm and static pressure, SP, of 9.0 inches of water column. In selecting the fan from the manufacturer's tables, let us assume we read 1832 RPM and 16.0 BHP. Restating the performance at standard conditions for the selected fan size and type: 10,000 cfm, SP = 9.0 inches, 1832 RPM, 16.0 BHP (standard conditions). . . .

Step 4. Now check performance when handling air at the actual operating conditions. Referring to the fan laws for variations in air density:

cfm: unchanged = 10,000 cfm
RPM: unchanged = 1832 RPM
SP_c   = $SP_s \times P'$
         = $9.0 \times .444 = 4.0$ in
BHP   = $BHP \times P'$
         = $16.0 \times .444 = 7.1$

where: BHP_c = brake horsepower at site conditions
BHP_s = brake horsepower at standard conditions

We have a choice of sizing the motor to drive the fan at start-up (ambient conditions 70° F) with a 20-HP motor or using a damper and/or a speed control device so

Table 13.2 Air Density Ratios of Various Altitudes and Air Temperatures*

Air Temp. in °F	Altitude in Feet above Sea Level								
	0	500	1,000	1,500	2,000	3,000	4,000	5,000	6,000
	Barometric Pressure in Inches								
	29.92	29.38	28.86	28.33	27.82	26.81	25.84	24.89	23.98
70	1.000	.981	.965	.947	.930	.896	.864	.832	.799
100	.946	.928	.913	.895	.880	.848	.817	.787	.756
150	.869	.852	.839	.823	.808	.779	.751	.723	.694
200	.803	.788	.775	.760	.747	.719	.694	.668	.642
250	.747	.733	.721	.707	.695	.669	.645	.622	.597
300	.697	.684	.673	.660	.648	.625	.602	.580	.557
350	.654	.642	.631	.619	.608	.586	.565	.544	.523
400	.616	.604	.594	.583	.573	.552	.532	.513	.492
450	.582	.571	.562	.551	.541	.521	.503	.484	.465
500	.552	.542	.533	.523	.513	.495	.477	.459	.441
550	.525	.515	.507	.497	.488	.470	.454	.437	.419
600	.500	.491	.483	.474	.465	.448	.432	.416	.400
650	.477	.468	.460	.452	.444	.427	.412	.397	.381
700	.457	.448	.441	.433	.425	.409	.395	.380	.365
750	.438	.430	.423	.415	.407	.392	.378	.364	.350
800	.421	.413	.406	.399	.392	.377	.364	.350	.336
850	.404	.396	.390	.383	.376	.362	.349	.336	.323
900	.389	.382	.375	.368	.362	.349	.336	.324	.311
950	.375	.369	.362	.355	.349	.336	.324	.312	.300
1,000	.363	.356	.350	.344	.338	.325	.314	.302	.290

*Unity basis standard air density of .075 lb. per cu. ft., which at sea level (29.2″ barometric pressure) is equivalent to dry air at 70°F.

as to reduce the motor size to the amount required under operating conditions (7.1 BHP) or a 7.5-HP motor.

Example 2: SCFM—Temperature Correction

Select a fan to move 20,000 SCFM at 400° F against a static pressure of 2.0 inches at 400° F and sea level:

Step 1. Since specifications indicate there will be a requirement for weight flow at 400° F, it is necessary to correct the SCFM specified to a yield cfm at 400° F delivering the same weight. If the SCFM did not indicate a temperature, it would be necessary to either question the specifications or to assume that the SCFM was at 70° F. Here are the corrections:

Step 2. $\text{cfm} = (\text{SCFM}) \, P_s/P_c$
$$= 20,000 \times 1/.616 = 32,500 \text{ cfm}$$

and correction of static pressure

$$SP_s = SP_c \, 1/P' = 2.0 \times 1/.616 = 3.25 \text{ in}$$

The fan unit would have to be selected from the manufacturer's rating tables for 32,500 cfm, 3.25 inches SP. Assuming the selected fan unit requires 700 RPM and 20.0 BHP, at standard conditions, we would determine the horsepower consumed at the actual operating conditions by multiplying by the density ratio factor:

Step 3. $\mathrm{HP_c} = \mathrm{HP_s} \times \mathrm{P'}$
$= 20.0 \times .616 = 12.32$ HP at actual conditions

Example 3: Wheel Diameter Correction

We require increased pressure and volume capability, but we are unable to increase the speed. Figure 13–15 illustrates the performance curve of an available 40-inch diameter centrifugal fan operating at 1000 RPM. Table 13.3 shows a functional data form listing operating points taken from the 40-inch diameter fan performance curve of Figure 13–15 and calculated data for plotting the 45 1/8-inch diameter fan performance curve at the same speed. The 40-inch diameter fan curve has been divided into several sections as indicated by the points "A," "B," "C," "D," etc. on the performance curves. By projecting down and across from these points, a flow (cfm), pressure, and horsepower can be determined for each point.

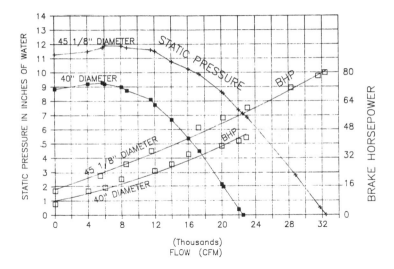

FIGURE 13–15   *Constant speed diameter performance curves for different diameter fans.*

Table 13.3 Fan Performance at a Series of Operating Points

Point	CFM	SP	BHP	CFM	SP	BHP
A	0	8.85	7.50	0	11.28	13.78
B	4,000	9.22	12.50	5,760	11.76	22.95
C	6,000	9.21	15.90	8,650	11.74	29.20
D	8,000	9.00	19.50	11,500	11.48	35.80
E	12,000	7.75	26.60	17,300	9.88	48.80
F	14,000	6.70	30.00	20,160	8.54	55.00
G	16,000	5.39	33.00	23,010	6.87	60.60
H	20,000	2.18	38.90	28,800	2.78	71.50
I	22,000	.40	41.80	31,700	.51	76.90
J	22,500	0	42.50	32,400	0	78.00

Abbreviations: CFM, cubic feet per minute; SP, static pressure; BHP, brake horsepower.

Fan Law Factors

$$\text{Wheel diameter factor} = \frac{45.125 \text{ in}}{40.000 \text{ in}} = 1.129$$

Flow factor	=	$(1.129)^3 = 1.44$
Pressure factor	=	$(1.129)^2 = 1.275$
Horsepower factor	=	$(1.129)^5 = 1.836$

From Table 13.3, we take Point "F" as a sample for applying the calculated fan law factors.

point	cfm	SP	BHP	cfm	SP	BHP
F	14,000	6.7	30	20,160	8.54	55.00

cfm   = $1.44 \times 14,000 = 20,160$
SP   = $1.275 \times 6.7 = 8.54$
BHP = $1.836 \times 30 = 55.00$

The 45 1/8-inch diameter fan performance curve at 1000 RPM has similar characteristics as the 40-inch diameter curve. In rating performance from one size fan to another, the rating process should always be accomplished from a base curve of the smaller fan.

Example 4: Speed Variation

A 30-inch radial bladed centrifugal fan handling 3000 cfm of clean air against a system resistance of 15.75 inch water column absorbs 12 BHP when running at 1780 RPM. For an uprated volume of 4000 cfm of air, the new speed, BHP and pressure would be calculated as follows:

$$\frac{\text{Original Volume}}{\text{New Volume}} = \frac{\text{Original speed}}{\text{New Speed}}$$

$$\frac{3000}{4000} = \frac{1780}{\text{New Speed}}$$

$$\text{New Speed} = \frac{4000 \times 1780}{3000}$$

$$= 2373.3$$

$$\frac{\text{New Pressure}}{15.75} = \frac{(2370)^2}{(1780)^2}$$

$$\text{New Pressure} = \frac{(2370)^2}{(1780)^2} \times 15.75$$

$$= 27.92 \text{ in Water Column}$$

$$\frac{\text{New BHP}}{\text{Original BHP}} = \frac{(\text{New Speed})^3}{(\text{Original Speed})^3}$$

$$\frac{\text{New BHP}}{12.0} = \frac{(2370)^3}{(1780)^3}$$

$$\text{New BHP} = \frac{(2370)^3}{(1780)^3} \times 12.0$$

$$= 28.32$$

CAPACITY CONTROL OF FANS

Centrifugal Fans

Control of the output capacity of a centrifugal fan is accomplished by adjusting either the fan characteristics or the system characteristics. Fan characteristics may be modified by changing the rotation or speed of the wheel or by altering the rotation or whirl of the inlet air. System characteristics may be altered by either increasing or decreasing resistance somewhere in the system.

A number of methods of capacity control are used in current practice. The relative importance assigned to various factors of the particular application governs the choice of method. Basically, the primary design function is to achieve maximum efficiency through control or maximum reduction in operating power to accompany the required reduction in capacity. Selection of a method of capacity control will result from evaluation of such factors as initial cost, operating costs, range of control required, speed of response, simplicity of operation, auxiliary controls, reliability, longevity, maintenance, etc.

The effects of the three primary methods of control (inlet vane control, discharge damper control, and speed control) in relation to fan performance are illustrated on the constant speed curves shown in Figures 13–16 through 13–18. The curves are identical in speed, and the performance characteristics are arbitrarily based on the nonoverloading airfoil-type fan.

Inlet Vane Control

Inlet vane control consists of a number of vanes or blades located at the inlet to the fan. The vanes may be adjusted to various positions so that the entering air is given a change in direction or a spin in the direction of wheel rotation as illustrated in Figure 13–16.

The initial spin modifies the basic characteristics of pressure output and power input, resulting in a new and reduced pressure and horsepower characteristic. This is graphically illustrated in the differential between the base curve and percent inlet guide vane closure curves shown in Figure 13–16. Adjustment of the vanes to various positions, thereby changing the extent of the initial spin, gives regulation to any required volumetric flow at only the pressure demanded by the system. A spin in either direction will reduce the capacity of the fan. A spin contrary to wheel rotation will often increase the shaft horsepower. A spin in the same direction as wheel rotation, as shown, will reduce shaft horsepower. With inlet guide vane control, the fan performance curve is repositioned as the vanes are moved from the wide open to the closed position, with the system resistance curve staying effectively the same.

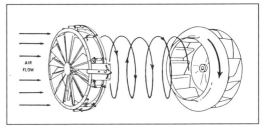

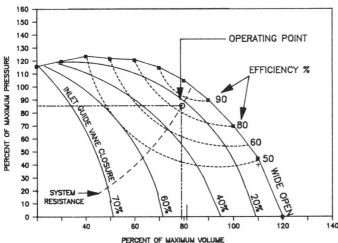

FIGURE 13-16 *Performance curves for inlet vane control at constant speed.*

Inlet vane control provides moderate first cost, excellent operating costs, wide-range regulation, simplicity in operation and auxiliary controls, low maintenance, and relatively long life. Of course, the vane structure must be sufficiently rigid to withstand significant aerodynamic forces, and the mechanism for moving the vanes should be such that all are positioned equally, with considerable degree of accuracy. Inlet vane control mechanisms are limited to use in relatively low temperature applications (up to 900 ° F) and may have limited life in abrasive and/or corrosive atmospheres.

It is often believed that inlet guide vane control acts like a multiple-leaf damper. This is quite wrong. A damper is independent of the fan and is not intended to affect fan performance in any way, although if a damper is located near the fan suction, it may do so to the detriment of fan performance. A damper is merely an adjustable resistance in the system intended to waste horsepower. Vane control, on the other hand, acts primarily to modify the fan performance so as to reduce the power required to attain a specific set of conditions. Such resistance as the inlet vane controller interposes is incidental and has no effect on its principal function of fan performance modification.

Damper Control

A damper consists of one or more pivoted blades acting as a valve in the duct whereby resistance to flow may be varied from wide open to virtually blocked tight.

Figure 13-17 illustrates the change in system characteristics achieved with damper control. The same figure also shows two different arrangements of the commonly used multivane damper normally mounted on the fan discharge or in the duct system

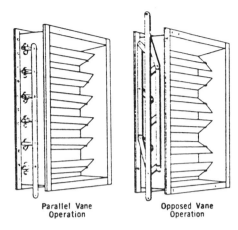

Parallel Vane
Operation

Opposed Vane
Operation

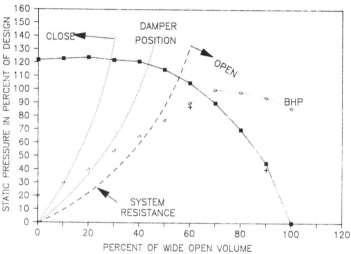

FIGURE 13–17 *Effect of damper control on system characteristics.*

downstream of the fan. Each of these arrangements has individual characteristics as to its effect on airflow; but, in general, both provide a variable resistance for the system.

The multiple vane damper, often called a louver damper, should be considered as an independent element interposed between the fan and the system. As previously mentioned, the system characteristics and not the fan characteristics are affected by the damper.

As shown in Figure 13–17, when the system resistance is altered, the operating point on the fan curve is shifted correspondingly. At reduced volume, the fan develops a higher pressure than the system requires. The curve is shifted to the left as the damper is moved from the wide open to the closed position, with the fan performance curve staying effectively the same.

A damper in a duct system controls the flow and dissipates energy by warming the flowing air. A considerable amount of the power input to the fan is converted by warming the air just slightly. Nevertheless, something must be done to control air flow, and the damper is often the most desirable method, even at the expense of wasted horsepower.

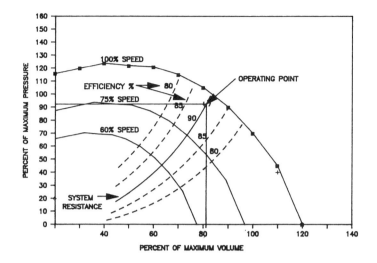

FIGURE 13-18   *Variable speed performance curves.*

The damper method of control provides minimal initial cost, moderate operating costs, wide-range regulation, simplicity of operation, and relatively long life. Dampers can be fabricated to withstand extremes in temperature and are capable of satisfactory operation in abrasive and/or corrosive atmospheres. Within design limitations, damper construction can be provided for gas-tight and insulated applications.

Speed Control

When a fan is controlled by varying its speed, there is no waste. At each volume, the fan develops a pressure just equal to the system resistance and no more. From the standpoint of the fan, variable speed control is most economically attained.

Variable speed control is predicated on the theory that the volume of air flowing is proportional to the fan speed; the pressure developed is proportional to the square of the speed; and the horsepower is proportional to the cube of the speed. Therefore, from a given constant speed performance curve, new curves at other speeds can be easily computed.

On reviewing Figure 13-18, it becomes obvious that each fan speed gives a complete and separate performance curve. Since the system resistance has not been physically altered, each reduction in fan speed will provide a new fan performance curve crossing the system resistance curve below the base curve, as illustrated. In like manner, fan speeds in excess of base curve speeds provide performance curves that cross the system resistance curve above the base curve.

Many methods of providing variable speed drives are available, each having its own specific desirable and undesirable features. Basically, variable speed control is accomplished by one of two methods: a variable speed prime mover connected directly to the fan shaft; or a constant speed prime mover with interconnecting adjustable means to vary the fan speed.

The turbine, direct current (DC) motor and variable frequency alternating current (AC) motor drive fall within the variable speed prime mover classification and may be considered as completely variable. The slip-ring AC motor and the two- and three-speed AC motors also fall within this classification but should be considered as semivariable and stepped.

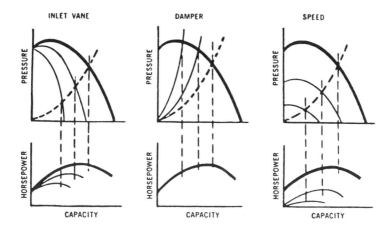

FIGURE 13–19 *Comparison of various fan control methods.*

The hydraulic and electrical coupling and the variable pitch drive are interconnecting adjustable mechanisms for variation of fan speed. For small fans, the variable pitch sheave and belt drive arrangement is by far the most popular today. Large fans are generally not equipped with belt drives.

The ultimate decision as to which control method or methods to use depends on the functional overall system requirements. Comparisons must also take into account differences in first cost, fixed charges, maintenance cost, use factor, environmental conditions, power demands, stability, etc.

A combination of two methods of control, such as a two-speed drive with vane control, will frequently provide the most suitable as well as the most economical method of controlling the fan. Also, a variable speed drive and a multivane damper can be combined to permit utilization of a much smaller motor in elevated temperature systems.

To ensure proper capacity control of a centrifugal fan, knowledge of the effects of each of the methods described above will be helpful. Figure 13–19 graphically compares the three basic methods.

Axial Fan Control

Although all of the previously mentioned control methods could be used, axial fan control is most often accomplished by adjusting the blade angle. This adjustment can normally be accomplished with the fan either shut down or in operation, and both pneumatic and hydraulic actuating mechanisms are available.

The pitch of all blades is changed simultaneously, and the actuating signal can be manual or automatic. Axial fan blade pitch controls have been configured to respond to changes in gas velocity, temperature, pressure, gas composition, and other parameters.

Blade angle adjustment allows efficiencies to be optimized under a wide range of pressure and flow conditions. Operation in the stall region of the performance curve can be effectively avoided by this highly flexible control method, which is shown in Figure 13–20.

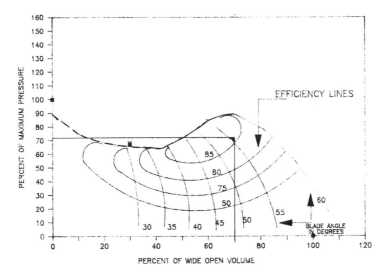

FIGURE 13–20   *Axial flow fan performance with variable pitch blade control.*

Fan Control Summary

Centrifugal fans can be controlled by the following:

- Movable inlet guide vanes that vary the pre-rotation angle of the incoming gas. The resulting change produces a shift in the performance curve, which now intersects with the system resistance curve at different pressures and flows, bringing about a different power requirement.
- Dampers, which throttle the flow either at the fan inlet or outlet and thus cause a shift in the system resistance curve. This shift results in a new point of intersection with the performance curve, and a new pressure-flow relationship is established. Damper control is both less efficient and lower in initial cost than movable inlet guide vane control.
- Speed changes that produce a parallel shift of the performance curves, which will now intersect the system resistance curve at different pressures and flows. This control method is more energy-efficient than either inlet guide vane or damper controls.

Axial fans can be most efficiently controlled by varying the blade pitch angle. Speed, damper, or stationary blade pitch angle controls are feasible, but they are usually less economically attractive.

Chapter 14

Reciprocating Compressors*

Reciprocating compressors are positive-displacement machines that compress and move gases by using a combination of rotational and linear (reciprocating) motion. Reciprocating compressors are used for a variety of industrial services. Their basic function is to raise the pressure level of the gas being compressed. Doing so is desirable for the following reasons:

1. Storage and transmission of energy, e.g., shop air compression.
2. Reduction of volume for the purpose of storage and transport, including gas lique-faction. Typical examples are bottled gases for industrial uses, natural gas storage in underground reservoirs, storage and transport of liquid natural gas (LNG) and liquid petroleum gas (LPG), and compression of natural gas as a fuel for automobiles.
3. Transmission. Examples include pipelines for natural gas, ethylene and other hydrocarbons, ammonia, oxygen, and nitrogen.
4. Process. Chemical reactions that take place at elevated pressures (e.g., ammonia synthesis at 5,000 pounds per square inch absolute [psia]) are illustrative.
5. Energy conversion. Mechanical to thermal energy conversion (refrigeration systems, heat pumps) is an example.

Reciprocating compressors have inherent advantages over other compressors in their ability to adapt to a wide range of load, speed, pressure conditions, and pressure ratios ($P_{discharge}/P_{suction}$). The load may be varied from 0 to 100 percent; the speed may have a wide range, depending on the driver. Pressures may vary from a few inches of mercury absolute suction pressure in the case of a vaccum pump to 50,000 psia or more discharge pressure for process-gas compressors. Pressure ratios may vary from slightly over 1, in the case of natural gas transmission pipeline service, to 8 or more in the case of shop air compressors. Several stages of compression are often used when the overall pressure ratio is high.

There are many prime movers suitable for driving reciprocating compressors. These include electric motors, turbines, natural gas engines, diesel engines, and dual-fuel engines. Many electric motors and reciprocating engines have rotative speeds similar to those of reciprocating compressors and can be directly connected, eliminating speed reduction (or increase) between the prime mover and the compressor.

A typical compressor and principal internal construction features are shown in Figures 14–1 and 14–2.

---

*Source: Sulzer-Burckhardt (Winterthur and Basel, Switzerland), unless otherwise noted. Reprinted and adapted by permission. Copyright retained by contributor.

FIGURE 14–1 *Single-stage dry-running compressor with two horizontally opposed cylinders for ethylene service. This machine is typical of small, skid-mounted equipment. (Source: Sulzer-Burckhardt, Winterthur and Basel, Switzerland.)*

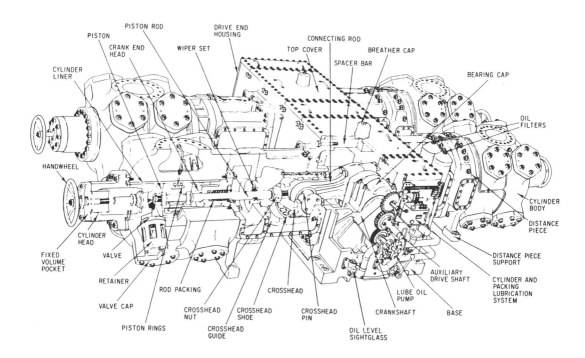

FIGURE 14–2 *Major internal parts of a large horizontally opposed reciprocating compressor. (Source: Transamerica De Laval, Trenton, NJ.)*

IDEAL COMPRESSOR CYCLE

Positive-displacement compressors are machines in which successive volumes of gas are confined within a closed space and elevated to a higher pressure. The reciprocating compressor is a special type of positive-displacement compressor that elevates the pressure of the trapped gas by decreasing the volume that the trapped gas occupies. A piston moving in a cylinder is used to reduce the volume of the trapped gas (Figure 14–3).

Compression

Referring to Figure 14–3, note that the cylinder has filled with gas at the suction pressure with the piston at position *a*. The piston moves from *a* toward *b*, compressing the gas isentropically (with no heat transfer and no turbulent or frictional losses) until the pressure within the cylinder reaches the discharge-line pressure.

Discharge

At this point, the discharge valve opens and permits gas to flow from the cylinder into the discharge line until the piston has reached the end of its stroke at point *c*.

Expansion

Since it is impossible to build a compressor with zero clearance volume, gas remains in the cylinder clearance volume at the end of the discharge stroke. The gas remaining expands isentropically to suction pressure as the piston moves from *c* to *d*.

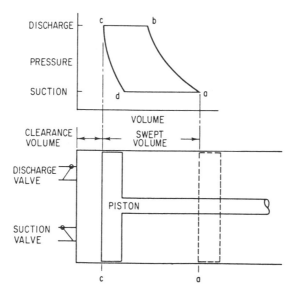

FIGURE 14–3   *Ideal compressor cycle. (Source: Transamerica De Laval, Trenton, NJ.)*

Suction

When the pressure within the cylinder reaches the suction pressure, the suction valve opens and permits gas at suction pressure to enter as the piston moves from *d* to *a*. Since points *b* and *d* are determined by the pressures during the cycle, the cycle is described as having a suction stroke (piston moves from *c* to *a*) and a discharge stroke (piston moves from *a* to *c*).

CLASSIFICATION OF RECIPROCATING COMPRESSORS

Reciprocating compressors can be categorized as follows:

I. Cylinder lubricated
   A. Trunk-piston type
II. Cylinder nonlubricated
   A. Dry running piston rings
   B. Ringless type (labyrinth-piston type)

Other criteria for classification include arrangement of cylinder:

- vertical in-line
- horizontal (balanced-opposed)
- V-, L-, or W-arrangement
- integral with internal combustion engine

and cooling:

- water-cooled
- air-cooled

Whether to use a compressor with a lubricated or nonlubricated cylinder is, as a rule, dictated by process requirements and the gas to be compressed. Some chemical processes do not permit the use of lubricants in the cylinders, even though the oil can be removed almost entirely by means of separators, filters, and the like, because the slightest traces of lubricant may "poison" the catalyst. In this context, it must be remembered that since most separation methods are based on the difference between the specific weights of the gas and the lubricant, oil separation becomes more difficult with higher gas pressures. Lubricating oil cannot be used in compressor parts making contact with such gases as oxygen and chlorine.

Compressors with Cylinder Lubrication

Except for diaphragm compressors, machines with cylinder lubrication are the only reciprocating compressors generally available for pressures in excess of approximately 4300 psia.

Small compressors are, as a rule, trunk-piston machines. Their construction details (see Figure 14–4) resemble automotive engines. For power inputs above approximately 100 kilowatts (kW), crosshead-type compressors are used. In contrast to the less costly trunk-piston types, compressors with crossheads permit the use of double-acting pistons. In other words, compression takes place during both forward stroke and reverse stroke of the piston.

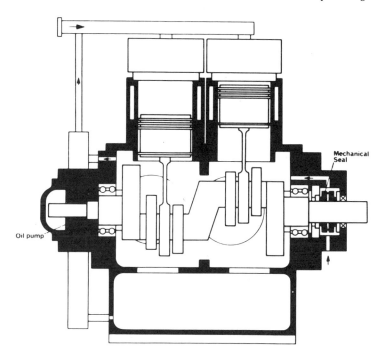

FIGURE 14–4 *Internal construction of a small trunk-piston reciprocating compressor. (Source: Sulzer-Burckhardt, Winterthur and Basel, Switzerland.)*

Where contamination of the gas by lubricating oil cannot be accepted, compressors with water-lubricated cylinders may be used. The water can then be removed by means of a dryer. These compressors are used mainly for filling high-pressure cylinders, where small volumes of gas have to be compressed to relatively high pressures. Typical usage ranges are suction volumes from 12 to 160 actual cubic feet per minute (acfm), and discharge pressures from 2000 to 3600 psia.

Compressors with Nonlubricated Cylinders

Compressors with nonlubricated cylinders are now used wherever possible. The occasional claim that if process permits, it is better to have a lubricated compressor, is not always borne out by the latest experience. Although tremendous progress has been made on nonlubricated compressors, where there is a choice between the oil-lubricated and nonlubricated type, the following factors must be considered:

- Nonlubricated compressors cost more than oil-lubricated machines.
- Nonlubricated compressors require more power.
- With the possible exception of frictionless compressors, nonlubricated compressors require more maintenance.

While there may be no ready answer to the question of lubricated versus nonlubricated compressor selection, the user should have little difficulty choosing the most applicable machine. A detailed investigation of service experience and maintenance frequency in a given service will be of great help to the specifying engineer.

COMPRESSOR ARRANGEMENT OVERVIEW

As mentioned earlier, reciprocating compressors can also be categorized according to the type and arrangement of their cylinders. Small compressors usually have *single-acting, trunk-piston cylinders* (see Figure 14–5). Most of the higher horsepower units are *double-acting, crosshead units* (see Figure 14–6).

Double-acting compressors have pistons that compress gas on both ends so that one end is on its suction stroke while the other end is on its discharge stroke. The force resulting from the pressure and area differential across the piston is referred to as the *piston-rod load.* Reciprocating compressors are rated in terms of their rod-load capability rather than by horsepower. Rod-load ratings range up to 175,000 lb. Higher horsepower compressors are built with a basic frame, and a wide range of cylinders that are interchangeable on the frame. The cylinders range from small-diameter high-pressure cylinders to large-diameter low-pressure cylinders. A line of cylinders is usually designed so that each cylinder matches the rod-load capability of the compressor frame at the maximum working pressure of the cylinder and the expected pressure ratio of the applications for which the cylinder is intended.

Double-acting compressors with cylinders of the type described above are designed as vertical in-line, vertical or horizontal straight-line, horizontal balanced-opposed, or "V"- or "L"-machines. Other cylinder arrangements are less common. Multi-cylinder compressors with horizontal cylinders are very often designed as *balanced-opposed type* (see Figure 14–7).

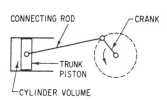

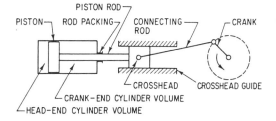

FIGURE 14–5   *Schematic view of a single-acting, trunk-piston machine. (Source: Transamerica De Laval, Trenton, NJ.)*

FIGURE 14–6   *Schematic view of a double-acting compressor design. (Source: Transamerica De Laval, Trenton, NJ.)*

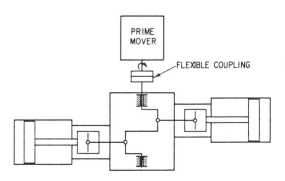

FIGURE 14–7   *Schematic view of a balanced-opposed compressor. (Source: Transamerica De Laval, Trenton, NJ.)*

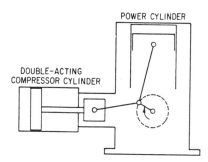

FIGURE 14–8 *Schematic view of an integral-type gas engine reciprocating compressor. (Source: Transamerica De Laval, Trenton, NJ.)*

The balanced-opposed frame is characterized by adjacent pairs of crank throws 180° out of phase and separated by crank webs only. With this configuration, the inertia forces are balanced if the reciprocating weights of opposing throws are balanced. The balanced-opposed design is a separable frame, thus the basic compressor can be driven by any number of prime movers, including diesel, natural gas and dual-fuel engines, gas and steam turbines, and electric motors.

Compressors driven by an internal combustion engine (e.g., diesel engine, gas engine) or by a steam engine can be designed as *integral engine-compressor units,* (Figure 14–8; see also Chapter 4). The integral compressor has compressor cylinders and power cylinders mounted on the same frame and driven by the same crankshaft. Some integrals have in-line power cylinders mounted vertically and compressor cylinders extending from one or both sides in the horizontal plane. Others are V engines with compressor cylinders extending from one or both sides. For a more detailed description of gas engines, refer to Chapter 4.

Cylinder Heads and Valves

The design of the cylinder heads and valves is dictated by the fact that these parts have to withstand, for years, a pressure that fluctuates considerably, e.g., between 900 and 2500 bar at a frequency of 3 to 4 Hz. Modern methods of investigation led to designs such as that illustrated in Figure 14–9, where inadmissible changes of combined strains could be kept within tolerable limits.

With large cylinders, combined suction discharge valves as illustrated in Figure 14–9 are used for very high pressures. This valve is fitted with multiple suction and delivery poppets in order to reduce the moving masses. It is interesting to note that poppet valves, which were used in the last century, have experienced a comeback in compressors that are at the other end of a more than 130-year-old development of reciprocating compressor design.

Capacity Range of Reciprocating Compressors versus Other Compressors

The pressure-flow relationship of reciprocating compressors relative to other compressor types is illustrated in Figure 14–10. Application ranges are identified as follows:

- A1: Reciprocating compressors with lubricated and nonlubricated cylinders
- A2: Reciprocating compressors for high and very high pressures with lubricated cylinders

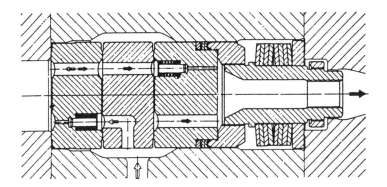

FIGURE 14-9   *Combination suction and discharge poppet valve for a secondary compressor. (Source: Sulzer-Burckhardt, Winterthur and Basel, Switzerland.)*

- B: Helical- or spiral-lobe compressors (rotary screw compressors) with dry or oil-flooded rotors
- C: Liquid ring compressors (also used as vacuum pumps)
- D: Two-impeller straight-lobe rotary compressors, oil-free (also used as vacuum pumps)
- E: Centrifugal turbocompressors
- F: Axial turbocompressors
- G: Diaphragm compressors

The most frequently used combinations of two different compressor types are identified in three fields:

- A+G: Oil-free reciprocating compressor followed by a diaphragm compressor
- E+A: Centrifugal turbocompressor followed by an oil-free reciprocating compressor
- F+E: Axial turbocompressor followed by a centrifugal turbocompressor

The above list is never quite complete in either design or application, since compressor innovations enter the marketplace rather frequently.

For moderate pressures, *rotary screw compressors* (Chapter 15) compete with reciprocating compressors. Oil-free screw compressors are built for compression ratios up to 4.5 per stage. Two-stage units with atmospheric intake are available for discharge pressures up to approximately 195 psia and three-stage units are available with pressures up to approximately 300 psia. The oil-flooded screw compressor is available for a compression ratio up to 13 in single-stage and up to 21 in two-stage configurations. Booster screw compressors with elevated suction pressure are available for a variety of conditions, with pressure limits in the neighborhood of 260 psia unless special designs and materials are used. Rotor deflection limits the maximum attainable pressure difference.

Unlike reciprocating compressors, rotary screw compressors have a fixed built-in pressure ratio. This means that in some cases, higher compression than required by the process occurs inside the compressor, thus causing higher discharge temperatures. With fluctuating intake and discharge pressures frequently occurring in chemical process cycles, it is not always possible to operate a compressor at its built-in compression ratio. Low efficiency and high power consumption may be the result.

Sliding-vane compressors are still widely used. Although they have limited discharge pressure capability (two-stage units are available with pressures up to 130 psia),

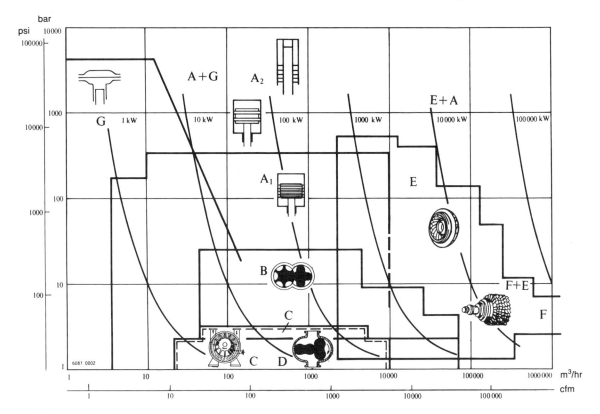

FIGURE 14–10 *Simplified capacity versus pressure diagram of the most widely used types of compressors in the CPI. Power consumption is approximate and is based on a suction pressure of 1 bar (14.5 psia). (Source: Sulzer-Burckhardt, Winterthur and Basel, Switzerland.)*

sliding vane compressors cover roughly the same capacity range as the screw compressors. Since they are not oil-free, they have lost ground to dry-running screw compressors and are not covered in this text.

There are no clearly defined limits for the application in series of two or more compressor types. However, in series applications, many factors and alternatives influencing compressor selection must be considered (see later).

Reciprocating compressors cover the range from the smallest capacity requirements through 10,000 m³/hour (6000 cfm) and more. The reason why larger suction flows are normally handled by turbocompressors is the relatively high price of very large reciprocating compressors in comparison with turbocompressors. In the chemical processing industry (CPI), reciprocating compressors are typically found with power inputs up to 2000 kW, although standard frames are available with power inputs exceeding 10,000 kW.

Most processes in the CPI require discharge pressures below 400 bar (5800 psia). The polymerization of ethylene to produce low-density polyethylene (LDPE) is an exception. This process requires discharge pressures between 1500 and 3500 bar (21,750 to 50,750 psia) and employs so-called secondary booster compressors with suction pressures between 100 and 300 bar (1450 to 4350 psia). Not only are these machines at the upper end of the pressure scale for chemical reactions, but they are also the most powerful reciprocating compressors built thus far. The power consumption of one existing unit is 15,200 kW!

Reciprocating compressors for discharge pressures of roughly 8000 bar (116,000 psia) have been used exclusively in high-pressure research (see heading "High Pressure Compressors," later).

Oil-free compressors can be built up to about 300 bar (4350 psia) discharge pressure (see heading "Labyrinth-Piston Compressors"). Beyond this limit, high-pressure reciprocating compressors have lubricated cylinders. For oil-free compression to even higher discharge pressures, an oil-free reciprocating compressor followed by a *diaphragm compressor* may represent the best solution. The diaphragm compressor is discussed at the end of this chapter.

Trunk-Piston Compressors

As was mentioned earlier, in trunk-piston machines, the pistons are also the crossheads. Since there are no piston rods, this design is only suited for single-acting pistons, where the gas is compressed on the cylinder head side of the piston only. These machines are used for relatively small power inputs, and ratings rarely exceed 200 kW.

A variety of cylinder arrangements is possible; two of these are represented in Figures 14–11 and 14–12. Most small compressors are built along this geometric arrangement. This design is highly suited for air and other noncorrosive gases compatible with the crankcase lubricating oil. Indeed, the gas blowing by the piston rings makes its way under the piston and from there into the crankcase. To prevent a gradual pressurization of the crankcase, the latter has to be connected to a system of adequately low pressure, in most cases the suction line, allowing a recycling of this blow-by gas without letting it escape to the atmosphere. In this case, it is necessary to design the crankcase as a gas-tight pressure vessel with mechanical crankshaft seals. One such machine is illustrated in Figure 14–4.

The gas intake and outlet connections of any given cylinder are generally concentrated in one location. This allows the use of combined suction and discharge valves. As shown in Figure 14–13, these concentrically designed valves incorporate circular spring-loaded plates.

The design concept represented by Figures 14–14 and 14–15 allows compressors to be built in up to five stages for discharge pressures up to 5000 psia and suction capacities between 3 and 280 acfm. Up to about 80 kW, these units are air-cooled, and above this limit, they are more often water-cooled.

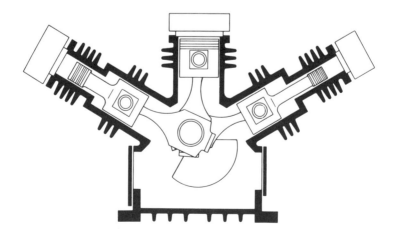

FIGURE 14–11   *Stepped cylinder arrangement, air-cooled. (Source: Sulzer-Burckhardt, Winterthur and Basel, Switzerland.)*

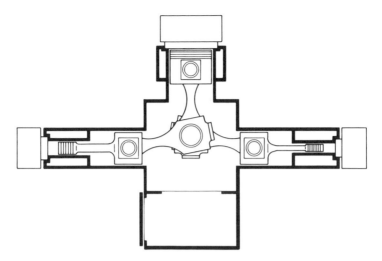

FIGURE 14–12 *Stepped cylinder arrangement, water-cooled. (Source: Sulzer-Burckhardt, Winterthur and Basel, Switzerland.)*

FIGURE 14–13 *Typical reciprocating compressor valve combining suction and discharge. (Source: Sulzer-Burckhardt, Winterthur and Basel, Switzerland.)*

Compressors of this size are supplied as compact packaged units on a sturdy steel frame. The assembly essentially includes the compressor, motor, V-belt drive, gas coolers, moisture separators, condensate receiving tank, filter, the piping for gas and, where required, for cooling water. The baseplate typically rests on vibration damping elements. The crankshaft is fitted with counterweights to reduce free unbalance forces to an acceptable minimum. Units of this kind are ready for use when placed on a substantial floor or simple foundation and connected to power, suction, and discharge gas lines and, if water cooled, to water lines.

Crosshead-Type Compressors With Cylinder Lubrication

Heavy-duty machines are of the crosshead type with entirely separate and well-controlled cylinder lubrication, water-cooled cylinders, and a relatively low operating

FIGURE 14–14 *Air-cooled trunk-piston compressor for 5000 psia service. (Source: Sulzer-Burckhardt, Winterthur and Basel, Switzerland.)*

FIGURE 14–15 *Water-cooled trunk-piston compressor for 4300 psia service. (Source: Sulzer-Burckhardt, Winterthur and Basel, Switzerland.)*

speed. Permanently mounted on a good foundation or isolated support system, they can be operated at full load for years with minimum attention.

While the crankcase is lubricated in the usual manner by a forced-feed lubricating pump driven by the crankshaft or by a separate electric motor, a high-pressure lubricat-

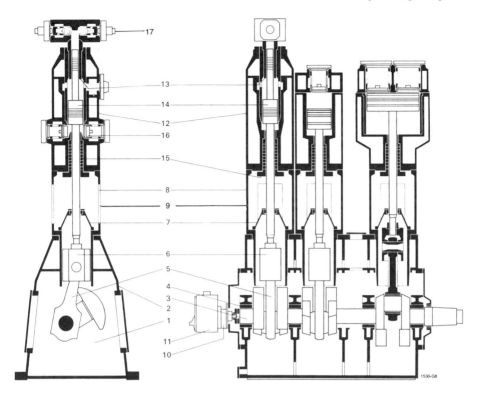

FIGURE 14-16 *Vertical, four-stage crosshead-type compressor with cylinder lubrication and single-compartment distance piece for nontoxic and nonflammable gases. 1 = Crankcase; 2 = frame; 3 = crankshaft; 4 = bearing; 5 = connecting rod; 6 = crosshead; 7 = cover; 8 = distance piece; 9 = purge chamber; 10 = lubricating pump for crankcase; 11 = cylinder lubricator; 12 = cylinder; 13 = cylinder liner; 14 = piston; 15 = piston-rod packing; 16 = valves; 17 = capacity control. (Source: Sulzer-Burckhardt, Winterthur and Basel, Switzerland.)*

ing oil pump, also driven by the crankshaft or by a separate electric motor, feeds the cylinders and piston-rod packings. Each high-pressure lubrication element has an adjustable metering device and visual flow control.

These machines have a separate crosshead, as shown in Figures 14-16 to 14-18, with a piston-rod connecting crosshead and piston.

Crosshead-type machines are the most widely used reciprocating compressors in the CPI. Their power input ranges from around 100 to approximately 10,000 kW. As was mentioned earlier, only secondary compressors for the very high pressures needed in the production of LDPE have a higher power input, i.e., up to 15,000 kW.

Heavy-duty reciprocating compressors typically use double-acting pistons and perhaps even step pistons for two, three, or even four compression stages on the same piston rod. Thus the gas forces can be equalized very well. The crankcase side of the piston is sealed off by means of piston-rod packing.

With some services, a short single-compartment distance piece is sufficient (Figure 14-16), while for flammable, hazardous, or toxic gases, a two-compartment distance piece as shown in Figure 14-17 is required.

Since process compressors are often custom-designed for a given duty, standardization, as a rule, is limited to the frame and running gear. There are two basic factors that guide the designer:

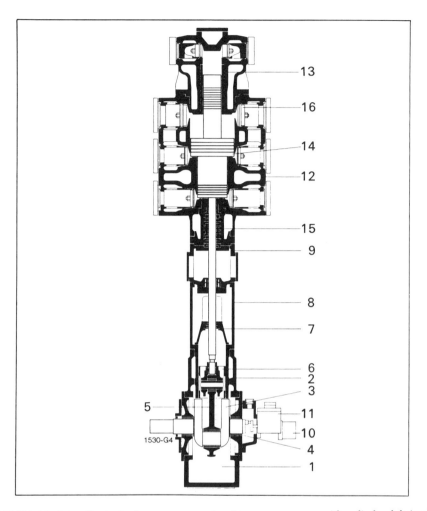

FIGURE 14–17   *Vertical, three-stage crosshead-type compressor with cylinder lubrication and two-compartment distance piece (8+9) for toxic and/or flammable gases. 1 = Crankcase; 2 = frame; 3 = crankshaft; 4 = bearing; 5 = connecting rod; 6 = crosshead; 7 = cover; 8 = distance piece; 9 = purge chamber; 10 = lubricating pump for crankcase; 11 = cylinder lubricator; 12 = cylinder; 13 = cylinder liner; 14 = piston; 15 = piston-rod packing; 16 = valves. (Source: Sulzer-Burckhardt, Winterthur and Basel, Switzerland.)*

1. The maximum power at a given speed that can be transmitted through the shaft and running gear to the pistons.
2. The load imposed on the piston rod, the so-called pin load.

There are other factors involved, for example, the number of cranks, the distance between them, and the piston stroke. These factors represent the working limits for design purposes. Other limits are set by the cylinders, for example, the maximum allowable pressure.

A typical example of a crosshead machine is represented in Figure 14–18.

To date, much has been published about the advantages of horizontal versus vertical designs and vice versa. Both designs have their merits, and in some cases, the selection is dictated by the kind of service intended (Figures 14–19 and 14–20).

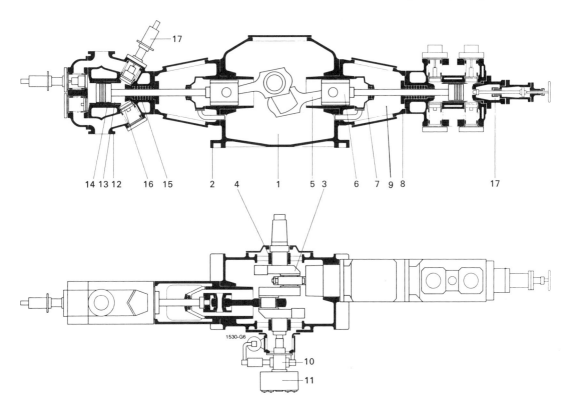

FIGURE 14–18 *Horizontal, two-stage crosshead-type compressor with cylinder lubrication. In addition to unloaders on all suction valves (17), the second stage cylinder on the right side is fitted with hand-operated clearance pocket control (17). 1 = Crankcase; 2 = frame; 3 = crankshaft; 4 = bearing; 5 = connecting rod; 6 = crosshead; 7 = cover; 8 = packing cartridge; 9 = purge chamber; 10 = lubricating pump for crankcase; 11 = cylinder lubricator; 12 = cylinder; 13 = cylinder liner; 14 = piston; 15 = piston-rod packing; 16 = valves; 17 = capacity control. (Source: Sulzer-Burckhardt, Winterthur and Basel, Switzerland.)*

It is claimed that a compressor with *horizontal* opposed cylinders is cheaper to build than one of the vertical type and that it is better balanced. The pipe connections between cylinders and coolers, being closer to floor level, are simpler and easier to support. This type of compressor can be built with an even number of cylinders only, which means that the very popular three-cylinder configuration is not possible. A prominent industry specification, API Standard 618, states that horizontal cylinders are required for handling saturated gases or for gases carrying injected flushing liquids. It also requires that horizontal cylinders have bottom discharge connections. To prevent liquid collecting in the cylinder, arrangements are normally made for delivery valve ports to be on the bottom face of the cylinders, so that liquid is automatically drained by the flow of gas through the cylinder. Horizontal compressors often experience unequal wear on crossheads and their guides, on piston-rod packings, and on piston rings. This may lead to uneven cylinder wear due to force of gravity action in the downward direction.

On the other hand, the *vertical* machine is more compact, requires less floor space, and has shorter interconnecting pipes. Also, with a direct-lifting crane over the machine, assembly and dismantling of cylinders and pistons is facilitated.

Figure 14–21 demonstrates that a very compact and space-saving arrangement of compressor, gas coolers, and piping is feasible with horizontal compressors as well. Such an arrangement, however, raises the question of accessibility. With this compres-

FIGURE 14–19   *Compact skid-mounted compressor package with a five-stage compressor with four horizontal opposed cylinders. Compression is from atmospheric to 2013 psia pressure, capacity is 1820 acfm, and power input is 690 kW at 705 RPM. (Source: Sulzer-Burckhardt, Winterthur and Basel, Switzerland.)*

FIGURE 14–20   *Vertical, three-stage compressor with a two-crank frame for sour gas re-injection in an oil field. Compression is from 70 to 2712 psia; the unit is designed in accordance with API Standard 618. (Source: Sulzer-Burckhardt, Winterthur and Basel, Switzerland.)*

FIGURE 14–21 *Three-stage compressor for service on an off-shore platform. Gas compressed is a hydrocarbon mixture containing hydrogen sulfide. Main data: 1680 acfm, 280 psia, 390 kW. Interstage cooling is by gas-to-air coolers. (Source: Sulzer-Burckhardt, Winterthur and Basel, Switzerland.)*

sor, the remaining free forces are taken up by the springs on which the unit is mounted. This ensures that only a fraction of the forces is transmitted to the substructure; there is no need for a heavy foundation.

Vertical compressors in atmospheric suction pressure services might have low-pressure cylinders with extremely large diameters. This might make it necessary to lengthen the distance between the cylinder centerlines, and the compressor frame could become too heavy. For atmospheric suction, a horizontal design may thus have advantages over a vertical design, particularly if only two cylinders are required.

Compressors With Water-Lubricated Cylinders

Until compressors with dry-running piston rings, labyrinth-piston compressors, and diaphragm compressors became available, compressors with water-lubricated cylinders were the only oil-free reciprocating compressors on the market. Before 1940, virtually all oxygen compressors were water-lubricated.

Generally speaking, water-lubricated machines are required whenever oil lubrication of the cylinders is not permitted and other oil-free compressors are not economically feasible. This is usually the case if a relatively small quantity of gas has to be compressed to a high pressure.

Three-stage compressors, as shown in Figures 14–22 and 14–23, are used for suction capacities to roughly 160 acfm and a maximum discharge pressure of about 3600 psia.

While the bearings of the running gear are lubricated in the conventional manner by means of a crankshaft-driven gear-type oil pump, the cylinders are lubricated with demineralized water. The water passes into the suction pipe at about atmospheric pressure and is carried through all three stages. It is recovered in a separator after the third

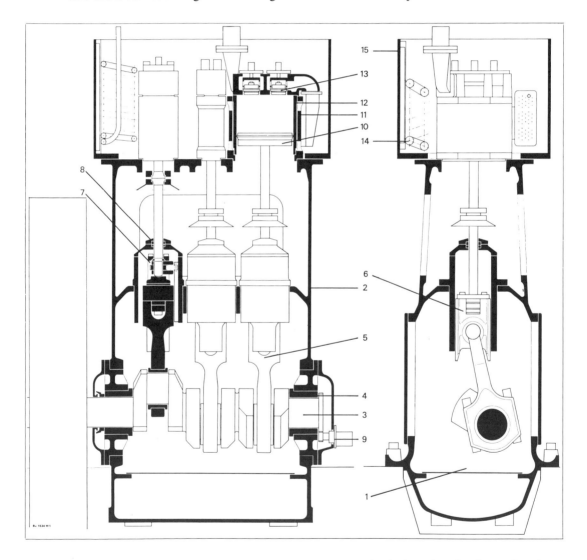

FIGURE 14–22   *Section views of a compressor with water-lubricated cylinders. 1 = Crankcase; 2 = frame; 3 = crankshaft; 4 = bearing; 5 = connecting rod; 6 = crosshead; 7 = ratchet drive for rotary motion of piston; 8 = cover; 9 = crankcase lubrication pump; 10 = piston; 11 = cylinder; 12 = cylinder liner; 13 = valves; 14 = gas cooler; 15 = cooling-water tank. (Source: Sulzer-Burckhardt, Winterthur and Basel, Switzerland.)*

FIGURE 14–23 *Three-stage compressor with water-lubricated cylinders. Note the lubrication-water tank on the wall above the compressor. Main data: Capacity at maximum speed, 270 RPM = 160 acfm; maximum discharge pressure = 3640 psia; piston stroke = 250 mm. (Source: Sulzer-Burckhardt, Winterthur and Basel, Switzerland.)*

stage, from where it is returned to the lubrication water tank. Since the latter is located above the compressor, the water is fed to the suction pipe by gravity. For compressors with elevated suction pressure, the water lubrication system can be pressurized.

The lubrication water tank is equipped with a cloth filter, which prevents solid particles from entering the compressor. Furthermore, it is fitted with a level gauge and a level switch and has connections for the feed, condensate return, and vent lines. A solenoid valve in the feed line is electrically interlocked with the driving motor, whereby the lubrication water circuit is automatically opened or shut off. A flow indicator with a needle valve allows the flow of lubrication water to be adjusted and regulated. Should the quantity of lubrication water not be sufficient, the drive motor is immediately switched off by means of a monitoring contact.

A ratched mechanism fitted to each crosshead rotates the piston stepwise at each stroke, thus preventing scoring of the mirror-polished running surfaces of the cylinder liners by the piston packing rings. The latter are made of leather in the first- and of fiber in the second- and third-stage cylinder.

All cylinders and associated coil-type gas coolers are submerged in a common cooling water tank. This efficient cooling and water lubrication of the piston packings allows compression to take place without any significant temperature rise.

These compressors are extremely well suited for compressing oxygen, hydrogen, helium, nitrogen, air, nitrous oxide gas, etc. The moisture content in the gas leaving a water-lubricated compressor is no higher than in an oil-lubricated or nonlubricated compressor handling humid gas, since the gas is saturated at the outlet of the after-cooler, whether the compressor is water-lubricated or not. If moisture in the gas is not permitted, the water can easily be removed by means of an absorption or refrigeration dryer.

Water-lubricated compressors are low-speed machines. Thanks to their efficient cooling, they allow a relatively high-pressure ratio per stage even in oxygen service. Like diaphragm compressors, they are very well suited for compressing small quantities of very light gases such as hydrogen and helium to relatively high pressures. They are, however, less costly than diaphragm compressors. Both hydrogen and helium are particularly "slippery" gases and have a pronounced tendency to slip past the piston rings during compression in a dry-running ring-type compressor and past the labyrinths in a labyrinth-piston–type compressor. This causes low efficiency and high discharge temperatures.

The gas handled by a water-lubricated compressor must be free of dust. Dust causes rapid wear of the cylinder liners and the piston packing rings.

Acetylene Compressors

Among the gases for which specially designed compressors with lubricated cylinders are required, acetylene is a typical example. Acetylene is produced from calcium carbide and water. Under certain temperatures and pressures and also through shock, the gas can decompose violently into its elements hydrogen and carbon. If decomposition occurs in a pressure vessel, the gas can reach a pressure about eleven times that which existed prior to decomposition. The higher the pressure, the smaller the initial force required to cause an explosion. Furthermore, acetylene is highly flammable.

This dangerous propensity is countered by shipping acetylene dissolved in acetone. One volume of acetone dissolves 300 volumes of acetylene at 175 psia. Acetylene cylinders are partly filled with a porous material that holds the acetone, thus preventing instability of the gas. Full cylinder pressure is 250 psia at 70 °C.

Most acetylene compressors are used for filling cylinders. In these machines, the explosion hazard must be minimized by other measures. In Europe and other countries, the design of equipment handling acetylene is governed by the Technical Rules for Acetylene Plants and Calcium Carbide Storage (TRAC) rules. These rules, however, do not apply to acetylene compressors used in a chemical process for which other safety rules exist.

Finally, the reader may be interested in reviewing the step-piston construction employed for the acetylene compressor shown in Figure 14–24.

Compressors for the Production of Low-Density Polyethylene

A characteristic feature of the high-pressure ethylene polymerization process is that a very large difference in pressure is necessary between the inlet gas entering the reactor and the outlet of the recycled gas. The recirculators, generally called secondary compressors (Figures 14–25 and 14–26), work between two limits, i.e., 100 to 300 bar on the suction side and 1500 to 3500 bar on the delivery side, for most of the existing processes. As the coefficient of reaction lies between 16 and 30 percent, the secondary compressors have to handle three to six times the quantity of ethylene that is polymerized, being thus by far the most powerful machines in the production loop. Their unit capacities, which when the industrial expansion first began were of 4 to 5 ton/hour, now lie between 15 and 118 ton/hour, and their power requirement per unit has been increased from 400 kW up to some 15,000 kW.

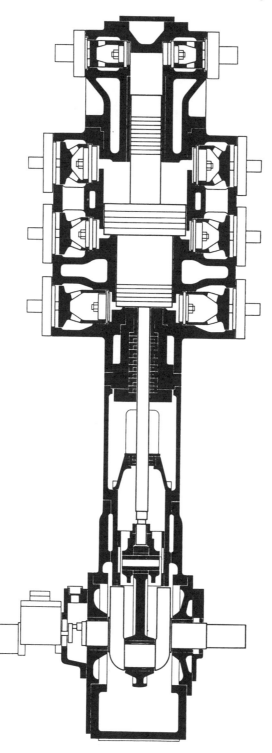

FIGURE 14–24 *Section view of acetylene compressor. (Source: Sulzer-Burckhardt, Winter-thur and Basel, Switzerland.)*

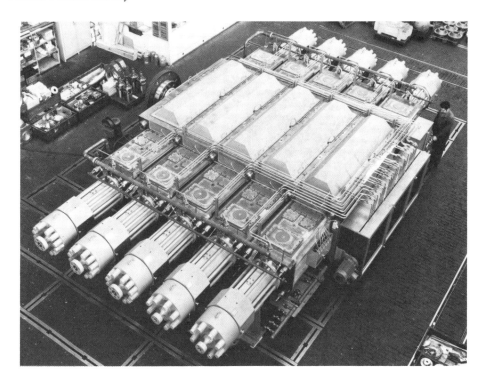

FIGURE 14–25   *Large secondary compressor (hyper compressor) (Source: Sulzer-Burck-hardt, Winterthur and Basel, Switzerland.)*

As the whole operating range of these secondary compressors is placed well above the critical point of ethylene, the thermodynamic behavior of the fluid lies somewhere between that of a gas and that of a liquid. This peculiar condition has two main effects. The first is a very small reduction of the specific volume with increasing pressure; for instance, at a temperature of 25 °C the specific volume is 3 dm^3/kg at 100 bar, 2 dm^3/kg at 700 bar, and 1.5 dm^3/kg at 4500 bar. The second effect is a very moderate rise of adiabatic temperature with increasing pressure; for instance, with suction conditions of 200 bar and 20 °C, the delivery temperature will reach only 100 °C at 2000 bar.

These particular thermodynamic conditions greatly influence the design of high-pressure ethylene compressors. Compared with the classical reciprocating compressor, the compression ratio is of little practical significance, the important factor being the final compression temperature, which should not exceed 80 to 120 °C, depending on process, gas purity, catalyst, etc., in order to avoid premature polymerization. The influence of the cylinder head clearance on the volumetric efficiency is slight because of the small reduction of specific volume, and very high compression ratios are therefore possible with quite acceptable efficiency. In addition, the stability of intermediate pressures depends chiefly on the accuracy of temperatures; for instance, in the case of two-stage compression from a suction pressure of 200 bar to a delivery pressure of 2500 bar, a drop in the first-stage suction temperature from 40 to 20 °C will cause the interstage pressure to rise from 1000 to almost 1600 bar.

For these reasons, a secondary compressor is required, which has only one or two stages, despite the very large pressure differences involved. However, this compels the designer to face extremely high mechanical strains due to the high amplitude of pressure fluctuation in the cylinders. Finally, an additional and sometimes disturbing fea-

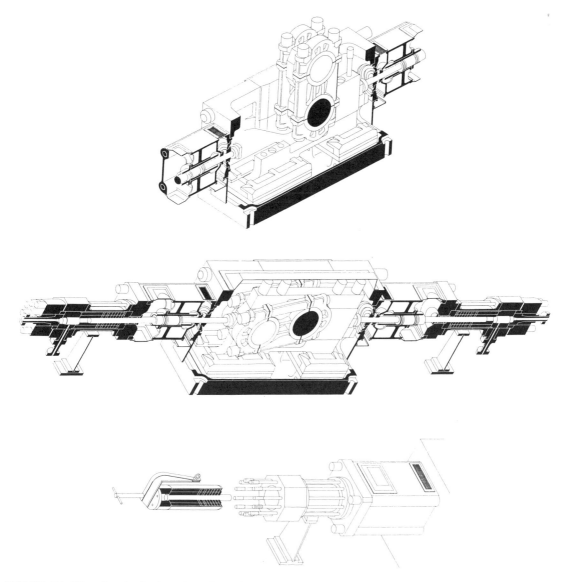

FIGURE 14–26 *Longitudinal sections through secondary compressor. The design of this crank mechanism is based on a moving frame surrounding the crankshaft and connecting the crosshead to the piston rods on either side. Thus the crosshead pin is loaded with the difference between the piston-rod loads only. (Source: Sulzer-Burckhardt, Winterthur and Basel, Switzerland.)*

ture of ethylene must be mentioned. If the gas reaches a very high pressure and a high temperature simultaneously (which can easily occur in a blocked delivery port owing to the very low compressibility), it will decompose into carbon black and hydrogen in an exothermic reaction of explosive character.

Cylinders and Piston Seals

Sealing of the high-pressure compression chamber is a major problem, and this could be solved by avoiding friction between moving and stationary parts. This has been realized for laboratory equipment and small-scale pilot plants by the use of either metallic diaphragms or mercury-based sealants in U-tubes, and such arrangements are still in use for research purposes. In addition, they have the advantage of avoiding any contamination of the compressed gas by lubricants. Unfortunately, chiefly for economic reasons, they proved to be impracticable for industrial compressors, at least in the present state of techniques. Thus, as labyrinth seals are out of the question for very high pressures, friction seals have to be accepted; in fact, two solutions are currently used— moving and stationary seals.

At first, metallic piston rings were the only sort of moving seals used in the large reciprocating type of compressor for very high pressures. They were generally made in three pieces; two sealing rings, each covering the slots of the other, and an expander ring behind both of them, which also sealed the gaps in the radial direction. The materials used were special grade cast iron, bronze, or a combination of both, with cast iron or steel for the expander. The piston, of built-up design, comprised a series of supporting and intermediate rings with a guide ring on top of them and a through-going bolt (two different designs are shown in Figure 14–27). All parts of the piston were made of high tensile steel and particular care had to be given to the design and to the stress calculation of the central bolt, which was subjected to severe stress fluctuations.

The use of piston rings allowed for a simple cylinder design, the main part of which was a liner that had been thermally shrunk to withstand the high variations of the internal pressure (Figure 14–28). The inner sleeve, which had previously been made of nitrided steel, was later produced of massive sintered material like tungsten carbide. The use of this expensive material was justified by two beneficial qualities: it possesses

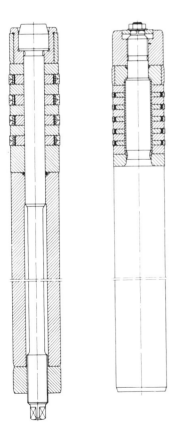

FIGURE 14–27 *Built-up piston designs for secondary compressors. (Source: Sulzer-Burckhardt, Winterthur and Basel, Switzerland.)*

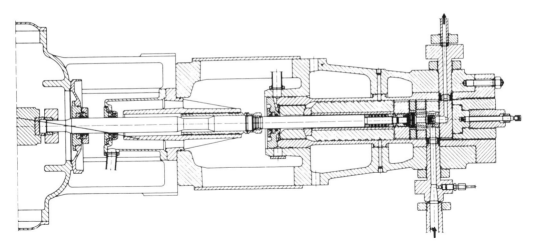

FIGURE 14–28   *Medium pressure cylinder for secondary compressor (hyper compressor). (Source: Sulzer-Burckhardt, Winterthur and Basel, Switzerland.)*

an extremely hard surface and has a high modulus of elasticity. The first considerably improves the conditions of friction and greatly reduces the danger of seizure. Owing to the high modulus of elasticity, the amplitude of the "breathing" movement during internal pressure fluctuations is much smaller than with steel, and thus the stress variations in the expanded outer sleeves is appreciably reduced. However, as these sintered materials have a very poor tensile strength, care must be taken to ensure that the inner sleeve is always under compression, even if the temperature increases. This is the main purpose of the external cooling of the liner and not, as is usual, to dissipate the heat of compression.

At present, plungers with packings of the self-adjusting type, as shown in Figure 14–29, are most widely used. The packings are usually assembled in pairs, the actual sealing ring tangentially split into three or six pieces being covered by a three-piece radially cut section. Both are usually made of bronze, kept closed by surrounding garter springs, and held in place by locating and supporting steel plates. The sealing rings are pressed against the plunger by gas pressure, which corresponds to the pressure difference across the sealing elements. The supporting steel plates must also be thermally shrunk to resist the high variations in internal pressure. Unfortunately, the use of sintered hard materials is restricted by the fact that the supporting plates are subjected, in the axial direction, to heavy bending and shearing forces, which these materials generally cannot withstand.

The plungers for medium pressures are made of steel and plated with tungsten carbide. For very high pressures, the use of solid bars of hard metal is the best wear-resistant solution for plungers. The disadvantage of the packed plunger design lies in the much larger joint diameters of the static cylinder parts, which require two to three times higher closing forces than the piston ring design. Large cylinders, such as the one shown in Figure 14–29, need a pretensioning of the cylinder bolts to about ten times the maximum plunger load. This ratio is higher for smaller cylinders.

For piston rings and packed plungers, the optimum number of sealing elements appears to be four or five. In both solutions, it is essential that the piston be accurately centered if the seals are to be effective; this is the reason for the guiding ring within or near the cylinder and for the additional guide at the connection between the piston and driving rod. At the base of the cylinder an additional low-pressure gland allows gas leaks to be collected and the plunger to be flushed and cooled.

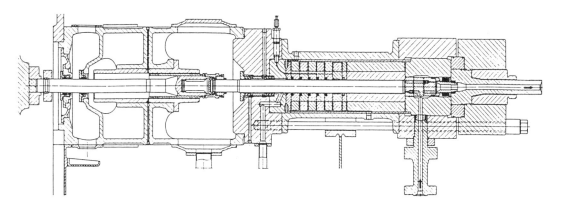

FIGURE 14–29   *Modern design cylinder with self-adjusting packing for secondary compressor. (Source: Sulzer-Burckhardt, Winterthur and Basel, Switzerland.)*

Other separate glands positioned on the rod connecting the piston to the drive (Figure 14–28) prevent the cylinder lubricant from mixing with the crankcase oil, and as the intermediate space is open to atmosphere, it is impossible for gas to enter the working parts.

From the point of view of design and maintenance, piston rings would appear to be the most adequate solution, and they are currently used for pressures up to 2000 bar or in some circumstances up to 3000 bar. The choice between them and the packed plungers depends largely on the process and type of lubricant used. One difficulty is that normal mineral oils are dissolved by ethylene under high pressure to such an extent that they no longer have any lubricating power. The glycerine used in earlier machines has been widely replaced by paraffin oil, either pure or with wax additives, which is much less diluted by the gas than other mineral oils. However, it is a rather poor lubricant and is inferior to the various types of new synthetic lubricants, which are generally based on hydrocarbons.

Since part of the LDPE is used for the manufacture of plastic films and bags for the foodstuff industry, the ethylene to be compressed must not be contaminated by a toxic lubricant. At present, synthetic oils without toxic ingredients are used almost exclusively.

The basic difference between piston rings and plunger packings is that the latter may be lubricated by direct injection, while piston rings are lubricated indirectly. This may be an advantage, since the low polymers carried by the return gas back from the reactor are reasonably good lubricants. However, too large an amount of low polymers causes the rings to stick in their grooves, and some kinds of catalyst carriers also brought back by the gas are excellent solvents for lubricants. Thus, the most convenient solution has to be selected for each specific case. In general, for higher delivery pressures (above 2000 to 2500 bar), better results are obtained with the use of packed plungers.

COMPRESSING DIFFICULT GASES

Difficult gases are those that require special attention by the compressor designer due to their specific properties. Very often standard compressors cannot be used for this service and the selection of the proper compressor type must be made very carefully in order to avoid misapplication and other costly consequences.

The following—incomplete—list demonstrates for which phenomena the compressor designer must be prepared when difficult gases are to be compressed.

These gases may

- assume the liquid state in intercoolers
- decompose at high temperature
- polymerize
- dimerize
- produce the chemical reaction for which they are used already in the compressor cylinder
- form explosive acetylides in the presence of materials of construction promoting such a reaction
- attack the materials of construction
- cause hydrogen embrittlement
- dissolve materials used for gaskets
- dilute the lubricating oil and cause oil foam
- form an explosive and self-igniting mixture with air
- cause a fire (oxygen)

The compressor designer is normally confronted with two or even more of the above phenomena requiring precautions that very often contradict each other. This may explain why difficult gases are also referred to as nasty or unpleasant gases.

The answer to all these problems is a purpose-oriented design that is based on previous and occasionally costly experience. While with some gases the problem can be solved by selecting a suitable lubricant, other gases prohibit the use of any lubricant at all and oil-free compressors have to be used.

Typical unpleasant gases include the following:

- acetylene (see previous paragraph)
- chlorine
- hydrogen sulfide
- sour gas (i.e., wet gases containing hydrogen sulfide)
- hydrogen chloride
- sulfur dioxide
- carbon monoxide

While these gases are normally inert to metals and do not attack the commonly used structural metals under normal conditions of use, they will corrode most normal materials of construction in the presence of moisture. Corrosion is accelerated at higher temperatures and pressures. Even if the gas is absolutely dry, corrosion may occur if atmospheric air is not prevented from entering the compressor. This means that the whole system has to be inerted carefully before and after overhaul services.

Normally, special materials have to be used for the gas-wetted parts of machines compressing difficult gases, and special precautions have to be taken in order to avoid leakage and to keep the gas temperature below the critical limit fixed in codes and standards for handling corrosive gases.*

For chlorine, which is normally compressed to 145 to 175 psia in order to liquefy it, special compressors are available. Since this gas forms hydrochloric acid with oil and water, oil lubrication of the cylinders is out of the question. Compressors with sulfuric acid lubrication used earlier have now been replaced by dry-running machines with piston rings and rod-packing glands of carbon or plastic. Clean and dry chlorine has lubricating qualities, and the wear on the piston rings is astonishingly small. It is,

---

*For wet gases containing hydrogen sulfide (sour gas), see National Association of Corrosion Engineers Standard MR-01-75, Sulfide Stress Cracking Resistant Metallic Material for Oil Field Equipment, published by the National Association of Corrosion Engineers, Houston.

however, essential that the chlorine be pure to prevent sealing rings sticking, and it is highly recommended to wash the chlorine gas with liquid chlorine from the liquefier before suction. Care must also be taken to prevent atmospheric moisture from entering the compressor in order to keep the gas dry and to prevent corrosion. The distance piece of these compressors has to be vented by means of dry air or preferably by dry nitrogen under a pressure slightly above atmospheric. The free length of the rod in the venting chamber is greater than the piston stroke, so that traces of moisture adhering to the rod can never come in contact with chlorine. In order to keep the delivery temperature below approximately 180 °F, chlorine compressors are usually made in three-stage designs for liquefaction at 175 psia.

COMPRESSORS WITH DRY-RUNNING PISTON RINGS

In contrast to the compressors described thus far, machines with dry-running piston rings and piston-rod–packing rings utilize no liquid lubricant, neither of a petroleum, synthetic, or other type, nor substitutes (for example, water) within the compression chamber. They belong to the group of nonlubricated compressors.

Basically, there are two different types of nonlubricated reciprocating compressors:

- dry-running compressors with piston rings and rod packing rings, which do not require lubrication
- frictionless, ringless labyrinth-piston–type compressors, which will be described later

The need for nonlubricated reciprocating compressors was first established in the United States by the brewing industry during the immediate postprohibition period. This industry was in the market for machines supplying uncontaminated compressed air. To meet these requirements, it was decided to build a carbon piston ring compressor based on previous experience with carbon brushes on commutator and collector rings in the electrical industry, which indicated that a reasonable life might be expected. It is interesting to note that the first labyrinth-piston–type compressor, built in Switzerland at about the same time, was also designed upon request of a brewery for compressed air service (see later).

Since the first appearance of nonlubricated compressors, these machines have gained a considerable market share. There are many reasons why nonlubricated compressors are used, and only the most important ones will be mentioned:

- Some gases do not permit the use of lubricating oils for safety reasons; oxygen is a typical example.
- Some gases attack lubricating oil, for example, chlorine.
- Lubricant contaminates gas stream (for example, instrument air, gases used in the foodstuff industry, and air and carbon dioxide in breweries).
- Lubricant carry-over fouls heat exchangers. This is an important factor in cryogenic cycles.
- No suitable lubricant is available for very low and very high temperatures (for example, boil-off compressors for liquid natural gas storage, steam compressors).
- Lubricant carry-over "poisons" catalyst.

Except for plant air, where the presence of trace amounts of oil may occasionally be welcome, there are no compression duties where contamination of the gas by lubricating oil is desired. For this reason, nonlubricated compressors are now used whenever possible and are commercially acceptable. However, when using nonlubricated compressors, one should bear in mind that the gas contains no oil to coat piping, pressure vessels, and heat exchangers. If these components are made of carbon steel, corrosion may occur.

Carbon used for piston rings and packing rings has been largely superseded in current practice by composition PTFE material, a fluorocarbon resin or plastic together with filler materials, such as glass fibers, bronze, and carbon. PTFE is not a self-lubricated material. The value of this material rests solely in its low coefficient of friction. Although only PTFE is mentioned here, the same comments would apply to future plastics, where improvements seem most likely. The plastics industry has developed low-friction materials for almost every gas. However, currently there is no universal material or compound that gives optimum service under all conditions. Although new engineering plastics, which can withstand relatively high stresses and temperatures, have been developed, there are limits to be observed by the designer when using nonmetallic materials.

These limits are set by the following main factors, which may adversely influence the durability of PTFE piston rings:

- pressure
- temperature
- properties of the gas
- dirt

Compressors with piston rings and piston-rod–packing rings of plastic can normally be used for *discharge pressures* up to some 2900 psia. This, however, is not a fixed limit. With dry gases, excessive wear already occurs at much lower pressures. Some compressor manufacturers claim that they have built dry-running compressors for pressures as high as 4000 psia even for such exacting duties as bone-dry inert gases, e.g., argon and nitrogen. Unfortunately, they often neglect to publish the life or durability of the dry-runnning parts under these conditions.

The author knows of such a compressor compressing ammonia in three stages from 270 to 4000 psia where the piston rings of the third stage had to be replaced every 200 operating hours. For process compressors, where uninterrupted service of 8400 hours (one year with an availability of 96 percent) is required, a discharge pressure of 1500 psia should not be exceeded. For pressure between 1500 and 2900 psia, it is very doubtful whether 8400 hours of uninterrupted service can be achieved, and pressures above 2900 psia are in the experimental range for nonlubricated reciprocating compressors with PTFE piston rings.

A large number of nonlubricated compressors with a static suction pressure of around 5000 psia and a delivery pressure roughly 400 to 600 psi higher than the suction pressure—so-called recirculators—have been successfully built. In these compressors, however, carbon rings have been used because PTFE did not withstand the high temperature created by friction heat. With carbon, an average life of 7000 operating hours has been reached. In the meantime, plastic materials have been developed for higher temperatures; however, it seems that these reciprocating recirculators have been phased out by the CPI.

Discharge temperature for dry-running compressors with PTFE piston rings should be held to a maximum of 350°F in order to achieve acceptable durability of the piston rings. By properly staging a compressor, the temperature can be kept below the critical limit.

Other problems can arise from the gas itself. A bone-dry gas, such as nitrogen from a cryogenic air separation plant or boil-off gas from a liquid gas storage vessel, can cause severe ring wear. When compressing argon, a bone-dry inert gas with a specific heat ratio of 1.67, the wear problem is aggravated by the relatively high discharge temperature. For dry-running pistons, dirt is usually the most severe problem.

To overcome wear problems, some manufacturers propose so-called minilube cylinders. These machines belong to the category of lubricated compressors.

The *design* of nonlubricated compressors is basically the same as for compressors with cylinder lubrication, as described earlier, except for the high-pressure lubricators for cylinders and rod packings, which are not required.

In nonlubricated compressors, the compression chamber must receive no lubricant from any source, i.e., not even from the piston rod that normally traverses both crankcase and rod packing. There is a tendency for oil to creep along the piston rod despite the provision of scraper rings and the high pressure in the cylinder. The only way to properly combat this slight oil contamination is to incorporate an extra length distance piece between cylinder and crankcase, longer than the piston stroke, so that the oil-wetted portion of the piston rod does not travel into the rod packing. In addition to this provision, a slinger has to be installed on the portion of the piston rod that passes into neither the cylinder packing nor the frame packing. Difficult conditions—compression of highly explosive, toxic, or extremely flammable gases—require the provision of a two-compartment distance piece with vent and purge connections.

Reciprocating compressors with dry-running piston rings and gland-packing rings can be built for power inputs up to several thousands kilowatts.

LABYRINTH-PISTON COMPRESSORS

Design Philosophy

In contrast to dry-running compressors of conventional design, as described earlier, the distinctive feature of the labyrinth-piston compressor is that no friction occurs in its gas-swept parts. Instead of piston rings, the labyrinth-piston compressor is provided with a large number of grooves producing a labyrinth-sealing effect against the cylinder wall, which is grooved as well. The piston moves with sufficient clearance so that no contact occurs between the latter and the cylinder wall. The same labyrinth-seal principle is used to seal the piston rod, so lubrication of the gland is unnecessary.

Figure 14–30 shows a standard machine in cross section, where the oil-free side of the compressor is marked "A," while part "B" is lubricated in the usual manner. This is achieved by using a crankshaft-driven gear-type pump that supplies the main bearings, the connecting-rod bearings, and the crosshead guide with pressurized oil. The piston-rod guide bearings are splash lubricated.

Oil scrapers mounted above the piston-rod guide bearings remove the oil from the piston rod, and a slinger on the piston rod (not shown in Figure 14–30) prevents the remaining oil film and oil droplets from entering the piston-rod gland. The distance between the crank gear and the gland has been so selected that the oil-wetted portion of the piston rod cannot reach the oil-free gland.

Labyrinth-Piston Compressors versus Compressors
With Dry-Running Piston Rings

Labyrinth-piston compressors have certain advantages but also disadvantages as compared with the oil-free compressors with dry-running sealing elements. As usual, the advantages of one compressor type are the disadvantages of the other one and vice versa.

The successful operation of filled PTFE rings in a dry-running compressor depends on the ability of the PTFE to coat the mating surface of cylinder bore and piston rod. Once the coating process is completed, the PTFE rings ride a film of the same material on the mating surface, minimizing ring wear. Until the PTFE coating is established on

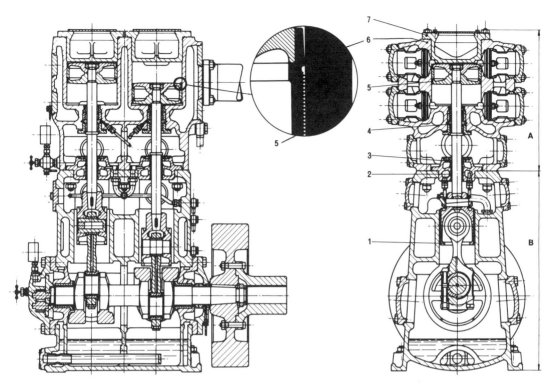

FIGURE 14–30 *Sections through a single-stage labyrinth-piston compressor. 1 = Crosshead; 2 = guide bearing; 3 = oil scraper; 4 = gland with labyrinth seal; 5 = piston with labyrinth grooves; 6 = cylinder with labyrinth grooves; 7 = cylinder head. (Source: Sulzer-Burckhardt, Winterthur and Basel, Switzerland.)*

the mating surface of piston rod and cylinder bore, however, excessive wear of PTFE rings and packing will occur. Oil carryover by the oil scraper rings will tend to remove the PTFE coating from the piston rod and cylinder bore. This is why periodic inspection of such a compressor is necessary.

Another hazard is the possibility that pipe scale or welding beads will be drawn into the inlet. They too remove the PTFE coating. If the compressor is to be idle for some time, condensation can occur in the cylinder because of the temperature difference between the wet gas and the cylinder jacket water, and rust will eventually form on the bore. Even if the cylinder liner is made of stainless steel in order to prevent the formation of rust inside the compressor, in most cases rust can not be avoided in the gas pipes, heat exchangers, and pressure vessels unless these are made of nonrusting materials. Rust and other solid particles are not only a cause for removal of the PTFE coating but have a tendency to embed in the surface of the PTFE rings, which leads to rapid wear of the cylinder liner. Solid particles stem not only from rusty surfaces but in many cases from the catalyst. Below a certain size it is very difficult to remove them by means of suction filters.

This problem can be overcome in labyrinth-piston compressors, which have no friction in the gas-swept part. Consequently, they are not particularly sensitive to damage from dust-laden gases. A slight formation of rust in the cylinder can be tolerated, since the proper functioning of the compressor is not dependent on a smooth cylinder surface. This also means that inexpensive materials, such as carbon steel and cast iron, can be used.

One of the problems with dry-running reciprocating compressors stems from friction in conjunction with the materials used for piston rings. They set a limit to the *pressure difference* and *discharge temperature* of each stage. This problem, too, can be overcome with the contactless and frictionless piston, which does not depend on the use of nonmetallic parts. With labyrinth-piston compressors, pressures up to 4300 psia can be attained regardless of the temperature and properties of the gas. Even higher pressures are possible. Until 1988, the highest pressure attained with a labyrinth-piston compressor in industrial service was 6980 psia. The medium compressed was ammonia synthesis gas.

Apart from the rod-packing rings, which are made of graphite, no nonmetallic parts, which limit the *temperature* in the cylinder, are used. Since graphite is a high-temperature–resisting material, relatively high discharge temperatures can be tolerated. The temperature limit is then set by the gas itself, which may decompose, polymerize, or dimerize, or which may become corrosive above a certain temperature. With suitable cylinder materials and a purpose-oriented design, which keeps heat expansion within tolerable limits, very high and very low gas temperatures can be handled.

The *disadvantages* of a labyrinth-piston compressor as compared with a machine with piston rings stem from the labyrinth principle itself. High discharge pressures lead to small pistons in the final compression stage, with a corresponding unfavorable ratio of the ring gap surface area between piston and cylinder-to-piston area. Where light gases, such as hydrogen and helium, are being compressed, this ratio imposes a limit on the application of the labyrinth-piston compressor at significantly lower pressures than mentioned earlier. Whereas energy losses in the labyrinth can be negligible for the majority of gases, such losses can be considerable when light gases are being compressed, particularly to higher pressures. However, this does not entirely exclude labyrinth-piston compressors from hydrogen and helium service. A substantial number of these machines have been installed in cryogenic cycles with helium as a refrigerant.

Due to the labyrinth losses between piston and cylinder, it is natural that labyrinth-piston compressors cannot be miniaturized at will. Their capacity range is consequently limited, so the lower limit of a single-stage compressor is approximately 10 acfm at atmospheric suction conditions, while that of the multistage machines is considerably higher. For most gases, the minimum suction capacity at atmospheric suction pressure of a four-stage machine for a maximum discharge pressure of 3570 psia is around 100 acfm. These figures are quite different for a light gas, such as helium. A four-stage machine has been built for helium service, and for a suction capacity of 290 acfm at atmospheric pressure, a discharge pressure of 870 psia was attained.

These few examples demonstrate what is meant by the quote, "the labyrinth-piston compressor cannot be miniaturized at will." Smaller capacities, however, are not normally required by the CPI. Supercritical gas extraction processes may open in the future a field of application for small high-pressure labyrinth-piston compressors. For these machines, a different design will be required. The upper limit of the capacity range is governed by economic considerations; it is around 6500 acfm.

Labyrinth-piston compressors are available as standardized and custom-designed units. More than 30 frame sizes cover a power range from 20 to 2,000 kW.

Labyrinth-piston compressors generally have a shorter *piston stroke* and are designed for higher speeds than compressors with piston rings. The reason why compressors with friction on the reciprocating parts have longer strokes and lower rotational speeds is obvious: the more strokes per minute, the more wear occurs at the points where the reciprocating parts change direction. On the other hand, a high rotational speed has a favorable effect on the labyrinth leakage in compressors with labyrinth pistons.

As mentioned earlier, the labyrinth sealing system has its price: labyrinth losses. It has been claimed that labyrinth-piston compressors require more power than non-lubricated compressors with piston rings. This raises the question as to whether the power loss caused by mechanical friction of the piston rings and gland rings is smaller than the power loss caused by labyrinth leakage between piston and cylinder and in the piston-rod gland. Experience gained with both compressor types shows that for medium to large swept volumes per unit time and for gases that are not unusually light, the energy loss due to labyrinth leakage is about equal or even less than that caused by friction.

In order to obtain more accurate results, comparative tests on air were made with a vertical single-throw standard crank machine at a rated speed of 750 RPM and two single-stage double-acting cylinders and piston sets. The two cylinders had exactly the same dimensions. The only difference was the surface finish of the cylinder walls. Whereas one piston was of standard labyrinth design, the other was fitted with three plastic piston rings.

In both instances, exactly the same compressor valves, pipe work, measuring instruments, and drive elements were used. In this way, the differences in the operating characteristics could be reliably established without interference from side effects.

The comparison was made between the two series of tests on the basis of the following efficiencies:

$$\eta_{ad} = \frac{P \text{ adiabatic}}{P \text{ effective}}$$

$$\eta_{is} = \frac{P \text{ isothermal}}{P \text{ effective}}$$

The test results are plotted in Figure 14–31 as a function of the pressure ratio at a speed of 600 RPM.

Piston-Rod Glands

As mentioned previously, the piston-rod glands of labyrinth-piston compressors are also based on the labyrinth sealing principle. The sealing elements are of the floating-ring type. These rings have labyrinth grooves inside, facing the reciprocating piston rod. They are made of special grades of graphite. Their good dry-running ability, chemical inertness, and low thermal expansion coefficient are almost ideal for most gases to be compressed in the CPI. Frictionless packings cannot "run hot."

The gland leakage is normally collected in a ring chamber at the lower end of the gland and vented back to the compressor intake. It does not represent a real loss of process gas (Figures 14–32 and 14–33).

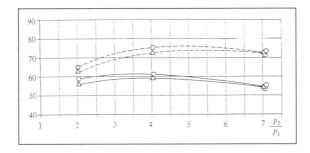

FIGURE 14–31 *Efficiency curves for a single-stage nonlubricated piston compressor.* (———) = isothermal; (— — —) = adiabatic; O = piston with three plastic rings, single-stage, double-acting, 150 mm stroke; △ = labyrinth piston; p_1 = intake pressure of 14.2 psia; p_2 = discharge pressure. (Source: Sulzer-Burckhardt, Winterthur and Basel, Switzerland.)

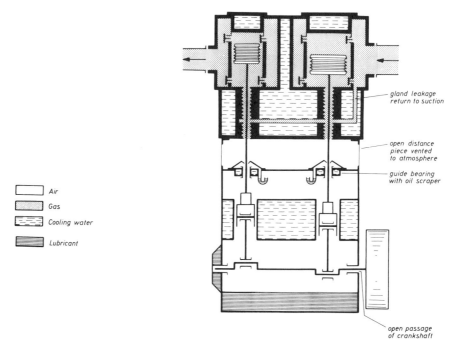

FIGURE 14–32   *Labyrinth-piston compressor, open type. (Source: Sulzer-Burckhardt, Winterthur and Basel, Switzerland.)*

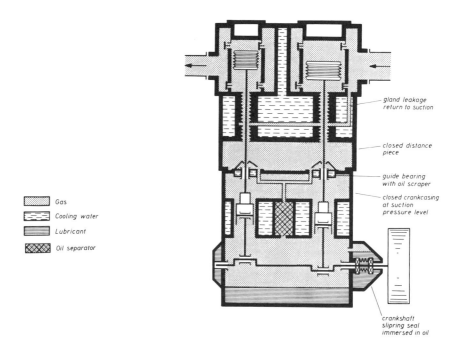

FIGURE 14–33   *Labyrinth-piston compressor with closed pressurized frame. (Source: Sulzer-Burckhardt, Winterthur and Basel, Switzerland.)*

In compressors handling air, nitrogen, oxygen, carbon dioxide, and other nontoxic and nonflammable gases, small gas losses to the surrounding atmosphere occurring at over-atmospheric suction pressure may be tolerated. They are vented to the atmosphere. In order to minimize this leakage, the lowest gland ring may be of the rubbing three-segment type with two garter springs around it and a smooth inner surface. Hot running of the lowest gland ring does not occur provided that the differential pressure across it is relatively small. For higher suction pressure, special gland designs are available. Labyrinth piston compressors as shown in Figure 14-32 are used in these services.

In compressors handling toxic, flammable, or very expensive gases, such as carbon monoxide, hydrocarbons, hydrogen, helium, and argon, even very slight gas losses to the atmosphere cannot be accepted. For these services, hermetically closed compressors as illustrated in Figure 14-33 are used. The only inevitable opening to the outside, which is the penetration of the crankshaft, is sealed hermetically in both stopped and running condition by means of a double mechanical seal immersed in an internal oil bath. This type of seal is effective even if the crankcase is under vacuum. The crankcase and the distance piece are filled with process gas and are subjected to suction pressure.

However, steps are taken to prevent the gas from carrying oil into the process. Depending on the necessity, this may be achieved by an oil separator in the crankcase or by a molecular sieve built into the leakage return line outside the crankcase. Practical results with closed refrigeration cycles have shown that the system remains free of oil even after such a machine has been in operation for years.

Basically all gases or gas mixtures that are compatible with lubricating oil can be compressed with these gas-tight compressors. They are also used in closed refrigeration cycles, particularly where low evaporation temperatures preclude the use of compressors with cylinder lubrication. Refrigerants used for this purpose are halogenated hydrocarbons, ethane, ethylene, and—for cryogenic cycles—helium. In closed cycles, these compressors must be able to withstand standstill pressures occurring when the entire system is warmed up to ambient temperature. To date, compressors for a crankcase pressure of 300 psia have been built.

For toxic and corrosive gases that are not compatible with lubricating oil, neither the open-type nor the closed-type compressor can be used. Chlorine belongs to this class of gases. For such hardship cases, compressors with a strict separation between the oil-free cylinder and the lubricated crankcase have to be used. Since leakage to the environment is not allowed, special piston-rod glands have to be applied with vent and purge connections that can be flushed with a sealing gas, for example, dry nitrogen.

Piston-rod gland problems are often the decisive factor for the selection of the compressor type. It should be borne in mind that no piston-rod packing is entirely gas tight. As a rule, lubricated glands have less leakage than nonlubricated ones, since oil also acts as a sealant. Entirely encapsulated compressors with mechanical crankshaft seals give best results as far as tightness is concerned, since it is much easier to seal a rotating shaft than reciprocating piston rods.

Typical Applications For Labyrinth-Piston Compressors

Labyrinth-piston compressors are often the answer to problems occurring with other types of compressors. Figure 14-34 depicts a three-stage oxygen compressor with a capacity of 1670 acfm and a discharge pressure of 600 psia. Since rubbing friction represents a source of ignition, frictionless labyrinth-piston compressors have found worldwide application in oxygen service. Packaged plants of the type shown with power ratings as high as 500 kW have been supplied.

FIGURE 14–34   *Three-stage oxygen compressor, 1700 acfm range. (Source: Sulzer-Burck-hardt, Winterthur and Basel, Switzerland.)*

FIGURE 14–35   *Four-stage labyrinth-piston compressor in oxygen service. This machine compresses 153 acfm to a discharge pressure of 3570 psia. (Source: Sulzer-Burckhardt, Winterthur and Basel, Switzerland.)*

FIGURE 14–36   *Propylene compressors in a United States Gulf Coast petrochemical plant. (Source: Sulzer-Burckhardt, Winterthur and Basel, Switzerland.)*

Figure 14–35 represents a four-stage oxygen compressor with a capacity of 153 acfm and a maximum discharge pressure of 3570 psia. With this compressor, dry oxygen and other dry gases can be compressed. In contrast to water-lubricated compressors, which are still widely used for this service, the gas remains dry and no dryer is required downstream of the machine. Two out of six two-stage propylene compressors located in a U.S. petrochemical plant are shown in Figure 14–36. Each of these compressors has a capacity of 2270 acfm and produces a discharge pressure of 300 psia. The compression ratio exceeds 10. These machines have performed admirably in a fouling service. Two skid-mounted two-stage labyrinth-piston compressors with pressure-tight casing are shown in Figure 14–37. Compressors of this type have found worldwide application as boil-off compressors for the sea transport and storage of liquefied gases, such as hydrocarbons and ammonia. In most cases, the vapor is com-

FIGURE 14–37   *Skid-mounted two-stage labyrinth-piston compressor with suction valve unloaders. (Source: Sulzer-Burckhardt, Winterthur and Basel, Switzerland.)*

pressed in order to be reliquefied and then recycled to the transport or storage tank. All suction valves are fitted with unloaders for unloaded start and capacity control. Compressors of this design are also used in refrigeration cycles.

DIAPHRAGM COMPRESSORS*

Earlier in this chapter, Figure 14–10 had assigned diaphragm compressors to the high-pressure low-flow range of application. Modern diaphragm compressors are often a combination of two systems—a hydraulic system and a gas compression system. A metal diaphragm group is the isolating component between these two systems.

In the diaphragm compressor depicted in Figure 14–38, the gas compression system consists of three flat metal diaphragms, which are clamped between two precisely contoured concave cavities and process gas inlet and outlet check valves. The hydraulic system includes a motor-driven crankshaft that reciprocates a piston in the hydraulic fluid medium. This positive-displacement piston pulses the hydraulic fluid against the lower side of the diaphragm group, causing it to sweep the cavity, displacing the process gas. Except for very small diaphragm compressors, the other components of the hydraulic system are the automatic hydraulic injection pump, the hydraulic fluid check valve, and the hydraulic relief valve. In addition to ensuring that the hydraulic system is always full for the compression cycle, the automatic injection pump provides for fast priming of the hydraulic system after extended compressor shutdowns. On smaller compressors, this same function is sometimes accomplished by a manual priming pump and a gravity oil-feed system. On the compression stroke of the main piston, the hydraulic check valve isolates the hydraulic system from the automatic injection pump

---

*Source: PPI Division, the Duriron Company, Inc., Warminster, Pennsylvania. Adapted by permission.

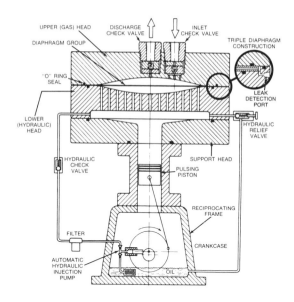

FIGURE 14–38 *Typical assembly of a diaphragm compressor. (Source: PPI Division of the Duriron Company, Inc., Warminster, PA.)*

or the manual priming pump so that an elevated pressure can be generated in the system against the hydraulic relief valve.

A full cycle of operation for the diaphragm compressor begins with the diaphragm group fully deflected to the bottom of the cavity by the gas suction pressure. The hydraulic piston is at bottom dead center, and the hydraulic system has just been filled by a single stroke of the automatic hydraulic fluid injection pump. On the process side of the diaphragm group, the cavity is now filled, at a given suction level, by the process gas that has entered through the inlet check valve. As the crankshaft rotates and the hydraulic piston moves from its bottom dead center position toward top dead center, the hydraulic pressure increases. This occurs because the hydraulic system injection and relief lines are blocked by the hydraulic check valve and relief valve, respectively.

As the piston continues toward top dead center, and as the hydraulic pressure reaches the pressure level of the process gas in the cavity, the diaphragm group begins to sweep the cavity toward its top dead center position, thereby compressing the process gas. When the pressure of the process gas in the cavity reaches the pressure level downstream of the discharge check valve, that check valve opens and the process gas is discharged from the cavity. Since the hydraulic system has slightly more displacement capacity than the gas system, the diaphragm group makes metal-to-metal contact with the process head cavity, assuring that all of the gas has been displaced. With the diaphragm group in this fully deflected discharge position, the piston still has a certain amount of travel required to reach its top dead center position. As the piston moves to top dead center, this additional hydraulic fluid volume is "overpumped" through the hydraulic relief valve, which is set at a pressure level above the desired process gas discharge pressure. At this point, the compression portion of the cycle has been completed.

The hydraulic piston now moves toward bottom dead center. As it does, the diaphragm group is deflected toward the bottom of the cavity by both the expansion of the residual gas contained in the clearance volume and by the additional process gas entering the cavity at suction pressure. During this suction stroke, a synchronized auxiliary eccentric cam on the crankshaft causes the hydraulic injection pump plunger to

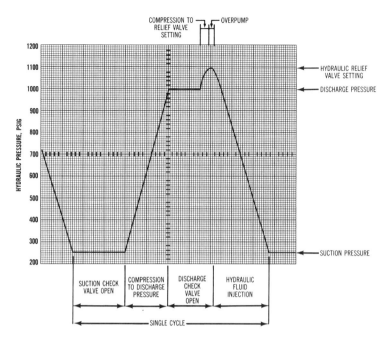

FIGURE 14-39　*Hydraulic system pressures acting over one operating cycle of typical diaphragm compressor. (Source: PPI Division of the Duriron Company, Inc., Warminster, PA.)*

stroke, injecting an amount of hydraulic fluid equal to that which was "overpumped" through the hydraulic relief valve at the end of the compression portion of the cycle. On small gravity oil-feed units, hydraulic fluid is drawn from the crankcase as the main piston travels to bottom dead center. When the hydraulic piston reaches its bottom dead center position, the hydraulic system is again filled. Since the diaphragm group is now in the bottom-most deflected position, a full gas cavity is assured. At this point, the cycle is complete.

The hydraulic pressure versus crank position has been plotted on Figure 14-39.

APPENDIX 14A

Performance Calculations

Performance calculations are made to determine the throughput of a reciprocating compressor and the horsepower absorbed by the process.

The *capacity throughput* is determined by the displacement rate of the compressor, i.e., the volume swept by piston or pistons, the actual suction volumetric efficiency, and the specific volume at suction conditions. The volume swept by the piston(s) is determined by the total piston area, the piston stroke, and the piston speed.

In the case of a multistage compressor, the calculations must be performed for the first stage only, and the pressure ratio of the first stage has to be used. However, this does not mean that the following formulas cannot be used for each individual stage as well.

The *actual volumetric efficiency* is defined as follows:

$$\lambda = \frac{\text{actual suction volume}}{\text{volume swept by piston(s)}} \qquad \%$$

In order to calculate this value, the *theoretical volumetric efficiency* based on ideal gas conditions and not considering efficiency losses of any kind must first be determined (Figure 14A-1).

$$\eta = 100 - V_o\left[\left(\frac{p_2}{p_1}\right)^{\frac{1}{k}} - 1\right] \qquad \%$$

where V_o = clearance volume (or space) expressed as percentage of the swept volume

p_2 = absolute discharge pressure of the first stage

p_1 = absolute suction pressure of the first stage

k = exponent of the re-expansion curve c–d

As a rule, the specific heat ratio is used for the sake of simplicity. The actual re-expansion curve deviates from the isentrope obtained in this way, and the difference has to be included in the coefficient of correction x (see below). This theoretical efficiency has to be corrected by x in order to obtain the *actual volumetric efficiency:*

$$\lambda = \eta - x \qquad \%$$

Correction x allows for all influences that decrease the volumetric efficiency, such as pressure losses across the valves increasing the pressure ratio, preheating of the gas in the suction chamber resulting in a decrease of its specific weight, deviation from ideal gas law (compressibility factors at both suction and discharge conditions), and internal gas leakages. Test results determine x. It is normally between 3 and 10 percent.

391

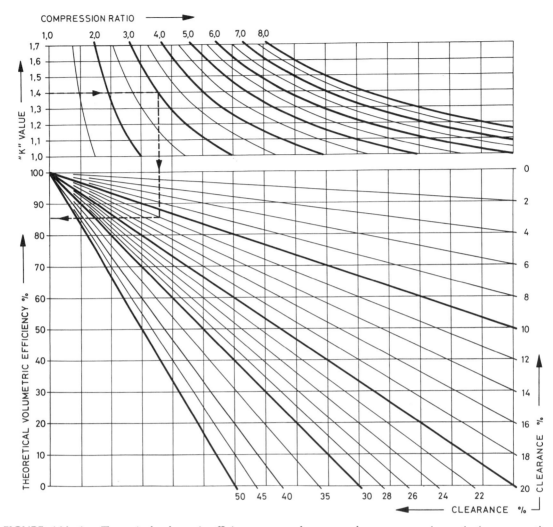

FIGURE 14A–1   *Theoretical volumetric efficiency versus clearance volume, compression ratio (pressure ratio P_1/P_2), and K value. Example: K = 1.4, compression ratio = 3, clearance volume = 12%, theoretical volumetric efficiency = 85.5%. (Source: Sulzer-Burckhardt, Winterthur and Basel, Switzerland.)*

Figure 14A–2 may be used to determine the theoretical volumetric efficiency. The *actual suction volume* at intake conditions can then be calculated as follows:

$$V_1 = Vs \times \lambda \qquad acfm$$

where Vs = volume swept by piston(s)   acfm

The *throughput* expressed as a mass flow is obtained by means of the following formula:

$$W = \frac{V_1}{\vartheta_1} \times 60 \qquad lb/hr$$

where ϑ_1 = specific volume of the gas at suction conditions, which is a $\dfrac{cu/ft}{lb}$
function of the pressure, temperature, gas composition, and
compressibility factor.

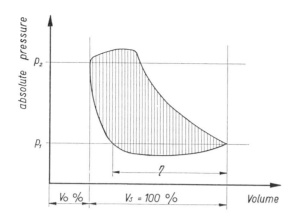

FIGURE 14A–2 *P-V indicator diagram, with shaded area representing work required to compress and deliver the gas. (Source: Sulzer-Burckhardt, Winterthur and Basel, Switzerland.)*

POWER REQUIREMENT

At a very early stage of each project in chemical engineering and in many cases before compressor manufacturers can be approached for quotations, the chemical engineer is confronted with at least three questions:

1. How big is the compressor I need?
2. What initial investment is required?
3. How much power is required?

Since the key to answering the first and second question is the power consumption, it may be useful to have a formula at hand, by means of which it is possible to work out at least an approximate figure for the power absorbed at the compressor shaft.

The basic theoretical single-stage horsepower formula is developed for the pv indicator diagram, the net area of which represents the work required to compress and deliver the gas. This is represented by the shaded area in Figure 14A–1. Unlike the diagram of the ideal cycle, the indicator diagram describes the real cycle and includes the pressure losses across the valves.

The actual power requirement is related to a theoretical cycle through an overall efficiency that has been determined by test on prior machines. The overall compressor efficiency is the ratio of theoretical to actual power on the compressor shaft.

As a rule, reciprocating compressors are compared with the isentropic (adiabatic) cycle. Since this leads to a somewhat complicated formula, a good many engineers still use the isothermal power as a basis, as in the formula below:

$$P_{is} = R \times T_1 \times \ln\frac{p_2}{p_1}$$

where R = gas constant
T_1 = absolute suction temperature

Since this is a general formula saying nothing about the type of compressor or the number of stages, it can also be used for multistage compressors provided that p_2/p_1 is the overall pressure ratio and not only the pressure ratio of one stage.

Using algebra, the above formula can be converted into the following ones, which are better suited for practical use:

Power absorbed in horsepower:

$$P = 10.05 \times 10^{-3} \times p_1 \times V_1 \times \log\frac{p_2}{p_1} \times \frac{1}{\eta_{is}} = HP$$

Power absorbed in kilowatts:

$$P = 7.50 \times 10^{-3} \times p_1 \times V_1 \times \log\frac{p_2}{p_1} \times \frac{1}{\eta_{is}} = kW$$

where p_1 = absolute suction pressure in psia
p_2 = absolute discharge pressure in psia
V_1 = actual suction volume in acfm
η_{is} = overall efficiency based on isothermal power

Overall efficiency based on isothermal power (η_{is}) is a figure that must be stated by the compressor manufacturer. In most cases, it is between 0.55 and 0.70 (55 to 70 percent). The value of η_{is} depends on many factors and its magnitude can be influenced only in part by the compressor manufacturer. It is mainly determined by the following:

- gas to be compressed (specific heat ratio, molecular weight, compressibility factors)
- size and type of compressor (as a rule, one 100 percent capacity compressor has a better efficiency than two 50 percent capacity machines)
- speed of the compressor (pressure losses in the gas stream)
- discharge pressure, pressure ratio
- number of stages
- intercooling with multistage compressors, i.e., the better the cooling effect, the better the efficiency

CONVERSION OF SUCTION CAPACITY

Compressor manufacturers state the actual *suction capacity* of their machines in acfm or cubic meters per hour, whereas chemical engineers prefer weight rates expressed in pounds per minute or pound moles per minute or flow rates expressed in standard cubic feet per minute (scfm).

The actual inlet volume may be calculated from the following formulas by using the proper inlet pressure and temperature and correcting for moisture content at these conditions and deviation from the ideal gas laws (compressibility):

From weight flow (W) to acfm (W lb/min, dry)

$$V_1 = 10.73 \times \frac{W}{M} \times \frac{T_1}{p_1} \times Z_1 \times C_1 = acfm$$

From scfm to acfm (V_o scfm at 14.7 psia, 60°F, dry)

$$V_1 = \frac{14.7}{520} \times V_o \times \frac{T_1}{p_1} \times Z_1 \times C_1 = acfm$$

From mole flow to acfm (N lb mole/min, dry)

$$V_1 = \frac{379 \times 14.7}{520} \times N \times \frac{T_1}{p_1} \times Z_1 \times C_1 = acfm$$

or

$$V_1 = 10.714 \times N \times \frac{T_1}{p_1} \times Z_1 \times C_1 = acfm$$

where M = molecular weight   dimensionless
       T_1 = absolute inlet temperature   degrees Rankine °R (/)
       p_1 = absolute inlet pressure   psia
       Z_1 = compressibility factor, at intake conditions   dimensionless
       C_1 = correction factor for the moisture at intake   dimensionless
              conditions

Z_1 depends on gas, pressure, and temperature. For ideal gases, Z_1 is 1. C_1 depends on moisture content, temperature, and pressure. For dry gases, C_1 is 1. For atmospheric pressure, 14.7 psia, 68 °F, and 100% relative humidity, it is 1.024. Use charts for Z_1 and C_1.

APPENDIX 14B

Capacity Control

There are two main reasons why compressor capacity regulation is used. The most prevalent one is to adjust the suction flow to match the process demand. The second reason is to save energy. As a rule, capacity control is determined by compressor discharge pressure. In cases where the system upstream of the compressor has to be protected against too low a suction pressure, the control point may be governed by intake pressure. Modern control technology permits using other parameters as control points; temperatures, flows, liquid levels are but a few examples.

Where changes in demand are infrequent and slow, the capacity control may be arranged for *manual operation*, either *directly* on the compressor or by means of *remote control*. Modern process plants in which changes in demand are rapid and not always predictable, or where compressors have to be operated without supervision by operating personnel, require *automatic control*.

There are a number of variations that can be grouped under two branches of capacity control, as shown in Figure 14B–1. The optimum capacity control is largely determined by the following parameters:

- the capacity range required
- how frequently changes in demand occur
- how long reduced capacity is required
- size of the compressor
- type of driver

Not all types of capacity controls can be used with a given compressor model, a specific pressure range, or a given gas. The process engineer involved in specifying the compressor should clearly describe the turndown requirements and work with the machine manufacturer in determining feasible capacity control strategies. It is sometimes necessary to combine two or more types of regulation for best efficiency, flexibility, and reliability. Table 14B.1 lists the main characteristics of the capacity control systems as described in Figure 14B–1.

Capacity control by means of an *overall bypass* can be applied without limitations to all compressors, provided that the gas recycled enters the suction line close to normal suction temperature. This means that an aftercooler or a bypass cooler may have to be used. In addition, a check valve in the discharge pipe is required to prevent the high-pressure gas from flowing back when the compressor is at standstill. Since this regulation is very uneconomical, it should only be used if the compressor has to be operated at reduced capacity for a short time or in combination with an energy-saving type of control.

In multistage compressors, a bypass around the first stage or a *partial bypass* can be used. The absolute power input can be reduced in this way, although the specific input is increased. In fixing the regulation range of a first-stage bypass, it must be

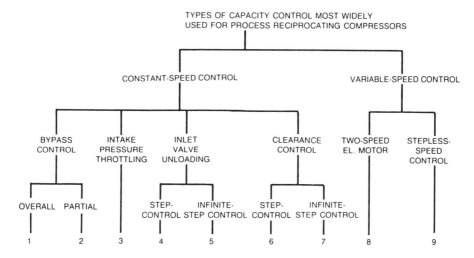

FIGURE 14B-1 *Types of capacity control most widely used for process reciprocating compressors. (Source: Sulzer-Burckhardt, Winterthur and Basel, Switzerland.)*

remembered that a reduction of the flow to the second stage causes a drop of all interstage pressures, and consequently it can lead to excessively high pressure ratios and discharge temperatures in the following stages. These pressure shifts can cause overload in the last stage. Therefore, the recycled flow has to be restricted by means of an orifice plate in the bypass pipe. The minimum capacity that can be attained depends on the number of compression stages. The more stages used for a given overall compression ratio, the wider the achievable control range. For oxygen service, a partial first-stage bypass is normally the only recommended means of constant-speed capacity control apart from an inefficient overall bypass. With three-stage compressors, the minimum attainable capacity is approximately 60 percent. If a lower turndown capacity is required with such a compressor, the problem can be solved by providing an additional compressor stage or cylinder.

Bypass regulations are very practical and easy to apply. Standardized components can be used.

Intake pressure throttling will provide an infinite number of steps between full load and reduced load. Capacity changes are achieved by reduction in both gas density and volumetric efficiency, since the latter depends on the pressure ratio. When this method is applied, it is again necessary to investigate beforehand the resulting changes in pressure, discharge temperatures, and load conditions in the upper stages. When compressing flammable gases, care must be taken not to create a vaccuum in the suction pipe in order to avoid the ingress of air.

Inlet valve unloading is the most widely used type of constant speed control. This consists simply of holding one or more inlet valves open during both suction and discharge strokes, so that the gas taken into the cylinder on the suction stroke is pushed back through the suction valves on the discharge stroke. With one double-acting cylinder, *two steps*—50 and 0 percent—are possible. With two first-stage double-acting cylinders running in parallel, 75, 50, 25, and 0 percent capacity can be attained.

With multistage compressors and control range down to 0 percent, all suction valves have to be fitted with unloaders. The cylinder shown in Figure 14B-2 is fitted with suction valve unloaders (A). The suction valve plates can be lifted either by means of threaded spindles with handwheels (not shown in the figure) or by means of servomotors, as represented in Figures 14B-3 and 14B-4. These are operated either by oil pressure or, less frequently, by compressed air, or by the process gas.

Table 1   Main Characteristics of Capacity Control Systems

NO.	TYPE OF CONTROL SYMBOL	POWER VS CAPACITY	RANGE %	ADVANTAGES	DISADVANTAGES	UNLOADED START
1		P / V_1	100 – 0	SIMPLE RESPONSIVE STEPLESS UNEXPENSIVE	WASTE OF POWER	NO
2		P / V_1	100 – ~60	STEPLESS INEXPENSIVE	CAN BE APPLIED TO MULTISTAGE COMPRESSORS ONLY INTERSTAGE PRESSURE SHIFT	PARTIAL
3		P / V_1	DEPENDS ON NUMBER OF STAGES	STEPLESS INEXPENSIVE	INCREASED PRESSURE RATIO HIGHER DISCHARGE TEMPERATURES	PARTIAL
4		P / V_1	100 ···· O	ECONOMICAL RESPONSIVE ONE STEP PER CYLINDER OR CYLINDER HALF	NOT FOR OXYGEN	YES
5		P / V_1	100 – 50 AND O	ECONOMICAL	NOT FOR OXYGEN NOT SUITED FOR MULTISTAGE COMPRESSORS	YES
6		P / V_1	100 ···~50	ECONOMICAL	NOT FOR OXYGEN CAN BE PROVIDED ON CYLINDER HEAD SIDE ONLY CAUSES DISCHARGE TEMPERATURE TO RISE	PARTIAL
7		P / V_1	100 – ~50	ECONOMICAL	SAME AS ABOVE	PARTIAL
8		P / V_1	100 AND 50 (100/67)	ECONOMICAL INEXPENSIVE	FOR ELECTRIC MOTORS BELOW 500KW ONLY	NO
9		P / V_1	100 – ~50	ECONOMICAL	WITH ELECTRIC MOTORS: EXPENSIVE	NO

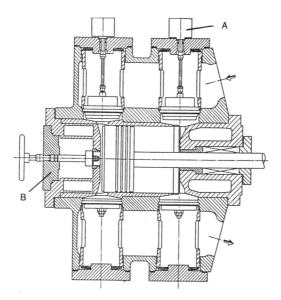

FIGURE 14B–2 *(A) Compressor cylinder with suction valve unloaders and (B) clearance control. (Source: Sulzer-Burckhardt, Winterthur and Basel, Switzerland.)*

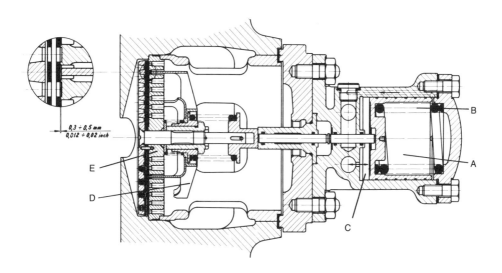

FIGURE 14B–3 *Suction valve unloader for reciprocating compressors. (A) Servomotor, (B) main spring, (C) oil under pressure, (D) bell, (E) suction valve. (Source: Sulzer-Burckhardt, Winterthur and Basel, Switzerland.)*

When the servomotor (A) is deactivated, a spring (B) exerts pressure on the spindle, thus keeping the valve plate in the open position. The oil pressure is, as a rule, generated by the crankshaft-driven gear-type lubricating oil pump, which also supplies oil to the compressor bearings. This means that the suction valves are kept open by spring force when the compressor is idle. As a consequence, a compressor equipped with hydraulic suction valve unloaders is started up without load and begins to deliver gas only when the oil pump generates sufficient oil pressure. When the compressor is

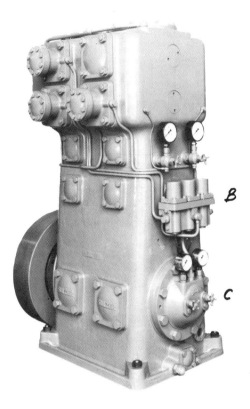

FIGURE 14B–4   *Single-stage labyrinth-piston compressor with two double-acting pistons working in parallel. One cylinder is fitted with suction valve unloaders (A) on both suction valves, while the other one has an unloader on the lower suction valve only. The unloaders are actuated by oil pressure supplied by the crankshaft-driven oil pump (C). The oil flow to and from the servomotors is controlled by three solenoid valves (B) so that capacity steps of 75, 50, and 25 percent can be achieved. (Source: Sulzer-Burckhardt, Winterthur and Basel, Switzerland.)*

running, its capacity is reduced in steps according to the number of suction valves kept open.

Infinite (stepless) control between 100 and approximately 50 percent of each individual suction valve and hence of each cylinder or cylinder half can be achieved by using springs between the servomotor spindle and the bell, which keeps the valve plate open. In this way it is possible to influence the time during which the gas is pushed back to the suction pipe at the discharge stroke. Valves controlled in this way are referred to as *timed suction valves*. By altering the spring force acting on the bell, the closing of the valve plate can be held up at any point between 0 and approximately 50 percent of the compression stroke, so that the flow can be adjusted continuously between 100 and approximately 50 percent.

The spring force depends on the position of the servomotor piston, which in turn depends on the oil pressure below the piston. Below 50 percent capacity, this control system jumps to the no-load position and the suction valve plates are then kept open permanently. This system also permits unloaded start of the compressor.

An example of a *one-step clearance control* is shown in Figure 14B–2 (B) and Figure 14B–5. Such a device is sometimes used in combination with valve unloading in steps. Figure 14B–5 illustrates cylinders of a two-stage labyrinth-piston compressor compressing ethylene from 280 to 1215 psia. The first-stage cylinder (right) is

FIGURE 14B–5   *One-step clearance controls on vertical compressor cylinder (right side, top). (Source: Sulzer-Burckhardt, Winterthur and Basel, Switzerland.)*

equipped with a one-step clearance control and with valve unloaders on the lower suction valves. The second-stage high-pressure cylinder has no capacity control. With this combination, steps of 73 and 60 percent can be achieved. Reducing the capacity in one cylinder only causes a drop of the interstage pressure, which could be tolerated in this specific case. The actuator for the clearance pocket control is operated by compressed air. The servomotors for the valve unloaders are actuated by oil pressure; however, the three-way valve controlling the oil flow is controlled pneumatically. The system is suited for manual as well as automatic remote control by means of a 60-psig compressed air signal.

All reciprocating machines have a clearance space, which is imposed by their design and which, particularly in compressors, is kept as small as possible. At the end of the compression stroke, this space is filled with the compressed gas. When the piston returns, the entrapped gas expands to the suction pressure before the suction valve can open and gas can flow into the cylinder. The larger the clearance space, the greater the loss of suction volume. Capacity control by additional clearance space is based on this fact. By artificially increasing the clearance space, the discharge quantity of the compressor can be regulated with only very slight losses. In its simplest form, this additional space is in the cylinder cover, as shown in Figure 14B–2, or is provided by a clearance bottle. By means of a valve, operated by hand, by air or oil pressure, or by a small electric motor, the additional clearance space can be added to the already existing one.

The practical application of this principle today consists in the inclusion of one, two, or more such additional spaces in the design of the cylinder cover, each fitted with suitable valves. With two additional spaces of different sizes, a and b, four stages of regulation—namely full load and the partial loads by connecting a, b, or a+b—can be attained. This method of regulation is reliable and economical, but it permits adjustment to fixed quantities only, so that it sometimes has to be combined with a bypass or suction valve control in order to bridge the intervals between the regulating steps.

Infinite stepless control by clearance space can be obtained if the additional clearance space is designed as a cylinder with a movable piston. Such a design, however, is far from simple and is generally expensive. This method of capacity control is therefore used very rarely, in spite of its technical advantage.

Permanent reduction of capacity by increasing clearance space is sometimes necessary if a compressor has to be operated at reduced flow for a relatively long period of time. In this case, capacity control systems are not the optimum solution. Clearance space increases are achieved by lifting cylinder cover and/or compressor valves by means of distance rings. If valves are lifted in this way, the lanterns have to be replaced by shorter ones. The higher the compression ratio, the greater the reduction in capacity that can be achieved. In most cases, a reduction of 10 percent or more is feasible at little cost and with practically no loss in power efficiency. When the full capacity is required again, the distance rings can be removed and the lanterns replaced by the initial ones.

VARIABLE-SPEED CONTROL

The optimum method of regulation is adapting the flow to process demand by changing the compressor speed. Variable-speed control is used whenever the driver is capable of operating at a speed commensurate with demand. Steam and gas turbines and internal combustion engines are in this class.

Most compressors today, however, are driven by electric motors. The simplest and least expensive method is to use a *two-speed motor*. However, two-speed operation may not be possible with large motors. The problem is aggravated by the fact that in contrast to dynamic compressors, reciprocating compressors have a constant torque over the full speed range. With direct current (DC) motors, a wide speed range can be attained, but DC drive systems can become very costly.

Recent developments in *thyristor technology* and electronic control made it possible to use *adjustable frequency* drives for large compressors, resulting in excellent overall efficiency. Adjustable frequency drives are very reliable and easier to maintain than other variable-speed electric motors. With adjustable frequency systems, both induction and synchronous motors may be used. A typical adjustable frequency drive system with electronic control consists of the following main equipment:

- a circuit breaker or motor starter
- a supply converter
- a DC link reactor
- a motor converter
- a brushless, alternating current (AC) synchronous motor

The use of variable-speed drivers creates some problems that deserve special attention:

- The flywheel effect decreases in proportion to the cube of the speed.
- Torsional and other vibrations could coincide with running speeds and cause damage to the machine.
- Gas pulsations in the pipe system must be kept under control over the full speed range.
- Poor lubrication of the compressor bearings occurs at minimum speed, particularly

if the lubricating oil pump is driven by the crankshaft. This problem can be solved by means of an independently driven pump.

- It may be necessary to add some additional mass to the reciprocating parts of the compressor in order to ensure piston-rod load reversal at the bottom speed. In most cases, this reversal is required to maintain proper lubrication between the crosshead pin and bushing.

The above list, which is not complete, demonstrates that in certain cases it is not possible to utilize the entire speed range of a driver.

Chapter 15

Rotating Positive
Displacement Compressors*

Rotary screw compressors and rotary piston blowers belong to the machinery group making up rotating positive-displacement compressors. Of these two machines, rotary screw compressors are primarily used in higher pressure air and process gas services, whereas the rotary piston blowers are more typically used in lower pressure, high-volume applications. Both machines can be used as dry or wet fluid movers.

Rotating positive-displacement machines offer the same advantage as reciprocating positive-displacement equipment with regard to flow versus pressure relationships, i.e., nearly constant inlet flow volume under varying discharge pressure conditions. Also, positive-displacement machines do not have a surge limitation, which is to say, there is no minimum throughput requirement for these compressors.

The rotor tip speeds on rotary screw and rotary piston blowers are low; this allows for liquid injection and handling of contaminated gases. By design, the rotors are self-cleaning during operation, which is a significant advantage in dirty-gas services.

ROTARY SCREW COMPRESSORS

Rotary screw compressors are available in oil-free or oil-flooded construction. Fields of application for oil-free machines include all processes that cannot tolerate contamination of the compressed gas or where the lubricating oil would be contaminated by the gas. Oil-flooded machines can achieve slightly higher efficiencies and utilize the oil for cooling as well.

Properly designed rotary screw compressors are constructed with no metallic contact whatsoever inside the compression chambers, either between the rotors themselves or between these and the walls of the housing.

Although originally intended for air compression, rotary screw compressors are now compressing a large number of process gases in the petrochemical and related industries. These include air separation plants, industrial refrigeration plants, evaporation plants, mining, and metallurgical plants.

Practically all gases can be compressed: ammonia, argon, ethylene, acetylene, butadine, chlorine gas, hydrochloric gas, natural gas, flare gas, blast furnace gas, swamp gas, helium, lime-kiln gas, coking-plant gas, carbon monoxide gas, all hydro-carbon combinations, town gas, air/methane gas, propane, propylene, flue gas, crude gas, sulphur dioxide, oxide of nitrogen, nitrogen, styrene gas, vinyl chloride gas, and hydrogen gas can be found on the reference tabulations of experienced manufactures.

---

*Source: Aerzener Maschinenfabrik, GmbH, D-3258 Aerzen, West Germany. Adapted with permission.

Application Limits for Rotary Screw

Application limits for rotary screw compressors are given by the pressure and temperature ranges and by the maximum allowable speed of the machines.

Oil-free rotary screw compressors can be mechanically loaded with pressure differences up to 12 bar, and oil-flooded compressors up to 20 bar. Higher pressure differences are possible in special cases.

The maximum allowable compression ratio for one screw compressor stage that will not cause the final compression temperature to rise above the permitted value of 250 °C will to a very large extent depend on the specific heat ratio c_p/c_v of the gas to be compressed. For example, where the specific heat ratio c_p/c_v equals 1.4, the maximum compression ratio would be approximately 4.5, and where the specific heat ratio c_p/c_v equals 1.2, the maximum compression ratio would be approximately 10 for one oil-free screw compressor stage.

Multistage (multicasing) arrangements are not uncommon and can result in pressure ranges from approximately 0.1 bar absolute to 40 bar. Interstage cooling is used in many of these applications.

Depending on compressor size, speeds from 2000 to 20,000 RPM can be encountered. The limiting factor is typically the circumferential speed of the male rotor, which typically ranges from 40 to approximately 120 m/sec, and up to a maximum of 150 m/sec for very light gases.

Flow volumes up to 60,000 m³/hr can be accommodated in these compressors.

Principal Construction Features

Rotary screw compressors (Figure 15–1) are dual-shaft rotary piston machines operating on the principle of positive displacement combined with internal compression. The gaseous medium moves from the suction port to the discharge port, entrapped in progressively decreasing spaces between the convolutions of the two helical rotors, being thus compressed up to the final pressure before it is discharged into the discharge nozzle. Figure 15–2 illustrates this process.

On small rotary screw compressors, the housing is vertically parted on the suction side. Cylinder and discharge side plate are frequently combined in one housing. The housings of larger machines are often parted horizontally for easy assembly.

Rotors and shafts are milled out of one piece of either forged or stainless steel. Some manufacturers provide rotors with synthetic coatings. Depending on service conditions, this may lead to a rapid drop in compressor efficiency due to loss of coating on the rotor edges.

Process gas machines incorporate direction of flow from the top to the bottom, thus facilitating liquid removal from the compression space whenever liquid is injected into the rotor chamber for cooling or cleaning during operation. On-stream cleaning is highly advantageous in services where gases are contaminated or tend to polymerize. The sealing area is equipped with connections for sealing medium supply and relief. In principle, it is possible to apply a cooling medium to the cylinder wall, but uncooled cylinder housings can be used as well.

The principal components of a large, two-stage rotary screw compressor are shown in Figure 15–3.

Figure 15–4 illustrates typical rotor combinations incorporating an asymmetrical rotor profile. The profile combination 4+6 means that the male rotor has four teeth and the female rotor, six. Due to this profile combination, the diameter of the rotor core is relatively thick. This allows for operation with large differential pressures.

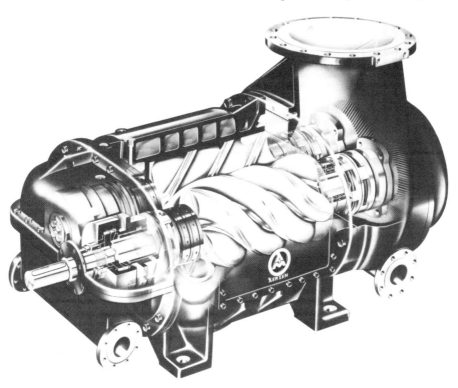

FIGURE 15–1 *Modern rotary screw compressor. (Source: Aerzener Maschinenfabrik GmbH, D-3258 Aerzen, West Germany.)*

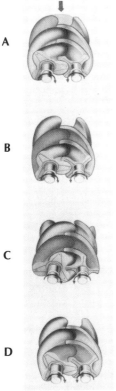

FIGURE 15–2 *Compression process in rotary screw compressor. (A) Suction intake: Gas enters through the intake aperture and flows into the helical grooves of the rotors, which are open. (B and C) Compression process: As rotation of the rotors proceeds, the air intake aperture closes, the volume diminishes, and the pressure rises. (D) Discharge: The compression process is completed, the final pressure is attained, the discharge commences. (Source: Aerzener Maschinenfabrik GmbH, D-3258 Aerzen, West Germany.)*

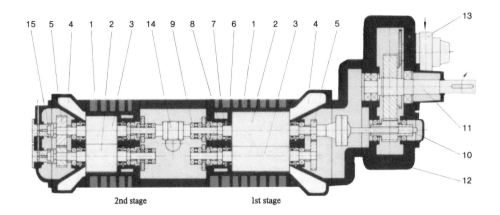

FIGURE 15–3   *Principal components of a two-stage rotary screw compressor. 1 = Housing; 2 = male rotor; 3 = female rotor; 4 = side plate on intake side; 5 = timing gears; 6 = graphite ring shaft-seal; 7 = oil seal; 8 = radial bearing; 9 = axial thrust bearing; 10 = torsion shaft; 11 = drive shaft; 12 = step-up gears; 13 = oil pump; 14 = coupling; 15 = compensating piston. (Source: Aerzener Maschinenfabrik GmbH, D-3258 Aerzen, West Germany.)*

Bearings

Although air machines are often equipped with rolling element bearings, the majority of compressors for process gas applications are furnished with journal bearings and thrust bearings of the type commonly found in centrifugal process gas compressors. The service life of these bearings is practically unlimited as long as proper lubricating and operating procedures are in force.

Seals

In many rotary screw compressor applications, it is necessary to provide a sealing barrier between the process gas and the bearings. A number of different seal types are feasible:

- carbon ring seals
- barrier water floating ring seals
- double-acting mechanical seals with stationary spring
- combined floating ring/mechanical seals

At the compressor input shaft, manufacturers often opt for:

- labyrinth seals, or
- double-acting mechanical seals with rotating springs

Carbon ring seals with connections for the injection and eduction of neutral, clean gases are used in cases where leakage gas, even in connection with sealing gas, may enter into the bearing areas or into the atmosphere. The gas pressure is relieved across floating carbon rings at the beginning of the seal chamber.

With barrier water floating ring seals, barrier water enters the seal chamber and a small amount of water reaches the compression space. Most of the water is returned to the barrier water system for cooling, filtration, and re-use. Barrier water seals are able to fully prevent gas leakage and can provide valuable cooling and scrubbing duties. Figure 15–5 depicts a flow diagram for a single-stage screw compressor with barrier water seals.

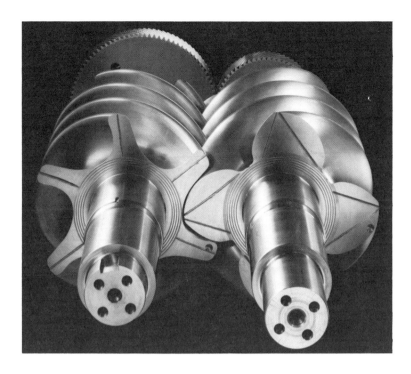

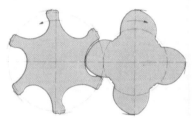

female rotor   male rotor

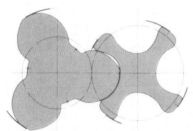

male rotor   female rotor

FIGURE 15–4   *Asymmetrical rotor sets for modern rotary screw compressors. (Source: Aerzener Maschinenfabrik GmbH, D-3258 Aerzen, West Germany.)*

A double-acting stationary spring mechanical seal and a combination mechanical and floating ring seal are primarily used for compression with high differential pressures.

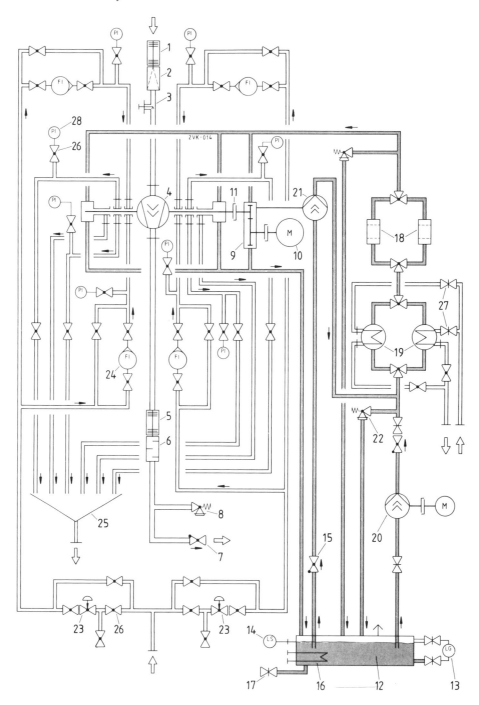

FIGURE 15–5   *Water sealing flow schematic. 1 = Lateral compensator ss*; 2 = starting strainer; 3 = water injection; 4 = screw compressor; 5 = lateral compensator ds*; 6 = discharge silencer; 7 = non-return valve; 8 = safety relief valve; 9 = gear box; 10 = drive motor; 11 = coupling; 12 = oil reservoir (base plate); 13 = oil sight glass; 14 = level controller (oil); 15 = non-return valve; 16 = oil heating; 17 = oil drain; 18 = twin oil filter; 19 = twin oil cooler; 20 = oil pump with motor; 21 = gear box oil pump; 22 = safety relief valve (oil); 23 = barrier water controller; 24 = flow indicator; 25 = barrier water return; 26 = valve; 27 = slide valve; 28 = manometer. (*ss = suction side; ds = discharge side). (Source: Aerzener Maschinenfabrik GmbH, D-3258 Aerzen, West Germany.)*

Operating Principles for Oil-Injected Compressors

Regardless of whether the screw compressor is executed for dry compression or oil injection, the gas is compressed in chambers progressively decreasing in size that are formed by the intermeshing action of the helical rotors and by the housing wall. However, oil-injected compressors do not incorporate timing gears. Instead, the driven male rotor interacts directly with the female rotor without use of timing gears. Oil injected into the compressor cavity provides intensive lubrication, and a large portion of the compression heat is absorbed. At the same time, the clearances between rotors and cylinder walls are filled with oil. This prevents the reverse flow of compressed gas and increases the overall efficiency of compression.

At the compressor discharge flange, gas and oil exit through a check valve to the oil reservoir where most of the oil is separated from the gas. The remaining oil is separated in a downstream separator, and residual oil amounts of typically 5 parts per million (ppm) continue to remain in the gas stream. Oil carryover can be further lowered by downstream cooling and final moisture separation. The oil separation unit has to be properly maintained and the pressure drop across the separator cartridges taken into account to determine the overall performance of the compressor package. It should also be recognized that the efficiency of oil separation depends on the degree of contamination of the separator elements.

The principle of oil separation is shown in Figure 15–6, and a typical oil-injected screw compressor is illustrated in Figure 15–7.

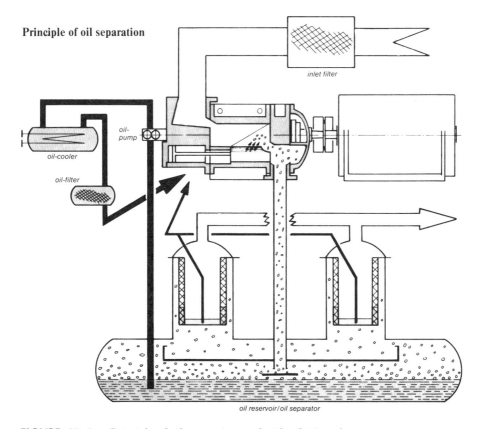

FIGURE 15–6   *Principle of oil separation used with oil-injected rotary screw compressors. (Source: Aerzener Maschinenfabrik GmbH, D-3258 Aerzen, West Germany.)*

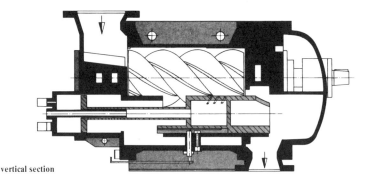

vertical section

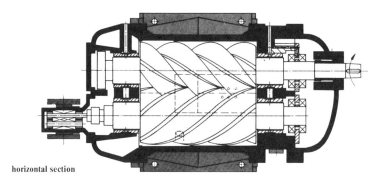

horizontal section

FIGURE 15–7   *Oil-injected rotary screw compressor cross section showing slide valve for capacity control. (Source: Aerzener Maschinenfabrik GmbH, D-3258 Aerzen, West Germany.)*

Compressor Volume Control

In principle, it is necessary to consider the problems of volume control for dry-running and for oil-injection–type screw compressors separately.

Dry Screw Compressors

Control by Variable Speed.   In consequence of the fact that screw compressors displace the medium positively, the most advantageous method of achieving volume control is that obtained by varying the speed. This may be done in any of the following ways:

- by variable speed electric motors
- by use of a torque converter
- by steam turbine drive

Speed may be reduced to about 50 percent of the maximum permissible speed. Induced flow volume and power transmitted through the coupling are in this manner reduced in approximately the same proportion. The allowable turn-down depends on the adequacy of bearing lubrication at low speed and compressor discharge temperature. More than 50 percent reduction is possible in special cases. As mentioned earlier, there is no surge limit for these positive-displacement machines.

FIGURE 15–8   *Slide valve inside housing in partial capacity position. (Source: Aerzener Maschinenfabrik GmbH, D-3258 Aerzen, West Germany.)*

Bypass.   Using this method, the surplus gas volume is allowed to flow back to the intake side by way of a compressor discharge pressure controller. An intermediate cooler brings the surplus gas volume down to intake temperature.

Full-Load/Idling-Speed Governor.   As soon as a predetermined final pressure is attained, a pressure controller operates a diaphragm valve that opens a bypass between the discharge and suction sides of the compressor. When this occurs, the compressor idles until pressure in the system drops to a predetermined minimum value. The valve will close once again on receiving an impulse from a pressure sensor. This brings the compressor back to full load.

Suction Throttle Control.   This method of control is suitable for air compressors only. As in the case of the full-load/idling-speed control method, a predetermined maximum pressure in the system, for example in a compressed air receiver, causes pressure on the discharge side to be relieved down to atmospheric pressure. Simultaneously, the suction side of the system is throttled down to about 0.15 bar absolute pressure. When pressure in the entire system has dropped to the permissible minimum value, full load is once again restored.

Screw Compressors Equipped With Oil Injection

Suction Throttle Control.   Since the final compression temperature is governed by the injected oil, a greater range of compression ratios, such as may arise when the induced volume is throttled down, can be safely coped with. This permits the main flow volume to be varied within wide limits.

Built-in Volume Governor.   Large compressors are frequently equipped with an internal volume-regulating device. By operating a slide valve (Figure 15–8) that is shaped to match the contours of the housing and that is built into the lower part of the housing, designed to move in a direction parallel to the rotors, the effective length of the rotors can be shortened. The range of this control mode is typically between about 10 percent and 100 percent. Compared with suction throttling, this type of control offers more efficient operation, as graphically represented in Figure 15–9.

Capacity

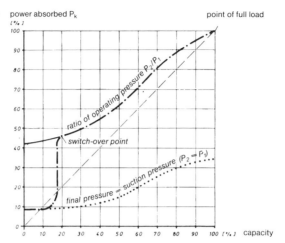

Fall-off in power absorbed P_k expressed as a percentage of the full-load performance when operating at partial capacity and with the constants:

- – – – – theoretical curve
- ———— curve obtained in actual practice
- ·—·—·—· curve obtained in actual practice by switching over from say 20 % capacity to idling speed ($p_1 = p_2 = 1$ bar)
- ········ curve obtained at idling speed ($p_1 = p_2 = 1$ bar).

FIGURE 15–9 *Capacity versus power curve pertaining to oil-injected rotary screw compressor. (Source: Aerzener Maschinenfabrik GmbH, D-3258 Aerzen, West Germany.)*

Calculation Procedures

As was discussed earlier, the spaces in which the gas is trapped and progressively advanced are those formed between the cylinder walls and the interlocking convolutions of the two helical rotors. The position of the edge of the outlet port determines the so-called "built-in volumetric ratio," v_i. The "built-in compression ratio," π_i, results from the equation $\pi_i = v_i^x$.* The compression process is shown in the theoretical p-v-diagram (Figure 15–10).

The induced flow volume may be calculated for any compression ratio, provided the data applicable to the particular compressor being considered are known. One revolution of the main helical rotor conveys the unit volume, q_0 (liter/rev). This gives us the theoretical induced flow volume, Q_0, at n revolutions:

$$Q_0 = \frac{n \cdot q_0}{1000} \, [\text{m}^3/\text{min}]$$

The actual induced flow volume, Q_1, is lower by the amount of gas, Q_v, flowing back through the very small clearances. Thus

$$Q_1 = Q_0 - Q_v \, [\text{m}^3/\text{min}]$$

---

*For a description of metric symbols used in this chapter, refer to Table 5.1 (page 420).

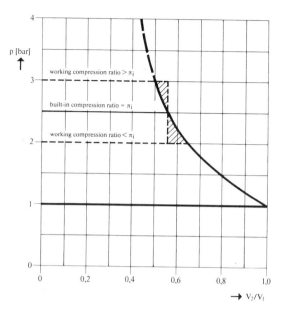

FIGURE 15–10 *Pressure-volume diagram for rotary screw compressor. (Source: Aerzener Maschinenfabrik GmbH, D-3258 Aerzen, West Germany.)*

The slip loss volume Q_v is mainly dependent on the following individual factors:

- total cross section of clearances
- density of medium handled
- compression ratio
- circumferential speed for rotor built-in volumetric ratio

Volumetric efficiency is

$$\eta_v = \frac{Q_1}{Q_0} = 1 - \frac{Q_v}{Q_0}$$

The theoretical power input required to compress the induced flow volume, Q_1, is

$$P_{th} = \frac{10^{-3}}{60} \cdot \rho_1 \cdot Q_0 \cdot h_{lad} \, [\text{kW}]$$

where $h_{lad} \left[\dfrac{J}{kg} \right]$ = the amount of energy required for the adiabatic compression of 1 kg of gas from p_1 to p_2

The theoretical power input requirement is increased by the dynamic flow loss, P_{dyn}, and by the mechanical losses, P_v. The latter consist of the losses in bearings, timing gears, and step-up gears. Thus, the power transmitted through the coupling is

$$P_k = P_{th} + P_{dyn} + P_v \, [\text{kW}]$$

Screw compressors are manufactured to American Petroleum Institute (API) 619 and/or VDI 2045 standards with a permissible tolerance of ± 5 percent in regard to power requirements and induced flow volume. These tolerances result from inaccuracies introduced during the manufacturing process. The final compression temperature is calculated for a dry-type compressor as follows:

$$t_{2\,th} = t_1 + \Delta t_{th} [^{\circ}C]$$

$$\Delta t_{th} = T_1 \left[\left(\frac{p_2}{p_1} \right)^{\frac{\chi - 1}{\chi}} - 1 \right] \frac{1}{\eta_v} [^{\circ}C]$$

When operating on the oil-free, dry-running principle, a screw compressor may attain a maximum final compression temperature of 250 °C. When air is the medium handled, this temperature (isentropic exponent $\chi = 1.4$) corresponds to a compression ratio of

$$\frac{p_2}{p_1} \approx 4.5$$

On the other hand, gases with a $\chi = 1.2$ will permit, within the temperature limits mentioned, a compression ratio as high as

$$\frac{p_2}{p_1} \approx 7$$

In the case of a screw compressor operating on the principle of oil injection, most of the drive energy applied to the machine is removed by the oil in the form of heat. The amount of oil injected is adjusted to ensure that final compression temperatures of approximately 90 °C are not exceeded. When taking in air under atmospheric pressure, compression ratios as high as $p_2/p_1 \approx 21$ are obtainable.

Typical Flow Schematic

Figure 15–11 represents a flow schematic for a two-stage rotary screw compression system.

ROTARY PISTON BLOWERS

Rotary piston, or lobe-type, blowers derive from the Roots compressor principle and have been built since 1864. They are used in a large variety of process plant applications, including pneumatic conveying of bulk materials, pressurized aeration of water at treatment plants, creation of vacuum, and gas movement in the petrochemical, pharmaceutical, and metallurgical industries. They range in size from fractional horsepower to literally hundreds of kilowatts.

Design and Construction

Rotary piston blowers are twin-shaft rotary machines. The two rotors are axially parallel to one another and located centrally inside the casing. The timing gears ensure that the rotors turn without contact. The rotors are supported on ball and roller bearings. The clearance between the rotors is kept to a minimum and selected for the expected pressure differential and thermal load under working conditions. Smaller rotors are adjusted in such a manner as to enable them to be run in either flow direction. Large blowers and gas blowers of special design are suitable for flow in one direction

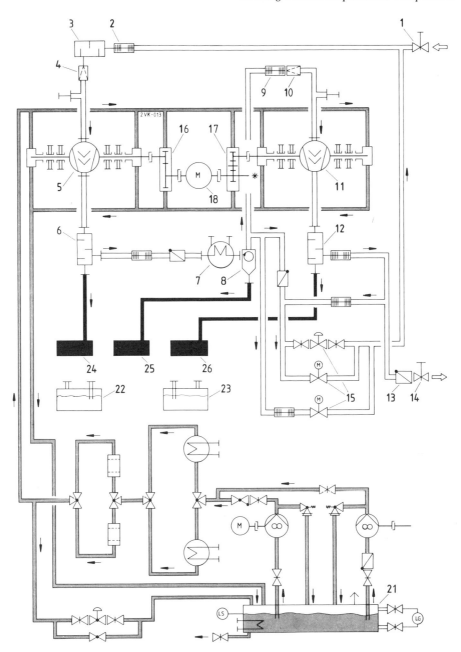

FIGURE 15–11   *Flow schematic for a two-stage rotary compression system. 1 = Slide valve; 2 = lateral compensator; 3 = intake silencer; 4 = starting strainer; 5 = screw compressor 1st stage; 6 = discharge silencer; 7 = gas cooler; 8 = separator; 9 = lateral compensator; 10 = starting strainer; 11 = screw compressor 2nd stage; 12 = discharge silencer; 13 = non-return valve; 14 = slide valve; 15 = control and shut-off devices; 16 = gear box 1st stage; 17 = gear box 2nd stage; 18 = drive motor; 19 = noise abatement hood (not shown); 20 = noise abatement hood (not shown); 21 = oil system; 22 = sealing water system; 23 = injecting water system; 24 = condensate tank 1; 25 = condensate tank 2; 26 = condensate tank 3. (Source: Aerzener Maschinenfabrik GmbH, D-3258 Aerzen, West Germany.)*

only, since in this case the clearances between the rotors and the casing must be kept larger on the low-pressure side to compensate for shaft deflection and bearing clearances. Axial thermal expansion of the rotors is compensated for by larger clearances between the rotors and the end plates at the free bearing end, i.e., at the bearing that is permitted to slide so as to accommodate thermal growth.

The shaft diameter is a very important factor in the evaluation of rotary lobe blower quality. It determines the amount of deflection and thus the magnitude of clearance and volumetric efficiency under load. Adequate bearing span provides space for proper sealing components between compression chamber and bearing housings. Needless to say, well-designed seals prevent contamination of gas by the lube oil and vice versa. This extends the life of bearings and gears.

The principal construction features of rotary piston blowers include driving and driven rotors, timing gears, bearings, and seals (Figure 15–12).

Method of Operation

Two symmetrical rotary pistons revolve in opposite directions timed to one another. The medium to be conveyed flows into the blower housing that surrounds the two rotors. From there, it is moved via positive displacement in the chambers formed between the rotors and the housing toward the discharge side. At the instant when one rotor tip passes the edge of the discharge port, the gas is compressed by the back flow from the discharge port as can be visualized from Figure 15–13. The final pressure adjusts itself to the pressure loss in the piping and in the plant equipment downstream from the blower.

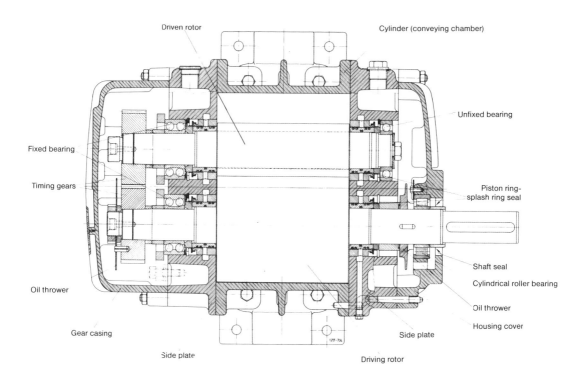

FIGURE 15–12   *Principal parts of a rotary piston blower. (Source: Aerzener Maschinenfabrik GmbH, D-3258 Aerzen, West Germany.)*

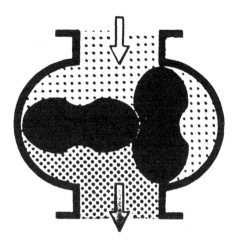

FIGURE 15-13 *Operating principle for rotary piston blower. (Source: Aerzener Maschinenfabrik GmbH, D-3258 Aerzen, West Germany.)*

The capacity of a given blower can be calculated for all types of gases and for every possible load condition. Each revolution of the rotors causes four separate volumes $q_0/4$ (liters/revolution) to be conveyed and compressed. The power transmitted through the blower coupling is

$$P = P_{th} + P_v (kW)$$

The main component, the theoretical power for compression, is thus independent of the type of gas involved and directly proportional to the working pressure differential and to the blower speed. Since no compression occurs internally, power absorbed when running under no-load conditions is nearly equal to the power loss, P_v. This will be approximately 3 to 5 percent of the power transmitted by the coupling when running on full load.

Rotary piston blowers are typically manufactured with a tolerance band of ± 5 percent, referred to the power consumption and the intake volume. These tolerances are composed of the sum of all manufacturing tolerances. This results in the theoretical capacity

$$Q_0 = \frac{n \cdot q_0}{1000} (m^3/min)$$

The actual capacity is obtained by taking the theoretical capacity and reducing it by the amount of gas, Q_v, flowing back through the clearances:

$$Q_1 = Q_o - Q_v (m^3/min)$$

The clearance losses depend on the specific density of the gas at intake, Δp, and on the total area, F, of the clearances. Volumetric efficiency is

$$\eta_v = \frac{Q_1}{Q_o} = 1 - \frac{Q_v}{Q_o}$$

Since the rotor clearances are extremely small, efficiency under working conditions is high. The capacity varies very little with changing loads (Figure 15-14). The amount of power needed to compress the capacity, Q_1, is theoretically

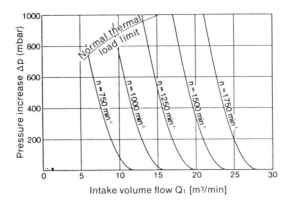

FIGURE 15–14   *Volumetric characteristics of a small rotary piston blower. (Source: Aerzener Maschinenfabrik GmbH, D-3258 Aerzen, West Germany.)*

Table 15.1   Nomenclature Typically Used for Rotating Positive Displacement Calculations

Symbol	Unit	Meaning
h	J/kg	Specific adiabatic work of compression
χ	—	Ratio of specific heats
n	min^{-1}	Speed of rotation
p_1	bar	Suction pressure
p_2	bar	Discharge pressure
p_e	bar	Compression gauge pressure
$-p_e$	bar	Negative vacuum pressure
P_{th}	kW	Theoretical power input
P_v	kW	Mechanical losses
P_{dyn}	kW	Dynamic losses
P_k	kW	Power transmitted through coupling
q_0	1/U	Unit volume
Q_0	m^3/min	Theoretical induced volume
Q_v	m^3/min	Slip loss volume
Q_1	m^3/min	Actual induced volume
t_1	°C	Inlet temperature
t_2	°C	Discharge temperature
Δt_{th}	°C	Theoretical increase in temperature
T_1	K	Absolute inlet temperature
v_i	—	Volumetric ratio
δ	kg/m^3	Density
η_v	—	Volumetric efficiency
π_i	—	Compression ratio

$$P_{th} = \frac{Q_o \cdot \Delta p}{600} \; (kW)$$

This power is in actual fact further increased by mechanical friction in bearings, timing gears, and sealing elements, and also by the dynamic losses occurring within the blower parts and in the compression chamber.

Capacity Control

The capacity of rotary piston blowers can be controlled by either varying the speed or by bypassing flow from discharge back to suction. Speed variation is, of course, the more efficient method, and turndown capacities of 70 percent can be reached in some cases. Power demand and pressure rise are almost directly proportional to blower speed.

Chapter 16

Mixers and Agitators*

Many process operations involve fluid mixing. Table 16.1 lists basic classifications, including liquid-solid mixing, gas-liquid mixing, liquid-liquid mixing, blending of miscible liquids, and fluid motion. In practice, most mixing operations involve several of these.

As shown in Table 16.1, process performance can be judged by physical uniformity, determined by taking samples and calculating the degree of uniformity produced, or by the time it takes to achieve a certain degree of uniformity.

When choosing mixing equipment, the viscosity of the fluids and mixtures is an important consideration, as is the density of materials and the resulting density of the mixture of materials involved. Mechanical mixing equipment can be categorized according to (1) the kinds of impellers used to suit specific process requirements; (2) the types of mechanical drives required to accommodate power, speed, and shaft length; (3) the necessity for sealing the tanks against high pressures imposed by certain processes. Another consideration is the requirement of stabilizing devices on the impellers as well as steady bearings in the tank. However, we will first take a look at some fluid mechanics involved in the mixing tank and some of the parameters that must be measured and observed—the mixers and agitators.

IMPELLER FLUID MECHANICS

A prime consideration is the power drawn by the mixer. In the average plant it is often necessary to measure horsepower, usually by means of electrical measurements for electric motor-driven equipment. Other types of drives are possible; these would include hydraulic motors, air motors, and steam or gas turbines, but electric motors are by far the most common. The most reliable way of measuring mixer power is by means of a recording wattmeter. Clamp-on ampmeters can be used to give an approximation, but they usually must be ratioed to full-load nameplate amperage and may not take into account voltage fluctuations that could exist in plant operation.

When taking a wattmeter reading, the electrical efficiency of the motor must be considered. Alternatively, a motor curve from the motor manufacturer would be helpful in relating amperage readings to the actual output horsepower. Motor manufacturers and mixer suppliers normally have these curves available for different horsepower ratings, different types of enclosures, and varying starting characteristics for motors. The curves would indicate average values expected for particular motor types. If a more accurate reading is desired, an individual motor curve should be considered. Most motors of 100 HP and over have had these electrical characteristics recorded, and motor manufacturers can often produce a motor curve for a particular motor on a given installation.

---

*Source: Mixing Equipment Co., Inc., Rochester, NY. Adapted by permission

Table 16.1 Mixing Processes

Physical Processing	Application Classes	Chemical Processing
Suspensions	Liquid-solid	Dissolving
Dispersions	Liquid-gas	Absorption
Emulsions	Immiscible liquids	Extraction
Blending	Miscible liquids	Reaction
Pumping	Fluid motion	Heat transfer

Source: Mixing Equipment Co., Inc., Rochester, NY.

In any event, a subtraction must now be made for mechanical efficiency in the gear reducer. Typically, gear reducer efficiencies for spiral bevel or helical gears are about two percent of rated horsepower per reduction. Another subtraction must be made if there is a stuffing box or mechanical seal involved in the equipment. This estimate usually must be obtained from mixer or seal manufacturers. The resulting quantity is the shaft horsepower, and all this horsepower is transformed into heat in the mixing vessel.

It is not recommended that no-load readings be taken as a subtractor for wattmeter amperage readings. Motor efficiencies are often very low at no-load readings, and these often give errors that are too large for accurate calculation of shaft horsepower.

All the power produced in the fluid is proportional to the pumping capacity (Q) and impeller velocity head (H). This can be expressed as $P \propto QH$. However, since the typical mixing system does not incorporate a casing around the impeller, the definition of pumping capacity must be through some arbitrarily chosen discharge area of the impeller. Considerable variation exists in practice in this regard, and a particular definition must be obtained when observing pumping capacity data from mixer manufacturers. In addition, the velocity head is not easily measured, so this too is a quantity that cannot be readily related to the shaft horsepower.

In concept, these two quantities are extremely important. The pumping capacity circulates fluid throughout the tank, and the process entrains other fluids with it; so the total flow of the tank may be anywhere from a few percent to as much as ten times higher than the actual flow from the impeller.

The velocity head is related to the fluid shear rate of the system. The fluid shear rate is a velocity gradient and is the only means by which particles get together in the mixing system. If it were not for shear rate, the particles would never meet each other; instead, each one would go around in its own velocity pattern, which is identical to the velocity pattern of every other particle.

Shear rate is an important factor in many processes, and process engineers must carefully consider the shear rate required and make sure that it is compatible with the process result desired.

Baffles

When low-viscosity media are mixed in tanks that are unbaffled, they tend to swirl and will produce a vortex, as shown in Figure 16-1. Sometimes this action is desirable, but usually it is not. Hence, four wall baffles are often used to produce the flow pattern shown in Figure 16-2. These baffles are usually 1/12 of the tank diameter in width. In calculating the strength for supporting the baffles, the mixing tank must resist all the torque applied by the mixer. Torque is the power divided by the impeller speed or consistent units

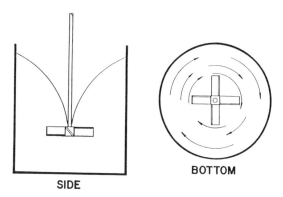

FIGURE 16–1 *Typical swirling condition of tank without baffles. (Source: Mixing Equipment Co., Inc., Rochester, NY.)*

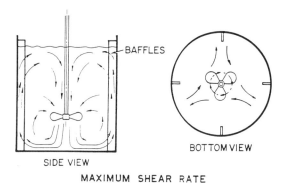

FIGURE 16–2 *Typical flow pattern with axial flow impeller tank with wall baffles. (Source: Mixing Equipment Co., Inc., Rochester, NY.)*

$$\text{Torque}_{(in-lb)} = \frac{HP(63,000)}{N(RPM)}$$

The baffles must resist this torque, allowing us to divide the torque by the number of baffles, and then by the tank radius to the center of the baffle. This gives the actual torque acting on each of the baffles. Although this torque is distributed over the length and width of the baffle, it is acceptable to assume that the torque operates in equal amounts at the location of the various impellers in the system. This concentrated load at the midpoint can be used to calculate the baffle thickness and the strength of the support arms required.

Tank Shape

Figure 16–3 depicts the nomenclature used in explaining mixer layout and design. Most tanks are cylindrical and typically have a liquid height over tank diameter (Z/T) ratio of about 1.0. On occasion, tanks have very low Z/T ratios of 0.2 to 0.4, which is typical of the large storage tanks in the petroleum industry. On the other hand, tanks can be very tall and slender. Multiple impellers must be used on large, tall tanks, while single

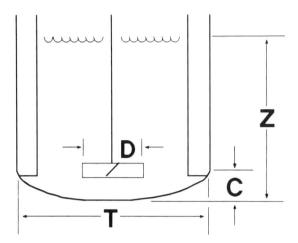

FIGURE 16-3   *Nomenclature employed in typical mixer applications. (Source: Mixing Equipment Co., Inc., Rochester, NY.)*

impellers are used on tanks with Z/T ratios of less than approximately 0.8. While the least power is usually required for tanks that have a Z/T ratio of 0.6, the specifying engineer must investigate some additional parameters in an effort to optimize the entire installation. He or she should be aware that tanks with a Z/T ratio of 0.6 are more costly to build than tanks having a Z/T ratio of 1.0, and this may offset the power savings.

Square tanks are often used, as are rectangular tanks. Up to a power level of about 1 HP/1000 gallons, these tanks are self-baffling and do not require additional baffles. Above that level, baffles are required, as shown in the drawing. Four baffles are typically used in square tanks and two baffles in rectangular tanks. There are also tanks with very complicated shapes, including elliptical heads, spherical heads, spherical tanks, and horizontal cylindrical tanks.

Impellers

There are two basic kinds of impellers, radial flow and axial flow. Figure 16-4 shows a spectrum of flow and shear rates for different impeller types. It will be noted that axial flow impellers are typically high-flow, low-shear devices, compared with the radial flow impellers. The axial flow device shown in Figure 16-5 has historically been the most common impeller used for axial flow situations. Axial flow impellers incorporate blade angles of about 45° to the horizontal, although this angle can vary between 25° and 60°. An impeller diameter over tank diameter (D/T) ratio of 0.3 to 0.5 is quite common, and these ratios are particularly applied to areas where high flow is needed, such as blending and solids suspension.

All impellers have a Reynolds number power curve, as shown in Figure 16-6. This allows the calculation of the Reynolds number when viscosity is known, the Reynolds number being the product of the impeller speed, the impeller diameter squared, and fluid density divided by the fluid viscosity. This has to be in consistent units. Having established the Reynolds number, Figure 16-6 can be used to obtain the power number, which is the power times g divided by density, speed cubed, and diameter to the fifth power. From this, the power can be calculated. These curves must be available for the particular geometry of the impeller, baffles, and tank configurations.

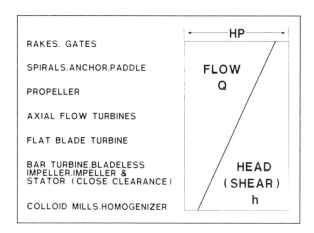

FIGURE 16–4 *Chart showing difference between flow and shear. (Source: Mixing Equipment Co., Inc., Rochester, NY.)*

FIGURE 16–5 *Typical axial flow turbine. (Source: Mixing Equipment Co., Inc., Rochester, NY.)*

The flat portion of the curve depicts the turbulent region, which is commonly for low-viscosity materials. There is also a viscous region in which the slope is minus 1, with a transition area in between.

Radial flow turbines are normally used where higher shear rates are required and where lower pumping capacity is needed. Radial flow turbines include the flat-blade turbine shown in Figure 16–7 and high-speed disc turbines for high-speed applications (Figure 16–8).

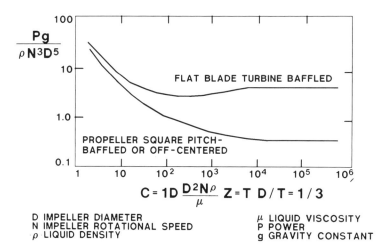

FIGURE 16–6   *Reynolds number-power number curve for two different impeller types, radial flow and axial flow. (Source: Mixing Equipment Co., Inc., Rochester, NY.)*

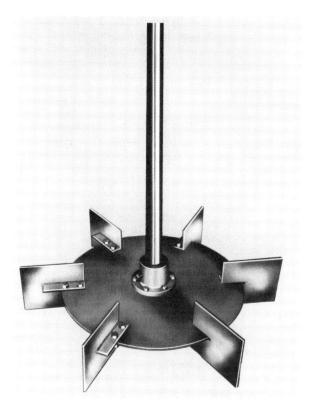

FIGURE 16–7   *Typical flat blade turbine. (Source: Mixing Equipment Co., Inc., Rochester, NY.)*

An axial flow airfoil impeller (Figure 16-9) has higher pumping capacity and lower shear rates than does the axial flow turbine and is particularly suited to blending and solids suspension. It is also particularly suited for large tanks, where often very poor blending conditions exist compared with flow patterns in small tanks in the pilot

FIGURE 16–8 *Typical high-speed disc-type radial flow high-shear impeller. (Source: Mixing Equipment Co., Inc., Rochester, NY.)*

plant. Axial flow airfoils are not particularly desirable for pilot plant operation, since blend times can appear to be adequate whereas full-scale counterparts may not be. For viscous materials, normally with Reynolds numbers of less than 10, the helical impellers shown in Figure 6–10 and anchor impellers (Figure 16–11) are typical. Both are very effective in providing visual blending throughout large-scale systems in viscous fluids and typically operate at speeds as low as 5 to 15 RPM. This is considerably lower than the speed of radial mixing impellers.

A recent innovation makes use of composite materials in which very exacting airfoil shapes can be produced. Figure 16–12 is one of these impellers that improve the pumping capacity, compared with the airfoil impeller made of metal shown in Figure 16–9 by approximately 10 to 30 percent.

Portable Mixers

Portable mixers are often provided with suitable clamping devices that permit mounting to the side of an open tank. Portable mixers range in horsepower up to about 3. Constant speed versions are usually either direct driven at 1150, 1450, or 1750 RPM or operate at either 280 or 350 RPM with a gear drive. The gear-driven portable mixer has a larger impeller and operates at slower speed than its direct-driven counterpart. It therefore develops more flow and less shear rate than the comparable direct-drive unit, which has a smaller impeller and higher speed. Using a ball and socket joint for

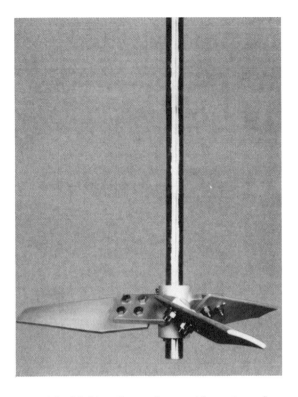

FIGURE 16–9   *Typical fluid foil impeller used to provide maximum flow and minimum shear rates. (Source: Mixing Equipment Co., Inc., Rochester, NY.)*

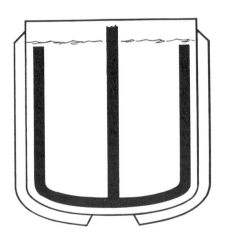

FIGURE 16–10   *Typical helical impeller for high-viscosity materials. (Source: Mixing Equipment Co., Inc., Rochester, NY.)*

FIGURE 16–11   *Anchor impeller for high-viscosity materials.*

a clamp allows the impeller to be skewed, as shown in Figure 16–13. Good flow patterns can thus be obtained without using baffles. These mixers can also be provided with a permanent angular mounting for open tanks. Portable mixers typically use the airfoil impeller (Figure 16–9), although other impellers may be used if called for by

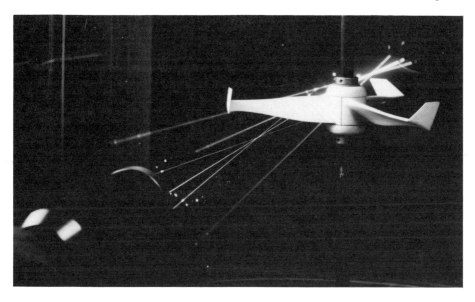

FIGURE 16–12   *Fluid foil impeller made from structural composite materials; also shown are laser measurements of flow characteristics. (Source: Mixing Equipment Co., Inc., Rochester, NY.)*

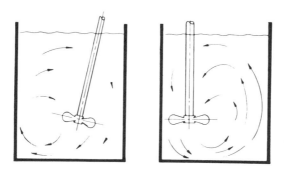

FIGURE 16–13   *Illustration of proper angle to achieve top-to-bottom turnover without wall baffles with portable mixers. (Source: Mixing Equipment Co., Inc., Rochester, NY.)*

special fluid conditions. The gear-driven portable mixer can be more expensive, since it requires a gear drive; this follows the general principle that mixers for low speed and high flow normally require higher capital outlays than mixers producing high shear and low flow. There are many other options available, such as hydraulic motors, air motors, and various kinds of variable-speed motors. Variable frequency, adjustable speed drive mechanisms represent another important option.

Heavy-Duty, Top-Entering Mixers

Industrial fluid mixers for large tanks usually incorporate a two- or three-stage gear speed reduction mechanism. These mixers go up from 2 HP to several thousand HP. The torque provided by the gear box can reach one million in-lb. These mixers are normally installed on the tank centerline. Wall baffles are typically used, and impellers may be either radial flow turbine, axial turbine, or airfoil type. Most gear drives are

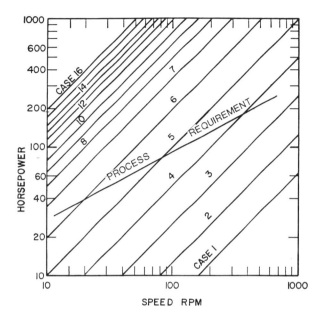

FIGURE 16–14   *Typical chart showing power versus output speed for various sizes of speed reducers. Most speed reducers are rated essentially at constant torque. (Source: Mixing Equipment Co., Inc., Rochester, NY.)*

rated at constant torque. Figure 16–14 illustrates the typical range of gear boxes; it indicates that for any given size, the mixer output is proportional to the output speed. Also shown on this curve is a typical process curve for a blending or solids suspension process. This confirms that for flow-controlled applications, there is a trade-off between operating costs and capital expenditures. For example, the choice of mixers can be between a 75-HP mixer, size five casing, or a 190-HP mixer, size four casing, which would cost less in capital dollars but would require more power. For high levels of power and high fluid shear rates, the radial flow flat blade turbine offers good mechanical stability. It has the ability to impart almost any power level that is required for the process.

The axial flow turbine, since it has a higher pumping capacity, is limited to lower power levels. These usually range from 0.5 to 5 HP/1000 gallons. Axial flow turbines are used in applications where circulating capacity is needed with a moderate amount of fluid shear rates. When the only process requirement is pumping capacity and minimum horsepower is needed, the airfoil impellers use power levels on the order of 0.1 and 2 HP/1000 gallons and are very adaptable to blending in solids suspension applications.

Typical D/T ratios for all three impeller types are usually between 0.3 and 0.5. When the D/T ratio gets much beyond 0.5, there is a tendency for additional entrainment of fluid to become very minimal. The advantage of large-diameter, slow-speed units to give more total pumping capacity is not realized.

One of the key design elements of top-entering mixers is that they must run below the first natural frequency of the mixer shaft. In designing a mixer of this type, the operating speed is selected for the particular diameter and power level desired for the process result. The operating speed is divided by the ratio of operating speed and critical speed desired, which most manufacturers set somewhere in the range of 0.7 to 0.8. The critical speed of the shaft can then be calculated.

The radius of the forces acting on the impeller must be calculated from the fluid mechanics and the fluid forces involved in these systems along with a suitable stress level for the material being used. The shaft diameter required at the maximum bending moment point is calculated. In addition, the diameter of the shaft is calculated based on torque requirements. The largest of these three diameters is the shaft required to satisfy all foreseeable conditions. As a general rule, there is some shaft diameter that will allow any length of the shaft to be run overhung with no steady bearing, operating at a suitable distance from the natural critical frequency. Using a steady bearing at the bottom of the tank allows the shaft diameter to be reduced. The cost of maintenance of the steady bearing must be balanced against the extra cost of a large diameter shaft without a steady bearing.

Light-Duty, Top-Entering Mixers

There is a gap between the range where typical portable applications stop and where the heavy-duty mixers described earlier must be used. This undefined range is usually in the vicinity of 5 to 15 HP with speeds on the order of 68 to 125 RPM. A variety of mixers are available in this area, all of which have somewhat differing basic design characteristics and which yield satisfactory performance at reasonable cost.

Side-Entering Mixers

Side-entering mixers must be properly positioned in the tank (Figure 16–15). Positioning about 10° off the tank diameter gives effective flow patterns for many types of blending applications. These mixers are quite commonly used in paperstock suspension blending, or in gasoline and crude oil tanks, as well as to prevent settling of solids in

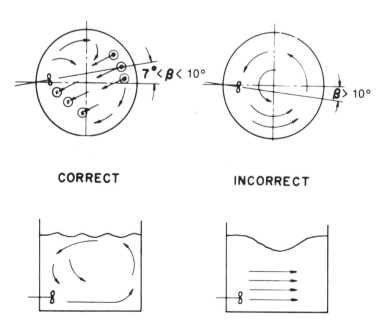

FIGURE 16–15 *Illustration of correct flow patterns for installation of side-entering mixers and large petroleum storage tanks. (Source: Mixing Equipment Co., Inc., Rochester, NY.)*

the sediment and water component of many oil pipelines. Side-entering mixers often produce quiescent zones 90° from the entry location. These zones can cause difficulty with solids settling if caution is not exercised in the process design. The major advantage of side-entering mixers is the low initial cost compared with that of top-entering mixers. Top-entering mixers must have a support structure, which adds considerably to the total cost of the installation.

Side-entering mixers will always require higher horsepower for the same application, since they run at higher speeds, typically 150 to 400 RPM. They have smaller diameter impellers than corresponding top-entering mixers and thus require more horsepower for the required circulating capacity. A relatively short shaft and cost savings on the gear box favor the economic evaluation. However, since the mixer shaft must penetrate the tank wall, a mechanical seal is needed. Materials that are very abrasive or corrosive can cause severe problems unless component configurations and materials of construction are carefully chosen and properly engineered.

Bottom-Entering Mixers

Bottom-entering mixers, as shown in Figure 16–16, are not as frequently used as top-entering mixers. They are chosen for certain reasons—one of them may be the desire to eliminate the shaft passing through the liquid surface, thus being subject to fouling due to collection of process material on the shaft. While it may be desirable to perform drive maintenance at floor level rather than on an elevated structure, it must be recognized that bottom-entering mixers require a mechanical seal at the bottom of the tank. Seal failure can cause the tank contents to discharge. Bottom-entering mixers allow the head space to be used for other types of piping connections. One disadvantage is that the high Z/T ratio is not readily handled, since the second and third impellers above the lower impeller are positioned on a shaft that is unsupported and may complicate the design.

Shaft Sealing Devices

There are two types of shaft sealing devices for mixers and agitators. One is the somewhat unsophisticated stuffing box that utilizes soft packing pressed against the shaft by an adjustable gland plate (Figure 16–17, upper half). The other sealing method employs mechanical face seals, as illustrated in the lower half of Figure 16–17.

Mechanical face seals are an indispensable component of high-pressure, toxic, flammable, or high-value mixer applications. Unlike soft packing, which must be adjusted so as to allow a small amount of leakage in order to provide a lubricating fluid film between shaft and packing, mechanical face seals can be arranged for virtually zero leakage. Moreover, mechanical seal housings can be designed to include stabilizing bearings (Figures 16–18 and 16–19) or emergency shutdown features (Figure 16–20).

Figures 16–18 and 16–19 illustrate double mechanical seals, which are typically surrounded by a buffer fluid that is maintained at about 10 to 25 psi over tank pressure. Should seal leakage occur, the buffer fluid captured between the two seals would either leak into the tank or toward the atmospheric (bearing housing) side. However, leakage rates are either extremely small and will be replenished from a makeup container, or else loss of buffer fluid can be detected by appropriate instrumentation and a safe shutdown be initiated without delay.

Figure 16–20 shows three of many shutdown or emergency seals for mixers and agitators. Upon loss of a main seal, the mixer will be automatically shut down and the emergency seal activated to contain the product in a safe fashion.

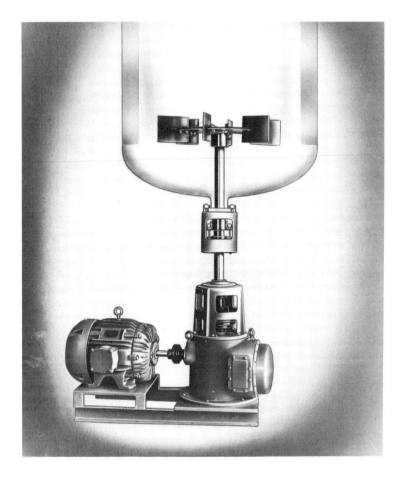

FIGURE 16–16 *Bottom-entering mixer. (Source: Mixing Equipment Co., Inc., Rochester, NY.)*

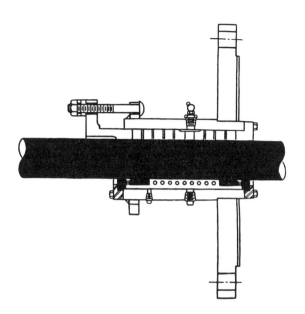

FIGURE 16–17 *Shaft sealing area for typical mixer-agitators. Soft packing is shown above center-line of shaft; a double mechanical seal is shown below the shaft centerline for comparison. (Source: Mixing Equipment Co., Inc., Rochester, NY.)*

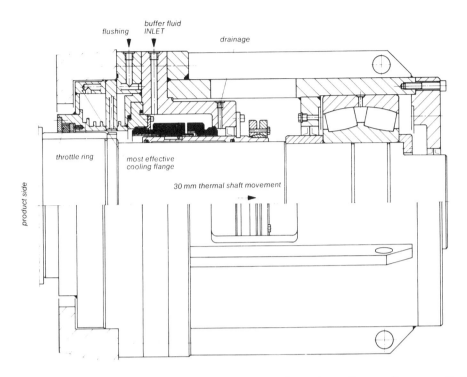

FIGURE 16–18   *Double mechanical seal for a modern mixer. (Source: Burgmann Seals America, Houston, TX.)*

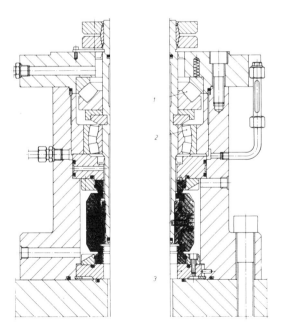

FIGURE 16–19   *Double mechanical seal for vertical mixer. Radial and thrust bearings are incorporated in the seal housing. (Source: Burgmann Seals America, Houston, TX.)*

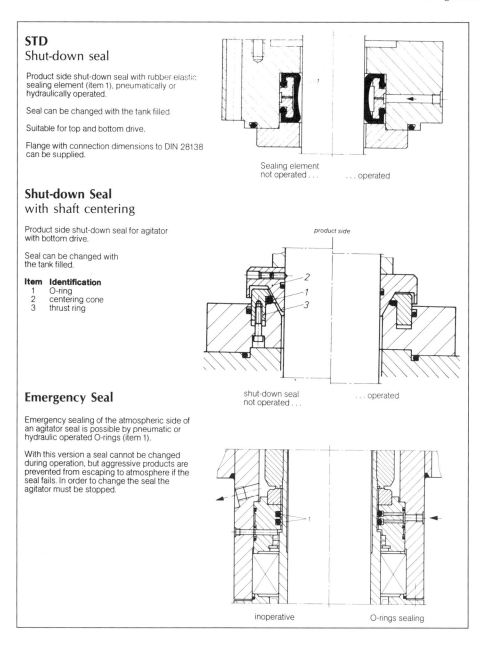

STD
Shut-down seal

Product side shut-down seal with rubber elastic sealing element (item 1), pneumatically or hydraulically operated.

Seal can be changed with the tank filled.

Suitable for top and bottom drive.

Flange with connection dimensions to DIN 28138 can be supplied.

Sealing element not operated . . .   . . . operated

Shut-down Seal
with shaft centering

Product side shut-down seal for agitator with bottom drive.

Seal can be changed with the tank filled.

Item	Identification
1	O-ring
2	centering cone
3	thrust ring

product side

shut-down seal not operated . . .   . . . operated

Emergency Seal

Emergency sealing of the atmospheric side of an agitator seal is possible by pneumatic or hydraulic operated O-rings (item 1).

With this version a seal cannot be changed during operation, but aggressive products are prevented from escaping to atmosphere if the seal fails. In order to change the seal the agitator must be stopped.

inoperative   O-rings sealing

FIGURE 16–20 *Shutdown or emergency seals for mixers and agitators. (Source: Burgmann Seals America, Houston, TX.)*

It is almost always a serious mistake to purchase the least expensive seal for a mixer or agitator application, as the resulting frequent seal failures will burden the user with high maintenance expenses. On the other hand, properly designed mechanical seals have extended the run time of modern mixers by orders of magnitude. The downtime and repair cost avoidance achieved with superior mechanical seals will often pay for the seals in the first few weeks of operation.

Chapter 17

Internal Mixers: Single- and Twin-Screw Extruders*

INTERNAL MIXERS

Internal (Banbury[†]-type) mixers have found widespread application in the rubber industry, where high-horsepower mixing is required for masticating and compounding; the technology has been successfully applied to plastics and chemicals for high- and low-viscosity systems:

Rubber Applications	Plastic Compounds	Other Applications
Tires, tubes	Polyvinylchloride (PVC)s (flexible, semirigid)	Adhesives, sealants
Molded articles, profiles		
Hoses, gaskets, seals	PVC scrap reclaim	Carbon electrodes
Shoe soles, heels	ABS (molding)	Ceramics
Foam, sponge rubber	Polyethylene, polypropylene, PVC, ABS color concentrates	Chewing gum
Packing, sealing, roofing	Phenolic	Dewatering, devolatilizing
Flooring (sheet)	Thermoplastic/rubber blends	Pharmaceuticals

The common goal is to mix solid and/or liquid additives into a rubber or plastic-type matrix. Additives (agglomerated particles or droplets bound by surface tension) must be separated, reduced in size, and uniformly distributed within the matrix. Two types of mixing phenomena are involved: extensive mixing and intensive mixing. Extensive (also known as distributive) mixing is responsible for spatial distribution of the individual particles within the polymeric matrix. Intensive (also known as dispersive) mixing is responsible for separating and reducing the particle size of the additives.

Principle of Operation

Internal mixers are designed to provide intensive and extensive mixing. Intensive mixing occurs in the narrow gap formed between the rotor tip and the mixing chamber

---

*Source: Werner & Pfleiderer Corporation, Ramsey, NJ. Adapted with permission.

†Banbury is a registered trademark identifying internal batch mixers made by Farrel Company and its predecessor companies since 1916.

wall. The mixture is repeatedly passed through this high shear field where fluid mechanical stresses separate and rupture agglomerates; dispersive forces are similar to those in a two-roll mill. Extensive mixing takes place between the rotors: the mixture is circulated from side to side and from one end of the mixer to the other after passing through the shear zone.

Energy dissipation from intensive mixing results in heating of the mixture. This heat is removed through the walls of the mixing chamber, the rotor bodies, and other contact parts (ram, discharge door, etc.) through cooling channels. The effective heat transfer of internal mixers can be the limiting factor to intensive work, since discharge temperatures cannot exceed the critical temperature of the mixture, which is determined, for example, by thermal breakdown of organic phase, onset of undesirable reactions (e.g., cross-linking), or the decrease of continuous-phase viscosity to a point where dispersion cannot proceed. Poor heat transfer will cause these temperature limits to be reached before dispersion is complete.

Design Features of Internal Mixers

Optimum dispersion is achieved through proper selection of machine type and process parameters. Internal mixers are available today with intermeshing and tangential (nonintermeshing) rotors. Rotor design is a critical factor in mixer performance, and some manufacturers offer various rotor configurations. (This topic is covered in detail later). Figure 17–1 shows the major components of internal mixers. A completely enclosed mixing chamber houses the spiral-shaped rotors, which rotate in opposite directions and at the same speed (intermeshing design) or at different speeds (tangential design) to keep the material circulating. The gap between the rotor tips and the chamber wall produces intensive shearing of the mixture. A hopper allows for loading ingredients, and an air-operated ram in the feeding neck confines the batch within the mixing chamber. A discharge door allows for quick and efficient unloading of the mixture at the end of the mixing cycle. The mixing chamber, rotors, and discharge door are all temperature controlled with steam and/or water.

Mixing Chamber

The mixing chamber body is usually of two-piece construction, split vertically. This allows for removal on-site without dismantling the entire mixer. The internal body halves are lined with wear-resistant materials to maintain rotor tip-to-wall clearances; the exterior is heavily reinforced for mechanical loading.

Temperature control of the mixing chamber is accomplished with steam or circulating water. The body halves are fitted with one of three types of chamber sides: cored, spray-type, or drilled.

Cored sides have passages arranged in a serpentine pattern running along the length of the sides of the chamber. These passages are usually formed in the casting, through which water or steam is circulated.

Spray-type sides use nozzles to spray water onto the sides for cooling effect; flood-type design is also available, without nozzles.

Drilled sides are the most common type of temperature control for the mixing chamber. Holes are drilled laterally in the body halves and provide a serpentine path for the flow of water. These holes are smaller and greater in number than the cored passages, as well as being closer in proximity to the chamber wall.

End Frames

The ends of the mixing chamber provide support for the mixer body. End frames carry the rotor bearing assemblies and dust-stop seals.

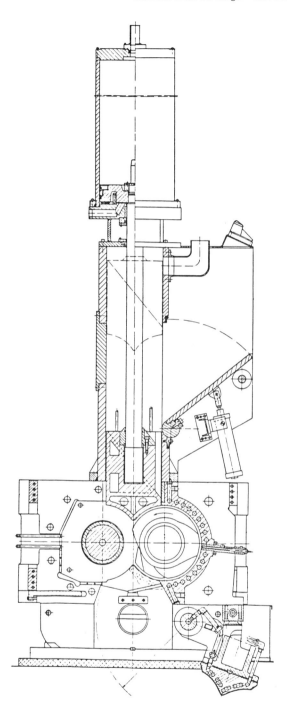

FIGURE 17–1 *Cross-sectional view of internal mixer with intermeshing rotors. (Source: Werner & Pfleiderer Corporation, Ramsey, NJ.)*

Self-aligning roller bearings are standard on most internal mixers. Double-acting axial thrust bearings can be used on large internal mixers to increase lifetime and efficiency of the rotor seals.

Dust stops are used to seal dust (carbon black, pigments, etc., used in the mix) within the mixing chamber. There are several designs available; one type relies on a hydraulically actuated yoke to apply pressure on the sealing rings. Self-sealing dust stops use the mechanical pressure of the mixture to maintain seals. Powders or dust that pass into the dust stops are flushed with process oil.

Discharge Doors

The discharge door of internal mixers is designed to provide quick and efficient dumping of the mixture. The door top must be hard-surfaced to withstand the mixing environment; some designs include a removable/replacement door top section. Most discharge doors are provided with passages for heating or cooling media; the door is usually cooled to prevent the mixture from sticking. Two types are available: drop door and sliding door.

The drop door type is pivoted on a shaft running through bearings located in the end frames. The door swings downward (135° to 180°) and away from the mixing chamber, providing a clear path for the material discharge.

The sliding door is mounted on an air-operated cylinder. Guides are necessary to provide clearances for door operation. Also available are mixing chambers that tilt over (up to 140°) to discharge the mixture through the top opening. Tilting mechanisms are electrically or hydraulically driven.

Feed Hopper

The feeding of materials into an internal mixer can actually take longer than the mixing cycle. Thus, efficient designs are utilized on the hopper assembly of internal mixers to facilitate quick charging of ingredients. The feed opening itself must be large to allow venting of air (and dust) as the batch is charged. The feed opening and throat on larger mixers are sized to accommodate rubber bales intact. Air-operated doors are provided in the hopper assembly to allow for loading of material into the mixer. Openings on the sides or in the rear of the hopper for charging of fillers, accelerators, curatives, and other dry powder components are normally connected to weigh hoppers or other means of feeding.

Ram Cylinder Assembly

A ram is used to confine the batch within the mixing chamber. Air pressure (10 to 120 psi) is applied to the cylinder, forcing the ram down into the mixing chamber. The bottom of the ram is usually shaped to conform to the gap between the rotors (V-bottom). The ram is fitted with a height indicator, used to gauge the state of the mix. Cavity cooling is used for temperature control of the ram bottom. The ram cylinder assembly is usually air-operated, but it can also be a hydraulic device. Ram pressure can have a significant impact on dispersion quality, and it is sometimes used as an operating parameter.

Drive Train

There are three mixer drive arrangements available: "standard," semi-unidrive, and unidrive. Standard drive trains use reduction gears mounted on the mixer base to drive the rotors; one rotor shaft is longer than the other, functioning as pinion shaft and reduction gear. Semi-unidrive systems use a separate reduction gearbox prior to the rotor shaft. One rotor shaft is longer than the other, carrying the pinion gear. Unidrive systems have speed reduction and dual output shafts within a single gearbox. Rotor shafts of equal length are coupled to the pinion shafts from the gearbox.

Rotor speed directly influences mixing quality, mixing time, and batch temperature. Optimum rotor speeds are chosen to process materials at their highest acceptable temperature within the shortest cycle time. A single-speed drive limits the number of formulations that can be processed optimally. Two-speed motors increase internal mixer flexibility. Several other alternatives are also available:

- constant speed motor with integral (or separate) two- or four-speed gearbox
- variable-speed DC or variable frequency AC motor (ultimate flexibility)

Bed Plate

The bed plate is the base frame that anchors the mixing chamber (and possibly gearbox). It is strengthened to withstand torque transfer and vibration of the rotors, as well as to evenly distribute the load of the mixer onto a foundation.

Auxiliary Systems

Operation of internal mixers requires several dedicated systems.

The *tempering system* supplies constant-temperature fluid to heat or cool the various mixer components. Steam/water, pressurized water, or heat-transfer oils can be used. Several separate circuits may be needed for efficient operation (e.g., ram and mixing chamber at one temperature, discharge door at another temperature, and rotors at yet a third temperature).

The *lubricating oil system* ensures adequate supply of lube oil to critical components in the drive train, rotor bearings, dust stops, etc.

The *process oil injection system* injects oil into the mixing chamber. The injection nozzle should be a self-sealing type to prevent fouling from mixture.

The *hydraulic or pneumatic system* operates discharge door (or tilting mechanism), ram cylinder, feed chute door, etc. It is controlled by means of solenoids.

The *temperature sensor,* strategically located, indicates batch temperature. It may be located in the feeding ram, in the discharge door, or through the end frames. Sensors should have intimate contact with the mixture to provide accurate readings.

Instrumentation and Controls

The degree of sophistication of internal mixer control systems can vary. They can

1. mix batch until desired temperature is reached
2. mix batch for predetermined time period
3. mix batch until predetermined energy is consumed
4. use various combinations of the above

Efficient operation of internal mixers is attained with automation of the mixing process. Increased productivity and consistency in product quality can be realized with computerized control of the mixing line from batch weighing through downstream processing.

Manual Control.   Standard internal mixer control systems provide interlocks and data acquisition for manual operation. Interlocks are installed on ram cylinders, feed hoppers, and discharge doors for operator safety and on drive components for overload protection. Operating data are usually recorded on strip-charts housed in a control panel for production monitoring of discharge temperature, power consumption, and cycle times. Stop/start push buttons are provided for drive motor and auxiliary equipment.

Automated Mixer Control Systems.   Application of a process computer system to an internal mixer significantly improves batch quality as well as quality consistency from batch to batch. Flow of raw materials, mixing control, downstream equipment, and production planning can all be integrated into a supervisory computer system.

Optimum mixer control is achieved with logic controls and/or combinations of mixing time, energy input, rotor speed, stock temperature, and chemical reaction parameters. Adaptive process control systems allow the mixing process to follow predefined energy and temperature curves that are stored in memory with each formulation.

Rotors

Mixing is achieved in internal mixers with the rotors, rotating toward each other at the same speed (intermeshing design) or at different speeds (tangential design). The rotors are designed to interact with each other via the rotor blades, called wings. Short wings and long wings are used in combination on each rotor. Mixers are available with either two-wing or four-wing rotors. Two-wing rotors have one short and one long wing, while four-wing rotors have two short and two long wings. The rotors are arranged in the mixer such that the long wing of one rotor interacts with the short wing of the other rotor (Figure 17–2). The edge of the blade is called the wing tip, which forms the shearing gap between the rotor blade and the chamber wall.

The rotors are temperature controlled with steam and/or water, which flows through the center of each shaft. Spray-type cooling or forced circulation is available.

Rotors and shafts can be manufactured as a one-piece steel casting or as two pieces: forged steel rotor shafts with rotor bodies shrunk onto the shafts. The entire rotor bodies are hard-surfaced for wear protection or are chrome plated with hardfacing only on the tips.

Rotor Design

Intensive mixing occurs where the material is compressed between the rotor wing tip and the mixing chamber wall. The width of the tip, the clearance between the wing tip and the chamber wall, and the leading/trailing angle affect dispersion. Wing length and helix angle influence the distribute mixing from rotor to rotor and from one end of the mixer to the other.

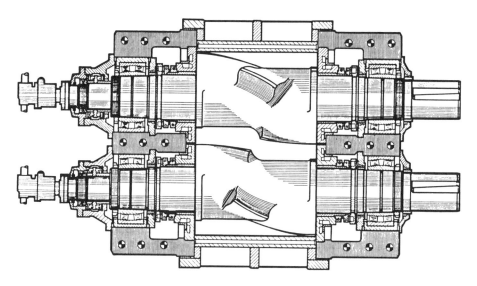

FIGURE 17–2   *Plan view of tangential four-wing rotors, rotor bearings, and dust stops. (Source: Werner & Pfleiderer Corporation, Ramsey, NJ.)*

Tangential Rotors

Conventional internal mixer designs use tangential rotors with two or four wings. The rotor diameter is equivalent to the center distance between the rotors. Tangential rotor mixers are mainly used for large-volume mixing, as in the tire industry.

Two-wing rotors are characterized by two flow regions due to different radial clearances between the rotors. Mixing intensity is directly influenced by ram pressure with a two-wing rotor system.

Four-wing rotors provide constant radial clearances between the rotors. Total wing length is greater than the two-wing system, producing higher specific energy input and better homogeneity.

The torque capacity of internal mixers supplied with four-wing rotors is approximately 30 percent greater than with the two-wing rotor system. Modified four-wing rotor designs have been developed with enhanced longitudinal (distributive) mixing.

Intermeshing Rotors

Manufacturers of internal mixers have developed intermeshing rotor systems in response to quality and productivity requirements of compounders. The diameter of intermeshing rotors is greater than the center distance between the rotors (Figure 17-3). A calendar effect is created by the intermeshing geometry of the wings, resulting in improved dispersion. Intermeshing rotor systems are capable of higher rates of energy input and better heat transfer than tangential designs. The number of mixing steps, as well as mixing time for each step, may be reduced by changing from tangential to intermeshing rotors. Internal mixers with intermeshing rotors are used mainly for high-quality mixing.

Operation of Internal Mixers

Several mixing techniques are commonly practiced using internal mixers: single-stage, masterbatch, and multistage mixing.

Single-stage mixing is used for materials that can tolerate relatively long mixing time at low rotor speed; temperature rise is the limiting factor. All ingredients can be charged at the same time or added sequentially while mixing. The sequence of addition of plasticizers and oils is critical to dispersion quality.

Masterbatching is an implied two-stage mixing process. Viscous components and fillers are mixed first. This stage can tolerate higher rotor speed and material temperature. Thus, short mix cycles produce complete dispersion. The batch is discharged and allowed to cool. A fraction of the first stage (masterbatch) is then loaded with the balance of ingredients that make up the total formulation. The predispersed masterbatch mixes efficiently and can produce higher quality dispersions in less total mixing time than a comparable single-stage mixing process.

Multistage mixing can include several masterbatching steps, a remilling stage to disperse the masterbatch, and final mixing.

Operating Parameters

Optimization of internal mixers requires knowledge of how operational parameters affect mixing quality. Generalized statements can be made as to the influence of fill factor, mixing sequence, ram pressure, rotor speed, mixing time, and temperature control.

Fill factor is an indication of the working volume of the mixer. Chamber volume is used to specify internal mixing capacity (in liters or cubic inches); empty volume is

FIGURE 17–3   *Tangential (top) and intermeshing (bottom) rotor geometries. (Source: Werner & Pfleiderer Corporation, Ramsey, NJ.)*

measured with the rotors installed. The working volume is the space occupied by the mixture (function of batch weight, specific gravity). The ratio of working to empty volume gives the fill factor.

Internal mixers are available with chamber volumes up to 650 liters (l), which can handle a 560-kilogram (kg) batch; laboratory-scale internal mixers used for research and development have net volumes as low as 0.4 l, requiring less than 0.3 kg per batch.

Fill factors vary for different mixer geometries (two-wing versus four-wing; intermeshing versus tangential) and different mixing tasks. Intermeshing rotor mixers generally use lower fill factors than tangential rotor systems, but they can achieve the same level of dispersion in a shorter mixing time. A compound that runs on a tangential mixer at 70 to 75 percent fill factor (four-wing rotor) would run between 60 and 65 percent fill factor on an intermeshing mixer.

Fill factor directly influences dispersion quality, specific energy input, and discharge temperature by providing empty volume for the mixture to circulate within.

Mixing sequence has a significant effect on dispersion quality, specific energy input, and discharge temperature when large amounts of oil are being processed. Oil or plasticizers are typically added later in the mixing cycle, after fillers have been dispersed.

Ram pressure is used as a process variable to influence the specific energy input and discharge temperature. Tangential rotor mixers are more sensitive to ram pressure than intermeshing rotor mixers. Increased ram pressure can shorten the mixing cycle by compressing the batch within the chamber to intensify mixing. Discharge temperature can be reduced by decreasing ram pressure.

Rotor speed is used to control the rate of specific energy input. Higher rotor speeds can reduce mixing time by dissipating more energy in a shorter time period. Rotor speed and ram pressure can be independently varied to achieve a target energy input, dispersion quality, or material temperature. Internal mixers supplied with a variable-speed motor have more flexibility in optimizing rotor speed for a given formulation than a single- or two-speed mixer. Large internal mixers operate at rotor speeds of 10 to 60 RPM; laboratory-scale internal mixers operate at speeds up to 110 RPM or higher.

Mixing time in conjunction with fill factor (batch weight) determines the throughput capacity of the mixer. Decreasing the mixing time (by increasing rotor speed or ram pressure) then increases mixer output. Mixing times vary widely depending on formulation and quality of mix. For example, single-stage mixing of one particular rubber formulation and oil, which takes eight minutes on a tangential mixer, can be processed in only five minutes on an intermeshing mixer. A typical tire formulation running on a tangential rotor system takes about three minutes for the masterbatch stage, three minutes for the remilling stage, and two minutes for final mixing.

Temperature control of mixer components (chamber body, rotors, ram, and discharge door) has several effects: adherance of material to rotors (rotor temperature), discharge temperature (cooling effect through chamber body), and efficiency of discharge (discharge door temperature). The mixer is usually started up hot to prevent slippage, and then cooling is applied when the material is in a fluxed state. Accurate temperature control of each part of the internal mixer helps maintain batch-to-batch uniformity.

Downstream Equipment

An internal mixer usually discharges directly into some type of shaping/forming equipment or onto conveyors. Downstream equipment is sized to provide a continuous process from the batch mixer; two mixers can also be arranged with alternating discharge.

Mixing Mill

Roll mills are used for cooling and shaping as well as for after-homogenizing and mixing. Cross-linking chemicals that cannot be added in the internal mixer due to temperature limitations in single-stage mixing can be added in the mixing mill. Mixing mills are built with fixed friction ratios (constant speed on rolls), or variable friction can be achieved with variable-speed control of each roll. Adjustment of the roll gap can be carried out under load with hydraulics.

Extruder

Single-screw extruders are used to produce pellets or sheet from the internal mixer. Pelletizing extruders are fitted with screens or strainers and die-face pelletizers or underwater pelletizers. Force-feeding devices (screws or rams) can also be installed. Sheet stock is produced from extruders equipped with a roller-die.

Maintenance of Internal Mixers

Routine maintenance of internal mixers requires attention to the various subsystems responsible for smooth operation (e.g., hydraulic, pneumatic, and temperature-control systems). Mixer manufacturers provide recommendations for preventive maintenance on rotor bearings and seals, lubricating oil changes, gearbox overhauls, etc. High on-stream factors are achieved when maintenance records are kept and factory recommendations are followed.

Major overhauls of internal mixers are required when the product specifications (dispersion quality, discharge temperature, etc.) are no longer acceptable. Abrasive wear on rotors, chamber body, and ram becomes evident in mixer performance: dispersion quality cannot be maintained as clearances increase between rotor tip and chamber wall; discharge temperature increases as a result of increased clearances and subsequent reduction of heat transfer. Manufacturers of internal mixers can provide a field inspection service to periodically document wear. Critical components (rotors, bearings, dust stops, etc.) are usually kept in stock for emergency delivery.

Rebuilding of internal mixers is a service performed by the manufacturers. Worn rotors and chamber bodies can be rebuilt to factory specifications. New dust-stop parts, door tops, and end frames are installed. Older mixer designs can be converted from cored or spray-type sides to drilled sides, from two-wing rotors to four-wing rotors, from standard gear to unidrive, etc.

SINGLE- AND TWIN-SCREW EXTRUDERS

Screw extruders were designed as continuous mixers for dispersing additives into a molten polymer. Intensive or extensive mixing takes place in the extruder as a function of screw geometry and operating conditions. Mixing time in continuous mixers cannot be arbitrarily chosen as in batch mixers, but is determined by operating conditions, screw geometry, and extruder length (residence time).

Evaluation of single- and twin-screw extruder designs for various process tasks requires a basic knowledge of key components: gearbox, drive motor, screw shafts, screws, and barrel (Figure 17–4).

Mechanical Description

Gearbox

The gearbox performs speed reduction, torque transfer, and thrust loading for the extruder. These functions can be handled separately with individual components or

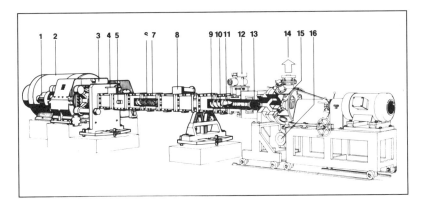

FIGURE 17–4 *Extruder components. 1 = Thrust bearing assembly; 2 = motor with overload safety clutch; 3 = gear box; 4 = feed barrel; 5 = electrical resistance heater shells; 6 = kneading elements; 7 = thermocouple for stock temperature; 8 = venting section; 9 = screw elements; 10 = stock pressure gauge; 11 = thermocouple for barrel temperature; 12 = start-up valve; 13 = screen pack changer; 14 = pellet/water discharge; 15 = UG (underground pelletizer); 16 = water inlet. (Source: Werner & Pfleiderer Corporation, Ramsey, NJ.)*

combined in a single unit. Drive motor speed is reduced to operable extruder screw speed in the gearbox. Various gear ratios are available to provide different output speed ranges from standard motor input speeds. Torque is transferred from the drive motor to a single output shaft (single-screw) or must be equally split for dual output shafts (twin-screw). Axial thrust loading (back-pressure) from the screw shaft(s) is taken up by thrust bearings.

Single-screw extruders are able to utilize large thrust bearings. Twin-screw extruders, however, must use alternative methods, since the close proximity of the screw shafts precludes the use of large bearings. Twin-screw extruder gearbox designs are inherently more complex than single-screw designs.

Drive Motor

Small single-screw (<8 in) and twin-screw (<130 mm) extruders use constant-torque variable-speed DC drive motors to provide process flexibility. Extruder systems can also use constant speed AC motors with mechanical variable-speed gearboxes, variable-frequency AC motors, or even hydraulic drives to achieve the same flexibility. Very large extruders are usually equipped with fixed- or two-speed drives, with an auxiliary drive for start up.

Drive motors can be direct-coupled to the gearbox or belt-driven. Mechanical slip clutches, positive disengagement couplings, or shear-pin devices are installed to provide overtorque protection.

Typical installed power for twin-screw extruders is shown in Figure 17–5.

Screw Shafts

The screw shaft(s) transfer the motor torque to the screw(s). Splines, keys, and other methods of torque transfer are used to mount the screws to the shafts. One end of the screw shaft is coupled to the output shaft of the gearbox; on some extruders the screw shaft is an extension of the pinion shaft and is not removable. The screws are fixed to the screw shaft with a screw tip that compresses the screw bushings together on the

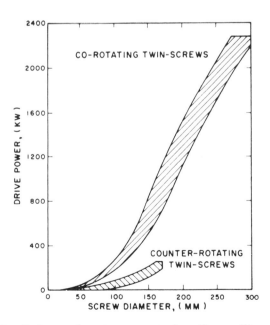

FIGURE 17–5 *Installed power for twin-screw extruders. (Source: Werner & Pfleiderer Corporation, Ramsey, NJ.)*

shaft. Screw shafts can be cored for flow of heat transfer fluids that can enter through the gearbox end or the discharge end.

Extruder Screws

The extruder screw(s) transfer the motor power into the material via viscous dissipation. Single screws are built as one piece or may be assembled from modular sections. Some twin screws are manufactured as one piece, but most are of modular design. The discrete screw elements offer two significant advantages:

1. Process flexibility to alter the screw geometry, thereby influencing shear, mixing, and residence time distribution.
2. As screw elements wear, only those pieces that are subject to abrasion need to be replaced, not the whole screw. Special materials of construction can be used for those particular screw pieces subject to abrasion or corrosion instead of for the entire screw.

Single screws typically have one or two flights; twin screws can have one, two, three, or more flights together on the same shaft (Figures 17–6 and 17–7). Screw elements can have various pitch angles, lengths, and number of flights, and can be forward or reverse conveying.

Co-rotating twin-screw extruders additionally use staggered screw discs (called kneading discs, mixing paddles, etc.) to provide additional mixing. The screw discs have various widths and stagger angles, as well as being forward or reverse conveying to tailor the amount and type of mixing required.

Extruder Barrel

As with screws, barrels can be built as one piece or of modular construction. The modular design provides the same advantages as for screws. Barrel sections are open for feeding and venting or closed for conveying and pressure build-up. The extruder barrel can have several openings for venting or feeding solids downstream. Access

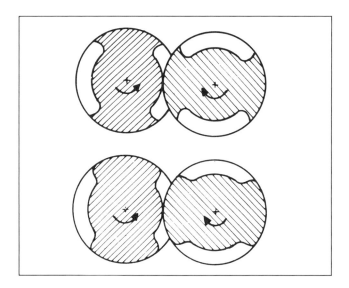

FIGURE 17–6   *Cross section of counter-rotating screws. (Source: Werner & Pfleiderer Corporation, Ramsey, NJ.)*

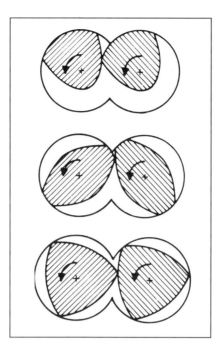

FIGURE 17–7   *Cross section of co-rotating screws: single-flight (top), two-flight (center), three-flight (bottom). (Source: Werner & Pfleiderer Corporation, Ramsey, NJ.)*

ports are provided to inject liquids or for installing melt thermocouples or pressure transducers.

Temperature control of the barrel can be accomplished in several ways: electric heating (with water or air cooling), steam heating, or fluid heating/cooling. Barrels either have drilled passages or are equipped with external jackets for flow of heating/cooling fluids. Maximum operating temperatures are usually limited by the materials of construction of the extruder barrel, screws, etc. Electric heating is provided with bolt-on resistance or induction heaters. Water or air is used to trim temperature. Several temperature zones can be installed, each with a dedicated controller. Steam heating requires mixing valves to control temperatures or flow control valves to control flow rate. Pressure ratings of barrel jackets or cored passages may limit steam pressure for high-temperature applications. Fluid heating/cooling provides very stable temperature control; commercial heat-transfer oils operate up to 650 ° F, while glycol solutions may be used for intensive cooling.

Extruder barrels are designed to provide access to the screws for cleaning, maintenance, etc. Clamshell designs are available, which open horizontally (top and bottom) or vertically (side-to-side). In other cases, the screw is pulled from the barrel, or the barrel housing can be pulled, leaving the screw attached to the gearbox.

Barrels can be manufactured with a replaceable liner for wear- or corrosion-resistance. The liners are press-fit or a split-barrel designed to bolt around the liner. Only those barrel sections where wear or corrosion is anticipated need to be built of special materials or replaced when worn.

Miscellaneous

A feeding device, located near the feed opening of the extruder, is a starting point common to single- and twin-screw extruders. Its purpose is to provide a continuous source of raw materials from a storage bin or mixer/surge system to the extruder.

Volumetric or gravimetric metering equipment is used to regulate throughput with twin-screw extruders; a simple gravity bin with bottom discharge can be used with single-screw extruders, since the extruder operates at 100 percent degree-of-fill in the feed section (flood feeding). Starve-fed extruders are very sensitive to feeding; disturbances or fluctuations may not be dampened out within the extruder. Separate feeding of several components to a twin-screw compounding extruder requires a gravimetric feeding system for accurate metering.

Force-feeding devices are used in conjunction with single- and twin-screw extruders for improved feeding of low-bulk-density materials. These are available in single- or twin-screw configurations for side feeding or vertical feeding (Figure 17–8).

Shear control mechanisms are designed to alter the shear intensity of the screw(s) while the machine is running. Several designs are available; most operate on the principle of a variable orifice restricting flow in the screw channels or at the screw tips.

Downstream equipment is used in conjunction with single- and twin-screw extruders to form, pelletize, and filter.

Extrusion dies are available to produce strands, sheet, film, fibers, and profiles. Heat- and shear-sensitive materials can be processed in a two-stage extrusion process: a second extruder or gear pump is used to pressurize the die, relieving the compounding extruder of this task.

Pelletizing equipment depends largely on the properties of the material. Molten strands can be pulled through a water bath (if the material has enough strength) and fed into a strand cutter. Underwater pelletizers are used if the material tends to smear when cut in air. Water-ring pelletizers and other types of die-face cutting are possible if the material can be cut in air and cooled in an air or water conveying system.

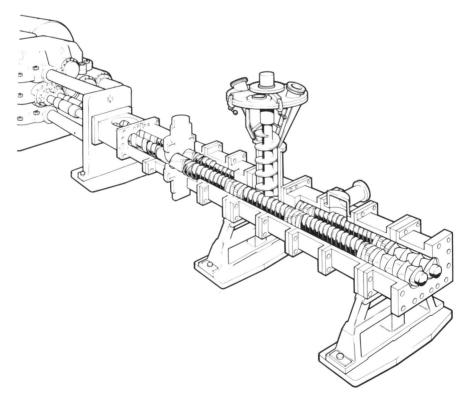

FIGURE 17–8   *Downstream feeding of abrasive resin extenders and additives reduces extruder wear and maintenance costs. (Source: Werner & Pfleiderer Corporation, Ramsey, NJ.)*

Screens are used to filter molten polymers for film and fiber extrusion. Fixed screens, screen-pack changers, or deep filter elements are installed between the extruder and the die assembly.

Process Task

In most thermoplastic extrusion operations, the basic function of the extruder is to accept a product, melt it through a combination of mechanical and thermal energy, and generate sufficient pressure to pass the molten polymer through a die. Sometimes the product is fed to the extruder in the form of a melt. In these cases, the task of the extruder is limited to providing additional energy input for homogenizing the melt with other ingredients and then to generate sufficient pressure to pass the molten material through a die. Other process tasks include

- devolatilization of residual monomers, solvents, water, etc.
- chemical reactions such as polymerization, polycondensation, polyaddition
- polymer alloying and blending
- incorporation of fibers (glass, carbon, etc.) for reinforcing

Screw extruders are applied in many areas outside of the plastics industry, for example:

- food processing (extrusion cooking of starch-based raw materials)
- energetic materials (mixing and forming of propellants)
- adhesives and sealants (compounding and feeding of coating equipment)
- chemicals and pharmaceuticals (reactive processing)
- hazardous waste treatment (encapsulating, solidification)

Design Feature—Single-Screw Extruders

A helical screw that rotates inside a cylindrical barrel housing can be referred to as a single-screw extruder. The screw consists of three sections: feeding, compression, and metering.

The feed section consists of deep screw flights designed to accommodate feed intake of low-bulk-density powders or pellets. Single-screw extruders usually operate with the feed section filled 100 percent (flood feeding); less than 100 percent fill in the feed section is called starve-feeding.

The compression section (sometimes referred to as the transition zone) is designed to input mechanical energy for melting, cooking, etc. *Compression ratio* is the term used to describe the ratio of flight depth or pitch in the feed zone to flight depth or pitch in the metering zone; typical compression ratios are from 1:1 to 5:1. There are two basic design approaches to achieve compression:

1. increase root diameter (decreasing flight depth)
2. decrease pitch, constant root diameter (constant flight depth)

The metering section is located before the die restriction. Most of the mechanical energy dissipation occurs in the metering section, producing the highest temperature in the extruder, high shear rates, and mixing. Enhanced mixing is achieved in the metering section with interrupted flights, mixing pins, screw barriers, etc.

The barrel housing may be grooved on the inner surface to improve frictional characteristics in the feed zone. Small grooves are cut into the barrel wall, either straight (axially down the barrel length) or spiral (helical grooves cut opposite to the conveying direction of the screw).

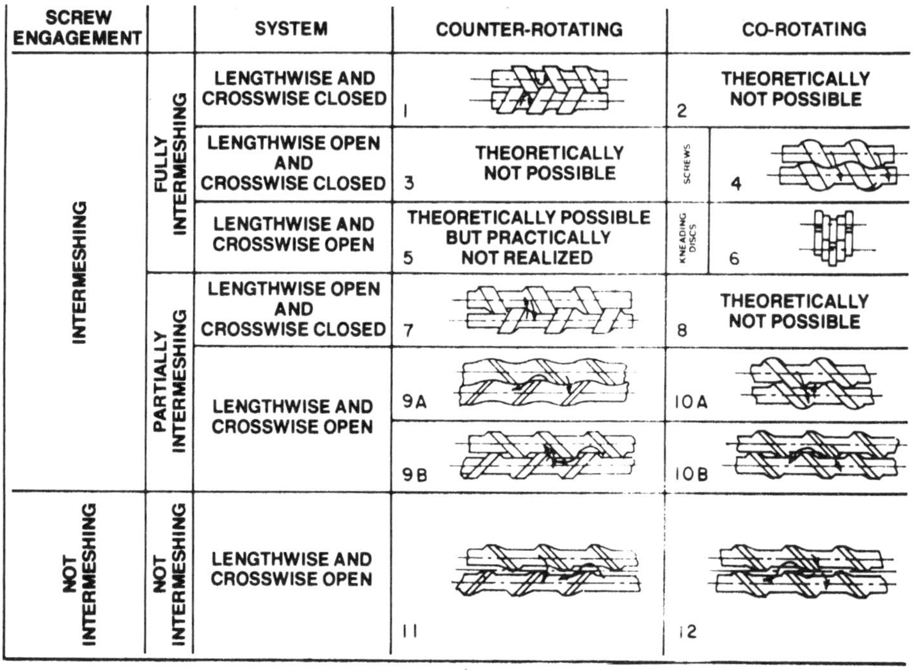

SCREW ENGAGEMENT		SYSTEM	COUNTER-ROTATING	CO-ROTATING
INTERMESHING	FULLY INTERMESHING	LENGTHWISE AND CROSSWISE CLOSED	1	THEORETICALLY NOT POSSIBLE 2
		LENGTHWISE OPEN AND CROSSWISE CLOSED	THEORETICALLY NOT POSSIBLE 3	SCREWS 4
		LENGTHWISE AND CROSSWISE OPEN	THEORETICALLY POSSIBLE BUT PRACTICALLY NOT REALIZED 5	KNEADING DISCS 6
	PARTIALLY INTERMESHING	LENGTHWISE OPEN AND CROSSWISE CLOSED	7	THEORETICALLY NOT POSSIBLE 8
		LENGTHWISE AND CROSSWISE OPEN	9A	10A
			9B	10B
NOT INTERMESHING	NOT INTERMESHING	LENGTHWISE AND CROSSWISE OPEN	11	12

FIGURE 17–9   *Screw meshes and configurations. (Source: Werner & Pfleiderer Corporation, Ramsey, NJ.)*

Several hybrid single-screw extruder designs are available; one incorporates a single screw that reciprocates as it rotates. Fixed pins on the barrel fit between interruptions in screw flights to promote mixing within the screw channel with each revolution. Some designs provide blades instead of pins, which can be adjusted to more or less restrict flow, depending on the angular position of the blade; shear intensity can be varied to suit the process application.

Design Features—Twin-Screw Extruders

Twin-screw extruders can be classified according to the direction of screw rotation and to the amount the screws intermesh with each other (Figure 17–9). A screw system that is lengthwise open (down channel) has a passage from the inlet to the outlet of the machine. Material exchange can take place lengthwise in the screw channel. In a lengthwise closed design, the screw flights in the longitudinal direction are closed at intervals. Crosswise open channels allow material exchange from one flight to another; in crosswise closed channels no material transfer is possible (neglecting mechanical clearances). Whether the screws are open lengthwise or crosswise or have a closed geometry has a direct effect on conveying characteristics, mixing, and pressure-generating capacity of the screw system.

Nonintermeshing twin-screws extruders are closest in principle to single-screw extruders. Two screws are arranged next to each other with the center line distance a little greater than the screw diameter; the screws can rotate at the same or different speeds. The interaction of the screws is limited to a random passage of material from one screw to the other (nonintermeshing screws are open both lengthwise and crosswise). Nonintermeshing systems are available with co-rotating or counter-rotating screws.

Fully intermeshing twin-screw systems can be co-rotating (both screws turn in the same direction) or counter-rotating (screws turn opposite, either toward or away from each other at the tip of the extruder). Fully intermeshing, counter-rotating twin-screw extruders are characterized as lengthwise and crosswise closed, similar to a screw pump. Fully intermeshing, co-rotating twin-screw extruders are lengthwise open and can be crosswise closed with conveying screws or crosswise open when staggered screw discs are used. Fully intermeshing twin-screws are also known as self-wiping.

Partially intermeshing twin-screw extruders are lengthwise open and can be crosswise closed (counter rotating) or crosswise open (co-rotating or counter-rotating). The degree of interaction between the screws is more pronounced than the nonintermeshing systems.

Twin-screw machines are now available that can operate both as a co-rotating and counter-rotating extruder using the same gearbox; a lever device is mounted on the gearbox to change rotation, and the appropriate screws are installed. Another twin-screw system that is not shown in Figure 17–9 is a fully intermeshing , counter-rotating conical screw system. The screw diameters decrease from the feed toward the discharge. This system provides a large free volume for feeding low-bulk-density materials and a large heat transfer surface area.

Single-Screw versus Co-Rotating versus Counter-Rotating

Feeding and Conveying of Solids

When fed with a solid, conveying is the first extrusion operation. Like any screw conveyor, an extruder operates on the principle of frictional relationships between a solid and the surrounding barrel walls. In order to analyze solids conveying, a force balance has to be made. The balance of the frictional forces determines the solid plug transport.

The driving force is the friction between the barrel surface and the plug and is proportional to the pressure exerted by the plug on the barrel. The retarding forces are friction forces on the flights and the root of the screw.

An increased coefficient of friction between the plug and the barrel wall (such as barrel grooves) and a lower coefficient of friction at the flights and the root of the screw (smooth screw) will result in better solids conveying. This conveying mechanism applies both to single- and twin-screw systems.

Single-screw extruders have deep-cut flights in the feed section to maximize volumetric capacity. Throughput is influenced by screw speed, since flood-feeding is commonly used. Barrel grooves improve frictional relationship for solids conveying.

Counter-rotating twin-screw extruders turn outward at the top, inward at the bottom (if the screws turned inward on top, the material would have to be pulled in by the calender gap formed between the screws). The material is conveyed to the lower wedge where it is partially compressed and conveyed as a unit volume. In order to utilize the largest possible free volume, multiflighted deep-cut screws are used. Since the conveying principle of the counter-rotating screws forms closed chamber unit volumes, this machine is ideal for feeding solids. The entire free volume can be utilized, assuming optimum pitch angle.

Co-rotating twin-screw extruders convey material from one screw toward the lower wedge, where the material is compressed and then picked up by the other screw and conveyed further. The flow path is then a figure-eight shape, material being continuously transferred from one screw to the other around the periphery as it moves downstream.

Melting Process

The melting process in all screw machines can be divided into two sections:

1. Compacting of the solids to eliminate air pockets. The air in most cases escapes through the feed opening. Simultaneously, a melt film is formed on the barrel walls; the material fuses due to heat and pressure.

2. Generation of melt through shearing of the material and mixing of the unmolten particles with the already formed melt. As a result, a large heat exchange surface is generated between the solid particles and the melt.

Single-screw extruders achieve compaction in the compression section of the screw where air is forced back to the feed opening. A melt film on the barrel wall is scraped off by the pushing side of the screw flank. The melt pool collects in front of the pushing flank.

Fully intermeshing, counter-rotating twin-screw extruders with tight clearances cannot efficiently compress the air out of the closed chambers toward the feed opening. Loose clearances are intentionally created for this reason. Solid particles are drawn into the wedge area and softened; the plasticized mass is taken in by the calendar gap and collected at the trailing flank.

Solids compression in co-rotating twin-screw extruders is produced through flow restriction. Reverse pitch screw elements or staggered screw discs are used to convey in the opposite direction, causing a pressure build-up upstream of the reverse pitch section. Air is allowed to flow back toward the feed, since the system is lengthwise open. Melting results from the backup length generated by the reverse pitch section.

Venting

Removal of volatiles, moisture, or entrapped air is commonly practiced in single- and twin-screw extruders. Screw geometry, vent port design, and process parameters will vary for venting on single-screw, co-rotating, and counter-rotating twin-screw extruders. Vent ports can be operated atmospherically or under vacuum, depending on the amount of volatiles to be removed.

Single-screw extruders use reduction of pitch or flight depth (increase in root diameter) prior to vent ports to produce a pressure drop for venting. The degree of fill under the vent is less than 100 percent, preventing material from being forced out of the screw. Downstream of the vent(s), the screw must have another compression zone and metering zone to overcome die pressure.

Counter-rotating twin-screw extruders have a reduced pitch prior to the vent opening to create a pressure drop under the vent where pitch may be increased again. Non-intermeshing (tangential) designs turn into one another at the top and drag material away from the vent opening. Lengthwise and crosswise closed channels on fully intermeshing counter-rotating extruders create melt sealing for vacuum venting. Fully intermeshing counter-rotating screws turning away from each other also prevent material from being forced out of the screw channels.

Co-rotating twin-screw extruders use reverse pitch screw elements or staggered screw discs to form a restriction in the screw. Conveying screws upstream of the restriction must overcome this resistance, resulting in a pressure drop in the vent area. Large pitch screws are installed in the vent section to provide a low degree of fill.

Several types of vent ports are used to keep the up-turning side of the screw(s) covered, preventing material from being forced out of the vent opening. Mechanical vent "stuffers" are available to keep material in the screw flights while under vacuum.

Pressure-Generating Capacity

The pressure-generating capacity of an extruder is a function of the screw geometry. Single-screw extruders have lengthwise open channels; pressure is generated by the metering section. Single-screw machines generally use the entire screw length to overcome die resistance (100 percent degree of fill throughout extruder).

Fully intermeshing, counter-rotating twin-screw extruders provide positive displacement through tight clearances. Closed chamber volumes intermittently releasing may lead to pressure fluctuations; enlarged clearances are sometimes used to overcome these pulsations, which also decreases positive-displacement capabilities.

Fully intermeshing, co-rotating twin-screw systems provide partial positive conveying through the wedge resistance. Material viscosity plays an important role in pressure build-up where a lengthwise open channel exists. Due to downstream restrictions, such as dies, the melt accumulates along the screw channel upstream, over several turns of the screw. Back-up length (zone of 100 percent degree of fill) is a function of material viscosity, screw geometry, and die resistance.

Process Variables

Process optimization on screw extruders requires special attention to the interaction of process variables and their influence on process parameters.

Length/diameter (L/D) ratio is used to characterize extruder size. Diameter refers to outside screw diameter, length refers to the effective barrel length. Single- and twin-screw extruders are available from L/D = 3 to L/D = 48. The L/D ratio is kept constant for geometric scale-up. Residence time is a function of extruder L/D ratio.

Screw geometry is a key factor in extruder performance. Screw configuration will influence residence time, mixing quality, specific energy input, discharge temperature, and degree of fill.

Single screw compression ratio, length of metering zone, and other geometric variables are well documented and predictable for thermoplastic processing. Barrier-type screws permit melt recirculation to improve melting and mixing performance.

Counter-rotating twin-screw extruders require selection of screw pitch, clearances between screws and between screw and barrel, and flight width. Interrupted flights are also used to enhance mixing.

Co-rotating twin-screw extruders offer different pitches and conveying directions (right or left hand); additional shear/mixing is provided with staggered screw discs (kneading discs). These mixing elements have various widths and stagger angles to influence mixing intensity and residence time distribution (Figure 17–10).

Screw speed has a direct influence on mixing quality, specific energy input, residence time, and degree of fill. Single-screw and co-rotating twin-screw extruders can operate at 300 to 500 RPM; counter-rotating twin-screw extruders and very large co-rotating twin-screw extruders run much slower due to mechanical limitations (Figure 17–11).

Throughput is used as an independent variable on twin-screw extruders; throughput is dependent on screw speed in single-screw extruders that are flood-fed. Throughput affects degree of fill, residence time, specific energy input, and mixing quality. Common throughputs for twin-screw machines are shown in Figure 17–12.

Barrel temperature has an effect on specific energy input and discharge temperature. The temperature profile can influence material viscosity, effectively thinning or building viscosity in the extruder screw. Extruders processing high-temperature polymers operate with barrel temperatures at 400 to 500 °C; extruders used for intensive cooling have glycol solution circulating through the barrel jackets and screw shafts at −15 °C.

Die pressure can have a significant impact on extruder performance. Open channel screws will have difficulties in generating high pressure with a very low viscosity product, where a closed screw channel will have no problem. Die pressure can be used as a variable in this way to influence degree of fill (called backup length), residence time, and specific energy input (open channel machines). Extrusion pressures for plastics processing are in the range of 500 to 3500 psi. Screw extruder designs are available for continuous service ratings of 9000 psi.

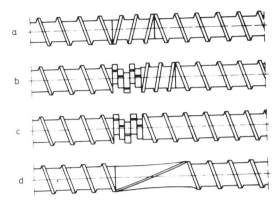

FIGURE 17–10   *Different screw arrangements in the plasticizing zone of a twin-screw, co-rotating extruder: (A) left-hand screw; (B) right-hand kneading block, left-hand screw; (C) right-hand kneading block; (D) large pitch left-hand screw. (Source: Werner & Pfleiderer Corporation, Ramsey, NJ.)*

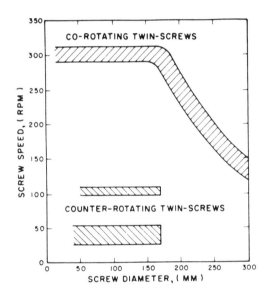

FIGURE 17–11   *Commercially common screw speeds for twin-screw machines. (Source: Werner & Pfleiderer Corporation, Ramsey, NJ.)*

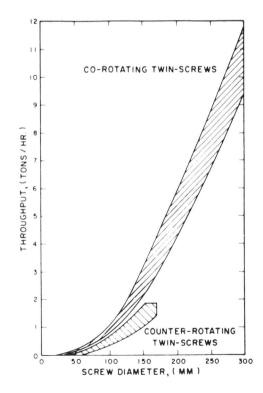

FIGURE 17–12   *Common throughputs for twin-screw machines. (Source: Werner & Pfleiderer Corporation, Ramsey, NJ.)*

Process Parameters

The combination of extruder process variables produces a unique set of process parameters (residence time distribution, specific energy input, heat transfer, mass transfer, and degree of fill) responsible for a given product quality. The objective in scale-up is to reproduce those unique process parameters at a higher rate with larger equipment.

Residence Time

Residence time distribution (RTD) in a screw extruder is a compromise between an ideal tubular reactor (plug flow) and a cascade of ideally mixed stirred tank reactors. The RTD of a screw extruder is more plug-like than a pipe of comparable L/D. Single-screw, non-intermeshing and partially intermeshing twin-screw extruders are not self-cleaning and exhibit long residence time tails. Fully intermeshing twin-screw extruders are self-cleaning; residence time distribution is sharply defined. Axial mixing in a longitudinal direction is shown in the width and length of the residence time distribution curves.

Average residence times can vary from less than ten seconds (short extruder, high speed) to more than ten minutes (long extruder, slow speed, low throughput).

Specific Energy

Specific energy input is defined as kilowatt-hours per kilogram (or horsepower-hours per pound) and describes the mechanical energy introduced into the product by the screw (specific mechanical energy) and thermal energy through the barrel jacket (specific thermal energy). Specific energy is a function of degree of fill, screw design, extruder length (L/D), and temperature.

Heat Transfer

Heat transfer in screw extruders occurs between the material in the screw channel and the barrel wall (or through the screw shaft for additional heat transfer surface area). Screw geometry and extruder design play a decisive role in achieving good heat transfer. Fully intermeshing co-rotating and counter-rotating twin-screw extruders provide constant renewal of material layers in the screw channel; multi-flighted screw systems provide shallow flight depths for better conductive heat transfer. Measurements with polymer melts on fully intermeshing, co-rotating twin-screw extruders have given heat transfer values of 200 to 500 $kcal/m^2$ hr °C.

Mass Transfer

Devolatilization of solvents, monomers, reaction products, or water is accomplished on screw extruders with the use of vent port openings in the barrel. Volatiles can be removed atmospherically or under vacuum, depending on the process requirements. Limiting factors for mass transfer in screw extruders are equilibrium volatile content, residence time in the vent, vapor velocity (vent open area), and film thickness in the screw channel. Devolatilizing extruders usually have several vents to provide staged vacuum strength, increased residence time, and maximum open area. Single-screw extruders use shallow flight depths, while twin-screw extruders use large screw pitch under the vents to produce thin film thickness (low degree of fill) for enhanced mass transfer.

Degree of Fill

The degree of fill in screw extruders is influenced by a combination of process variables. Single-screw extruders are typically 100 percent filled throughout the screw length (except for venting extruders), while twin-screw extruders operate at anywhere from 20 to 100 percent fill along the screw length. Degree of fill is manipulated by screw geometry (forward or reverse conveying, screw pitch), throughput, and screw speed. Low degree of fill is required for venting or downstream feeding and high degree of fill is necessary for pressure build-up.

Extruder Instrumentation and Controls

Extruders are equipped to varying degrees with instrumentation for obtaining process information. Analog or digital meters and strip-chart recorders are usually provided with the appropriate sensors:

- *Torque* indicates percent of available power that is transferred to the process material. Torque may be displayed as percent or motor amps used to indicate load (torque instrumentation must be calibrated to 100 percent of available power).
- *Kilowatts* may be displayed to provide an absolute value of energy consumption. The signal is the product of torque and speed signals.
- *Screw speed* is taken from a tach generator on the drive motor or a speed pickup in the gearbox.
- *Die pressure* is displayed as an absolute value (psi) from a pressure transducer placed in the extruder die, or locally indicated with pressure gauges. Thrust bearings in the extruder gearbox may be fitted with pressure transducers or strain gauges for pressure measurement.
- *Barrel temperatures* indicate the actual steel temperature detected by thermocouples placed along the extruder barrel.
- *Product temperature* is taken from a thermocouple placed in the barrel (or die) that actually contacts the material in the extruder.

Additionally, thermal (heater) power may be indicated with ampmeters or totalizers, flow rates of components from gravimetric feeders, pelletizer speed, and so on. Start/stop push buttons are provided to energize main drive, auxiliary pumps, feeders, and potentiometers used to control screw speed, pelletizer speed, etc.

Screw extruders are operated with open-loop (manual) or closed-loop (automatic) control systems. Microprocessor-based control systems can operate any number of extruders, including upstream and downstream equipment.

Open-Loop (Manual) Controls

Open-loop control is used to describe manual setting of feed rate, screw speed, barrel temperature, etc. Steady state conditions are reached when material temperature, torque, die pressure, and barrel temperatures are at equilibrium ("lined-out"). Thermal equilibrium takes the longest to achieve, since the mass of material within the screw is very small compared with the mass of metal in the barrel.

Several control system interlocks are typically used to prevent mechanical damage from occurring:

- *Underload protection* prevents the extruder screw(s) from running dry in the event of a feeder malfunction or loss of feed material. Typically, extruder drive shuts down when motor torque remains below a preset value for a specified time period.
- *Overload protection* prevents damage to screws, shaft(s), and gearbox. High motor torque shuts down the extruder main drive or causes the coupling to disengage.
- *High pressure interlock* for protecting the thrust bearings in the gearbox will shut down the main drive.

- *Lube oil pressure* and *temperature alarms* will shut down main drive for low lube oil pressure or high oil temperature to protect the gearbox from damage.
- Large extruders may have strategic bearings in the gearbox instrumented for *shock pulse measurements* (SPM) to warn of possible bearing failure. The SPM system is usually programmed to shut down the main drive when the signals reach preset values.

Closed-Loop (Automatic) Controls

Closed loop control of screw extruders provides the same interlocks as in an open-loop control system, and additionally may include automatic start-up and shutdown, data acquisition, and control of one or several process parameters.

Automatic start-up and shutdown is optimized for each product to reduce the amount of scrap material produced, resulting in increased productivity. Start-up and shutdown is reproducible; ramp functions control screw speed, feeders, etc.

Data acquisition requires constant monitoring of process variables. Production reports are generated from process data.

Automatic control of screw extruders provides a method for on-line quality measurement and control. Several control strategies are possible:

1. Constant melt temperature can be controlled by adjustable shear-control devices.
2. Constant pressure at the screw tips can be controlled by adjustable throttle devices.
3. Constant specific energy can be maintained by adjusting feed rate or screw speed.
4. Constant melt viscosity can be controlled by metering viscosity-controlling additive(s) (e.g., free radical chain scission reactants) or process variables (e.g., screw speed).

On-line methods of measuring various aspects of product quality (e.g., viscosity, color, composition, dispersion quality) are being developed to further increase productivity of screw extruders.

Scale-Up

Scaling up is an extrusion operation that requires monitoring of all process variables on laboratory-scale or pilot plant equipment. Scale-up factors are increasing as larger extrusion equipment becomes available; e.g., polyolefin compounding extruders are capable of 20,000 kg/hr, and larger capacity extruders are being designed. Determination of the limiting factor in scale-up places emphasis on accuracy of measurements and reproducibility of pilot plant experiments.

Scale-Up of Feeding and Conveying

A step-by-step analysis of extrusion technology can be accomplished by breaking the process down into its elementary unit operations (feeding, melting, mixing, conveying, venting, and pressure generation). Process parameters that influence scale-up can be isolated, including the effect of basic extruder design (free volume).

Small single screw extruders (3-in to 6-in screw diameter) are relatively easily controlled in terms of power input, feed rate, and conveying. In larger extruders that have much deeper screw channels, the frictional relationship between the material and barrel is very difficult to control. Figure 17–13 shows a typical solids-fed single-screw extruder with one vent port. In the feed section, the solids feed must be compacted to create sufficient friction on the inside barrel walls to assure efficient intake; careful temperature control of the feed barrel area is required. Axial or spiral grooves in the barrel may be necessary in addition to precise temperature control, especially when

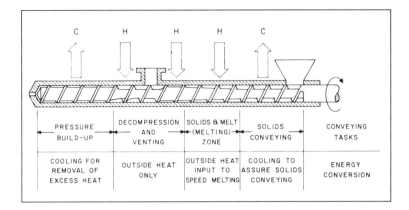

FIGURE 17–13   *Basic conveying functions of single-screw extruder and need for addition and withdrawal of energy. (Source: Werner & Pfleiderer Corporation, Ramsey, NJ.)*

materials with low coefficients of friction are fed (e.g., polymers with slip additives or lubricants).

The basic conveying capabilities of single-screw extruders depend on a number of interrelated factors, some of which are not easily controlled. To achieve proper conveyance in the solids, melt, mixing, and pressure build-up areas, screw design and screw operating conditions are all closely interdependent. This interrelationship becomes more critical with larger screw diameters and larger screw channels.

Twin-screw extruders have improved feeding and conveying characteristics due to the screw geometry. Feed intake and conveying of slippery or wet materials is predictable, dependent solely on free volume (function of third power of screw diameter) and screw speed. Twin-screw compounding extruders usually run faster and provide higher free volume/length than comparable single-screw machines and, therefore, have improved volumetric intake capability.

Scale-Up of Melting and Mixing

An extruder is an energy conversion device. In order to maintain a proper energy balance, it is generally necessary to add heat by conduction and to withdraw heat by cooling. The poor heat transfer characteristics of polymer melts are well known; problems associated with heat transfer become more severe with increasing screw channel depth and increased thicknesses of the polymer layer in the screw channel. The heat transfer area in an extruder increases with the second power of screw diameter, but the volume of material with a given residence time increases with the third power of screw diameter (assuming a constant L/D ratio and comparable screw geometry). Heat transfer scale-up in single-screw extruders becomes increasingly difficult with larger screw diameters; the ratio of inner barrel surface area to volume in twin-screw pilot plant extruders does not differ much from production size units.

Laminar flow patterns in the screw channel of single-screw extruders are disrupted with sufficient statistical frequency by leak-flow over the screw flights leading to changes of shear direction; this is essential for good mixing. As screw diameter increases, the statistical frequency (per volume of material) of shear direction changes decreases, often to a point where adequate mixing cannot be achieved. Additional shear and mixing devices are used as modifications of basic single-screw concepts (e.g., barrier-type screws, mixing pins, etc.) with good results in operation, but are difficult to scale-up.

Twin-screw extruders provide rearrangement of material layers with every rotation as material is transferred from one screw to the other. The direction of shear relative to the original shear planes can be changed with selection of screw pitch (counter-rotating) or use of staggered screw discs (co-rotating).

Scale-Up of Venting Operations

In removing large quantities of air or volatiles, the problems of foaming and entrainment exist. In the case of removal of volatiles in low concentrations, the problems are primarily caused by very slow diffusion speed of the volatiles in the molten polymer. Residence time, film thickness, and renewal of surfaces become a significant factor in venting. The thickness of the material layer controls diffusion rate; increasing screw channel depths on large-diameter screw extruders makes diffusion-limited scale-up difficult for single- and twin-screw machines. Vent open area and exposed screw surface area are scaled to produce similar vapor velocities in the vent ports.

Scale-Up of Pressure Generation

Screw extruders dissipate energy in generating sufficient pressure to pump molten material through a die. This energy results in temperature rise of the material being pumped. Controlling discharge temperature becomes increasingly difficult with deeper screw flights due to poor heat transfer. The pressure drop for a given die geometry can be calculated, given the rheological behavior of the material. The scale-up task is to predict the temperature rise in screw extruders as a function of pressure.

The "wetted" section of screw used to build up pressure is called the backup length (zone of 100 percent degree of fill).

In single-screw extruders, the entire screw is 100 percent filled (except for vent areas); the length of the metering zone with fully molten material is used to determine pressure-generating capability and temperature rise. The backup length in twin-screw extruders might only be a few turns of the screw. Because of improved pumping efficiency, viscous energy dissipation of intermeshing twin-screw extruders is greatly reduced.

Scale-up models can predict temperature rise in generating a specific pressure with a given screw geometry (single- or twin-screw) and rheological properties of the material; backup length (twin-screw) or length of the metering zone (single-screw) necessary to develop this pressure can also be calculated.

Materials of Construction

Screw extruders experience abrasion and/or corrosion depending on the process application; materials of construction for screw and barrels are selected for optimum service life. Most screw extruder components are manufactured from carbon steel and may be nitrided or through-hardened for wear protection. Modular designs of screw elements and barrels have distinct advantages for corrosion and wear protection. New machines only have to be equipped with wear-protected components in the area where wear or corrosion is expected. In a corrosive environment, the generation of a galvanic element by pairing of different materials has to be avoided. With varying barrel wear along the processing section, barrels can be exchanged and relocated so that all barrels are worn down uniformly to the point where a replacement becomes necessary. The same philosophy applies to screw elements.

Besides adhesive wear, which is experienced in some extruders due to contact between screw and barrel, the following wear mechanisms may be observed:

1. *abrasive wear* due to the material being processed or fillers contained in the material (e.g., glass-fiber-reinforced thermoplastics, mineral fillers, etc.)
2. *corrosion* due to aggressive acidic or basic components in the processed material (e.g., fluoropolymers, chemical reactions, etc.)

Designs for Wear-Protected Barrels

Several designs are available for manufacture of barrels for corrosion- and wear-protection. Exchangeable liners provide the most flexible and economic solution. Liners can be press-fit into barrels or split-barrels designed to bolt around the liner. The space between the liner and split-barrel is usually filled with heat-transfer cement to eliminate air gaps. Barrel liners are typically through-hardened, manufactured from castings, or machined from forgings. High chrome, high carbon alloys provide excellent wear resistance due to a high content of chromium carbides. Bimetallic liners may also be utilized, with two bimetallic tubes welded together for twin-screw applications. The bimetallic liner has high adhesive wear resistance (NiCrB or NiCrMo alloys with or without carbide reinforcement) with a ductile backing material.

Barrels that do not have liners may have surfaces welded with hardfacing materials and machined to original bore dimensions. Unlined barrels may also be machined to accept a liner.

Corrosion-resistant barrels or liners may be manufactured from hardenable chromium steel or CoMoCr alloys for extremely high corrosion resistance in addition to very good wear resistance.

Designs for Wear-Resistant Screws

Extruder screws are subject to the same corrosive and/or abrasive environment as barrels and are manufactured from material suited to the application. The crests of screw flights can be welded with hardened materials (e.g., stellite), screws can be through hardened (e.g., tool steel), spray-coated (e.g., carbide coatings), or chrome plated; corrosion-resistant materials (e.g., Inconel® -type, hardenable stainless steel) are also available.

Extruder manufacturers can provide recommendations for materials of construction based on field experience.

Extruder Maintenance

Routine maintenance of screw extruders can be divided into two sections: gearbox/drive motor and process section. Extruder manufacturers provide recommendations for preventive maintenance intervals based on experience in normal operations; modifications to these intervals will depend on the particular process application.

Gearbox/Drive Motor Maintenance

Scheduled maintenance on the extruder gearbox involves checking lubricating oil level, temperature, and pressure on a regular basis. Lube oil specifications are provided by the extruder manufacturer. Service life of gear oil depends on gearbox loading and environmental factors, but it typically should not exceed 12,000 operating hours or three years. Oil should be inspected every 2,000 operating hours (sample analysis).

Radial bearings have an expected lifetime as a function of average screw shaft speed and average load. This information is provided by the extruder manufacturer to schedule radial bearing replacement in the gearbox.

Thrust bearings also have an expected lifetime as a function of average screw shaft speed, average material pressure, and average load. As with the radial bearings, expected lifetimes are calculated for normal operating conditions; operation at low screw shaft speed, low material pressure, or low motor loading increases lifetime. In all cases, bearing overhaul is typically recommended within 64,000 operating hours.

Shock pulse metering (SPM) or acceleration-spike energy systems are available to monitor extruder bearing conditions, giving early warning of possible failure.

Gearbox overhauls are usually performed on site with or without supervision from the extruder manufacturer, or gearboxes can be returned to the manufacturer for overhaul.

Torque-limiting clutches that are not positive-disengagement-type should be checked if clutch slipping occurs; a clutch will disengage at a different setting as pads or shoes wear (recalibration may be necessary).

Maintenance of drive motors is generally performed under recommendations from the motor manufacturer.

Process Section Maintenance

Maintenance of extruder screws and barrels becomes necessary when machine performance deteriorates due to abrasive or corrosive wear; routine inspection of process section components is recommended.

Worn barrels and screws can be reconditioned to factory specifications depending on their condition and material of construction. New crests can be welded on worn screw flights and precision-ground to original dimensions.

Barrel cooling channels may become fouled or have scale deposits that restrict the flow of heating/cooling fluids. Plugged or fouled cooling bores are often repaired by the extruder manufacturer.

SUGGESTED READING

Internal Mixers

Funt J.M.: Principles of Mixing and Measurement of Dispersion. Rubber World, February 1986, pp. 21–32.
Giffin H.: Interlocking Rotor Developments. European Rubber Journal, May 1984, pp. 25–29.
Melotto M.A.: Rotor Design and Mixing Efficiency. Rubber World, February 1986, pp. 37–39.
Nevett R.F.: Mixing Practice. Rubber World, February 1986, pp. 33–36.
Schmid H.M.: Quality and Productivity Improvements Using Intermeshing Rotor System for Internal Mixers.

Extruders

Eise K., Herrmann H., Werner H., Burkhardt U.: An Analysis of Twin-Screw Extruder Mechanisms. Advances in Plastics Technology, Vol. 1, No. 2, April 1981, pp. 18–39.
Accuracy in Extruder Scale-Up. Polymer Processing News, Vol. 5, No. 1, Werner & Pfleiderer Corp.
An Analysis of Conveyance in Twin-Screw, Co-Rotating Extruders. Polymer Processing News, Vol. 11, No. 2, Werner & Pfleiderer Corp.
Twin-Screw vs. Single-Screw Extrusion. Polymer Processing News, Vol. 4, No. 2, Werner & Pfleiderer Corp.

Chapter 18

Conveyor-Based Processing Systems*

With numerous process plants employing conveyors of one type or another, it was felt that this text should give at least an introduction to this type of machinery by focusing on one of the more sophisticated executions: steel-belt conveyors.

The use of steel-belt conveyors has spread throughout the processing industries. Applications of steel-belt conveyors include cooling/solidification, drying, pressing, freezing, baking, and materials conveying.

The steel belt is made from flat strip steel from a rolling mill, prepared through special techniques that straighten, flatten, and make the ordinary strip suitable for welding into endless bands continuously running around two terminals. The conveyors based on this specialized technology are designed for the processing industries according to the needs of the product and the special needs of the steel belt.

Table 18.1 summarizes a wide variety of steel-belt applications and the important steel-belt properties that make the applications successful. The general categories that are shown in Table 18.1 are material handling, food processing, industrial processing, and presses for particle boards, plastics, and rubber. Table 18.1 also indicates the four major steel-belt grades that are in common use.

The following discussions describe the applications and processes for which steel-belt conveyors have been selected as the best of competing alternatives, including the types of materials used for conveyor belts. Throughout this chapter, the reader will find examples of industries and products where steel belts are being used.

BELT GRADES

There are numerous grades of steel that can be rolled and processed to be suitable for use as a conveyor belt. The selection of belt material is governed by the product to be conveyed and the process conditions.

For the large majority of applications, four grades of steel conveyor belts are used:

Austenitic Stainless Steels

American Iron and Steel Institute (AISI) type 301 is available in widths up to 61 inches. For conveyor belt applications, it is cold rolled to a hardness in the range of Vickers Hardness (HV) 380 resulting in a high tensile strength steel in the range of 160,000 psi. This steel has good corrosion resistance and remains stable at temperatures ranging

---

*Source: Sandvik Process System, Inc., Totowa, NJ. Adapted with permission.

Table 18.1 Areas of Steel Belt Application

Category	Steel Belt Installation Type	Strength—Ambient Temperature	Strength—High Temperature	Strength—Cryogenic Temperature	Stability against Temperature Difference	Corrosion Resistance	Wear Resistance	Hygienic	Flatness	Straightness	Surface Finish	Carbon (C)	Austenitic (SA)	Martensitic (SM)	Strength Martensitic (SM)
Material Handling	Conveyors, general	•	(•)				•					X			X
	Sorting systems	•	(•)			(•)	•		•			X			X
	Work tables, general					(•)	(•)		(•)			X	X	X	
Food Industry	Meat cutting tables					•		•	(•)				X		
	Bottle handling	•				(•)	•		•			X			X
	Bake ovens	(•)			(•)		•		•	(•)	(•)	X			
	Contact freezers			•		(•)		(•)	(•)				X		
	Belt coolers/food				(•)	(•)		(•)	(•)				X	X	
Industrial Process Applications	Belt coolers/chemicals				(•)	(•)		(•)	(•)				X	X	
	Steel belt dryers		(•)		(•)	(•)			•	(•)	(•)	X	X		
	Flow-through belt unit				(•)	(•)			(•)			X	X	X	
	Double belt presses	•			•	(•)			•		(•)	X	X		X
	Belt skimmers				•								X		
Particle Board Industry	Single opening presses				•	(•)	•		•	•		X			X
	Multi opening presses				•	(•)						X		X	
	Rotation presses	•			•	(•)			•	•			X		X
Rubber	Rotation presses	•							•	•	(•)		X		X
Plastics	Rotation presses	•					•		•	•	(•)		X		X

(•) = property sometimes of importance

 • = property always of importance

Source: Sandvik Process System, Inc., Totowa, NJ.

from $-200\,°C$ to $+400\,°C$. Due to physical properties (relatively low thermal conductivity and high thermal expansion of austenic stainless steels), the temperature must be kept uniform over the width to avoid thermal distortion, which could affect the product or process. This material has good welding properties; however, the weld must be work hardened to restore hardness and strength in the weld area.

AISI type 316 has a higher nickel content and is alloyed with molybdenum to obtain better corrosion resistance. This material is also cold rolled to increase strength, with maximum hardness in the range of HV 320. Tensile strength is about 130,000 psi at 167 °F. As type 316 is more difficult to produce and has lower mechanical properties, it generally is used only when 301 is unsuitable.

Hardened and Tempered Carbon Steel

This grade approximates AISI 1065 and is available in widths up to 48 inches. After cold rolling it is hardened and tempered through heat treatment to obtain a hardness of about HV 400 and tensile strength up to 180,000 psi at 120 °F. The hard springy nature of this steel makes it suitable for almost any application where the risk of corrosion is low. It has excellent heat transfer properties and a low coefficient of expansion, making it ideal for heating or baking ovens and many material handling applications where a hard, smooth, wear-resistant belt surface is desirable. Due to the carbon and manganese content and heat treatment of this steel, the welding procedure is rather complicated; however, with proper temperature control and heat treatment of the weld, very good results can be obtained.

Low Carbon Martensitic Stainless

This is a grade characterized by high strength with hardness of HV 350 and tensile strength of about 155,000 psi, with good corrosion resistance and excellent welding properties. This material is furnished in widths up to 61 inches in heat-treated condition. This material has a thermal conductivity slightly better than austenitic grades and a thermal expansion closer to carbon steel, making it more stable against thermal distortion than the austenitic grades. Strength of the weld is very close to the parent strength.

Precipitation-Hardened Stainless Steel

This material is characterized by extra-high tensile strength, high fatigue strength, good corrosion resistance, excellent weld strength, and good repairability. It is available in widths of up to 61 inches and has a hardness of HV 500 with tensile strength up to 240,000 psi. This grade is not suitable for use at low temperatures.

As stated earlier, there are numerous factors to consider in selecting the best belt grade for any given process or product. Table 18.2 gives some of the physical properties of steel-belt materials.

All of the grades mentioned can be longitudinally welded to obtain wider belts. Depending on the material, the longitudinal welds are made by the same procedure as transverse welds, with work hardening or heat treatment requirements identical. With existing equipment, it is possible to make up to three longitudinal welds with a maximum width of 180 inches. The length of any belt segment is limited to approximately 300 feet; however, these segments may be transverse welded to produce almost any length.

Other specialized treatments of steel belts include perforation, which is usually done to allow air to flow through the belt, but is also used in belt washers and for

Table 18.2 Selected Physical Properties of Steel Belt Materials

	Yield Strength 0.2% offset Vickers (ksi)		Tensile Strength Vickers (ksi)		Hardness (HV)		Thermal Conductivity (BTU/hr-ft°F)	Thermal Expansion (10⁻⁶/°F)
	Parent	Weld	Parent	Weld	Parent	Weld		
Austenitic stainless steel								
AISI 301	142	91	160	128	380	300	9.5	9.8
AISI 316	132	73	141	102	320	250	9.3	8.9
Carbon steel	174	128	186	144	400	350	22	6.2
Low carbon martensitic	145	138	157	152	350	350	11.8	6.0
Precipitation hardened stainless steel	236	217	239	232	500	500	9.2	6.1

Notes: Yield strength and tensile strength at room temperature; thermal conductivity at 212 °F; thermal expansion over range 68–212 °F.

Source: Sandvik Process System, Inc., Totowa, NJ.

baking. Polished or matter finishes may be supplied for special product surface effects. A variety of edging or vee-rope guides may be attached to belts for guiding or holding a liquid on the belts.

COOLER/FLAKERS

The most prevalent application of processing on steel belts is for cooling and solidifying molten chemicals. Some of the products presently being processed on steel-belt coolers are listed in Table 18.3.

Steel-belt coolers are also known as flakers. In the case of sulphur solidification, the steel-belt cooler is called a slater, for the 1/4- to 1/2-inch thick flat sulphur product known as slates.

The competitive device for solidifying chemicals is the drum flaker, a circular, internally cooled drum that has become relatively expensive compared with steel-belt flakers and therefore is not presently used very often for new installations.

Manufacturers of standard steel belt coolers have designs for a variety of belt widths (i.e., 500, 800, 1200, and 1500 mm). Zones or modules in the United States are eight, twelve, or sixteen feet long in the direction of the conveyor. Any number of these zones (within reason) may be combined to give the desired process and the desired capacity.

Cooling a Product on a Steel Belt

In the majority of steel-belt cooler applications, a coolant liquid is sprayed against the underside of the top strand of the conveyor. Temperature of the coolant may be specially selected depending on the nature of the product being carried on the belt, or in some cases, it may be simply whatever is available. Selection of the optimum coolant program is discussed below.

A typcial cross section showing coolant impinging on the underside of the belt is pictured in Figure 18–1. Volumetric flow rates are varied from 0.5 to 1.5 gpm per square foot of cooled belt, depending on temperature level, heat transfer from the hot product, and limitations on heating the coolant. The coolant film is scraped off as it

Table 18.3 Products Cooled or Flaked on Steel Belt Conveyors

Anti-oxidants	Phthalic anhydride
Hot melt adhesives	Asphalt and pitch
Resins of all types	Chocolate
Aluminum sulfate	Pesticides
Sulfur	Surfactants
Agar-agar	Atactic polypropylene
Maleic anhydride	Stearic acid
Animal fat	Waxes

Source: Sandvik Process System, Inc., Totowa, NJ.

passes over periodic cross bars, being immediately restored by the next set of spray nozzles.

Film coefficients of a variety of spray nozzle types were measured for this configuration as approximately 40 BTU/hr-ft^2 °F for design flow rates. For thin, thermally conductive products, the coolant film coefficient is relatively important for optimum flaker performance. Good spray nozzle operation must be maintained. On the other hand, thick insulating products cannot be cooled more rapidly by improvements in coolant film coefficient. The temperature of the product as carried down the conveyor is as shown in Figure 18–2. As a rule of thumb for typical belt cooler/flakers, one-half the heat load from the product is absorbed in the first third of the cooling length.

One of the advantages of processing on a conveyor is the ability to put a number of cooling zones or tanks in series. Each zone can be treated differently than the others, if desired.

If a product is placed on the belt in a thick layer and does not immediately "set up," retaining side strips on the belt may be required. Some products, placed molten on the cooling belt, do not remain as flat solids but tend to "curl" up at the edges or take a characteristic gull-wing shape. This tendency is probably due to nonuniform changes in the coefficient of expansion as the product is solidified and cooled. When one of these shapes occurs, the product pulls away from the belt, conductive heat transfer from product to belt ceases, and cooling of the product takes much longer. In many cases, curling of the product can be minimized by tempering the coolant water, i.e., by using high-temperature coolant in the initial zone and gradually reducing the coolant temperature in successive zones. Often, three different temperature levels are required. It is postulated that in the cases where tempering is effective, the product cools uniformly throughout its thickness, thus minimizing any differential contraction/expansion effects.

The opposite condition from a curling product is when the product sticks too rigidly to the belt. Some products tend to form a good adhesive bond to the belt when the belt is clean, dry, or heated. To avoid sticking, (1) the belt could be cooled before the product is placed on it, (2) a release agent such as silicone could be used, or (3) moisture on the belt could also act as a release agent. The challenge to the conveyor process engineer is to find the set of conditions whereby the product will initially adhere to the belt, thus promoting good heat transfer, but on reaching the discharge terminal will separate cleanly from the belt. A surprisingly large fraction of the products run on steel belts have these characteristics.

Without mechanical cooling from above, the product, cooling to the ambient air, still represents 10 to 15 percent of the total heat load. With properly designed fans and nozzles directed against the product, 50 percent of the heat load could be transferred above the belt. For thin, brittle products, high-velocity air may tend to blow the product off the belt and so must be used with appropriate caution.

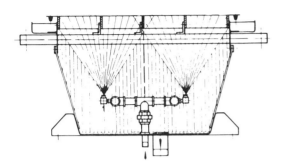

FIGURE 18–1   *Cross section showing conveyor cooling water being sprayed on bottom of belt. (Source: Sandvik Process System, Inc., Totowa, NJ.)*

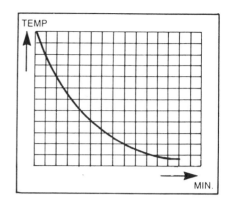

FIGURE 18–2   *Product temperature versus time for a cooling conveyor. (Source: Sandvik Process System, Inc., Totowa, NJ.)*

Supercooling Materials

Supercooling materials do not solidify as do "well-behaved" materials, but instead cool below their melting points without a transition to a hard, solid state. These supercooling materials may remain as cold, mushy pastes unsuitable for packaging or further processing. Typical examples of supercoolers are antioxidants for the rubber industry and aluminum sulphate.

When supercooling occurs, the techniques used for obtaining solidification include careful temperature control, addition of solid, granulated seed material, and energy addition through use of a scraped-surface heat exchanger before the product is deposited on the conveyor/cooler. After being formed into a sheet on the conveyor, cooling must be carefully controlled for solidification to a hard flake to take place. An additional curing conveyor is sometimes used to provide additional time for product hardening. Figure 18–3 shows the complete process requirements for aluminum sulphate.

Steel-Belt Accessories

Various feeding devices are used to form continuous sheets, strips, or droplets. The sheets or slabs of product are typically one to three millimeters thick, although there are certainly examples outside this range. Sheets are formed by (1) extruding through a die, (2) liquid flowing over a weir, and (3) a variety of special devices suited to the particular properties of the product.

Hot-melt adhesives, in particular, are formed in strips, cooled, and then cut into *slats* or *chicklets* by a multibladed cutter at the discharge end of the steel-belt cooler. Strip forming of brittle materials such as caprolactam is used to produce a more uniform particle size than if conventional flaking is used.

Maximum uniformity of solidified product is obtained with drop-forming or pastillating devices. In the past, these devices were reciprocating, required relatively slow belt speeds, and required a lot of maintenance for optimum product quality. Now, a rotary system is available that overcomes the previous objections. Producers are interested in forming drops that solidify to become an almost completely dust-free product in order to provide a marketing advantage over their competitors and because there is virtually no loss in production capacity for a given area of cooling belt. Some of the

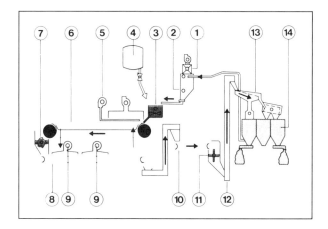

FIGURE 18–3 *Aluminum sulphate process with steel-belt solidification. 1 = Seeding material; 2 = vibrator; 3 = agitator; 4 = boiler; 5 = fan; 6 = steel belt cooler; 7 = coarse crusher; 8 = accumulation and recooling belt; 9 = fans; 10 = belt conveyor; 11 = mill; 12 = elevator; 13 = vibratory screen; 14 = silos. (Source: Sandvik Process System, Inc., Totowa, NJ.)*

products presently being formed in pastille shape are shown in Table 18.4. A photograph of a rotating drop-forming device on a steel belt is shown in Figure 18–4.

For producers not having a need for uniform, dust-free products, the steel-belt conveyor manufacturers provide rotating-bar breakers and crushers that reduce flake size for easy handling and bagging.

Double-Belt Cooler

In some specialized cases, an effective method of cooling is to place a second steel belt above the product, as shown in Figure 18–5. The product is loaded onto the upper surface of the unit's lower belt (1), which carries it into the cooling zone, where the pressure of the upper belt (2) ensures contact with both cooling surfaces. The upper belt is sprayed with cooling water (3) from above, and the lower belt is sprayed from below. Loading (4) and discharge methods (5) are simply arranged to suit individual products and can incorporate breaker equipment (6) at the discharge station if required.

The double-belt cooler is used for

- curling products that cannot be controlled otherwise
- minimizing oxidation of products while hot

Table 18.4 Products Being Drop-Formed on Steel Belts

Hydrocarbon resin	Wax
Bitumen mix	Acrylic resin
Antioxidants	Nickel catalyst
Pitch	Petroleum resin
Cobalt naphthenate	Hot melt adhesive
Cobalt stearate	Sulphur
Terpene resins	Epoxy resin
Rosin resins	Food sauce
Vegetable fat	Polyester resin

Source: Sandvik Process System, Inc., Totowa, NJ.

FIGURE 18–4   *Rotoformer formation of pastilles on a steel belt. (Source: Sandvik Process System, Inc., Totowa, NJ.)*

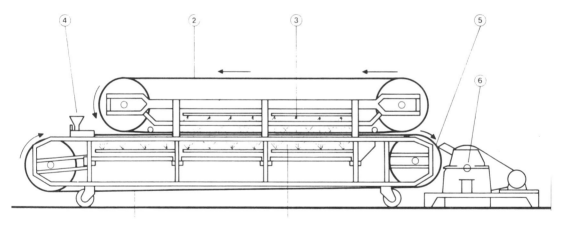

FIGURE 18–5   *Schematic of a double-belt cooler. (Source: Sandvik Process System, Inc., Totowa, NJ.)*

- increased capacity where space is a limitation
- more uniform discharge temperatures across width to eliminate downstream grinding problems

HEATING/DRYING APPLICATIONS

There are several prominent applications of steel-belt conveyors in which the product is heated, dried, or baked while on the belt. In each case, the unique configuration of the wide, flat, moving surface contributes substantially to the process. The types of applications for which steel belts provide the best solutions to the needs of manufacturers are

- film or sheet forming
- hot air drying or baking where sanitation or sequenced zones are important
- high temperature heating on a smooth belt

Examples of these applications will be discussed.

Film formation or casting a liquid on a steel belt uses the belt as part of the forming process and then uses the belt as the carrier during the drying or evaporation stage. Often the belt is polished. Starting with the standard 2B mill finish, belt polishing can be carried out to any degree up to a mirror finish. Product film properties will reflect the belt surface. Photographic film, filter membranes, polyvinyl alcohol film, and modified tobacco sheet are common examples of steel-belt processed products.

Dried tobacco sheet prepared to an approximate thickness of 0.022 inches and a moisture level of 11% is shown in Figure 18–6 being doctored off the discharge drum of a steel-belt conveyor moving at approximately 150 feet per minute. This sheet is formed from a high-moisture-containing slurry and requires careful atmospheric control of temperature and humidity to maintain the integrity of the drying sheet.

For another group of applications, direct heating of the belt by gas ribbon burners and/or indirect convective heating is used for baking cookies, drying cereals, and drying fruit such as apples. For the cookies, the carbon steel belt provides a surface on which a forming roll drops individual pieces. By passing through the continuous baking oven, the belt provides the proper thermal properties for this particular baking process.

FIGURE 18–6   *Tobacco sheet on a steel-belt conveyor dryer. (Source: Sandvik Process System, Inc., Totowa, NJ.)*

In drying some products such as apples, initial drying of the wet product is performed with a relatively thin bed and high temperature. When the surface water is evaporated, the temperature in the next zone in the dryer is reduced and the bed depth is increased.

Zones may be used in several ways in belt dryers to provide the optimum conditions required by the product. With a single belt, each module along the length of the belt may have a different temperature and humidity. With multiple belt systems, even more flexibility is possible.

Cereal preparation on a steel belt provides a sanitary surface that permits cutting and stamping operations. Sugar application is practical because the belt can be continuously washed if desired.

Many rubber profiles are cured on steel belts running in high temperature ovens. The belt provides a smooth, clean surface with belt temperature capability to 750°F.

DOUBLE BELT PRESS

The double belt press is based on the principle of combining and compressing products between two solid steel conveyor belts that are welded endless and are passing continuously through the press. As heating and/or cooling are generally involved in most press applications, the moving belts and belt support systems must have good heat transfer properties and must be capable of continuous operation at elevated temperatures and pressure. A smooth, uninterrupted surface is also required for a good surface finish on the product.

The smooth, flat, high-strength endless steel belt meets all of these requirements. Because the press can be built up in modules, it is possible to satisfy a great variety of process requirements with various combinations of pressing, heating, and cooling zones.

The double belt press is normally designed to operate in a pressure range of one to ten bar. If the belts were placed directly in contact with the platens, the sliding friction created between the belt and press platen would be too high for pulling the belts through the press. Therefore, a unique traveling roller support system between the belts and press platens is used to eliminate sliding friction, allowing the belts to glide easily through the pressing zones (Figures 18–7 and 18–8).

Figure 18–8 shows the overall press configuration with two terminal sections, a heating press zone and a heating or cooling press zone. Each belt is continuous and the product may be produced in any length. The roller chain assemblies called out in Figure

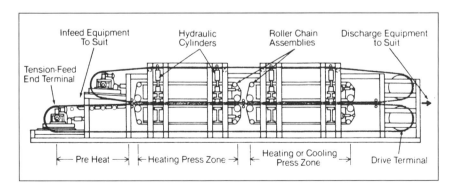

FIGURE 18–7   *Double-belt press general assembly. (Source: Sandvik Process System, Inc., Totowa, NJ.)*

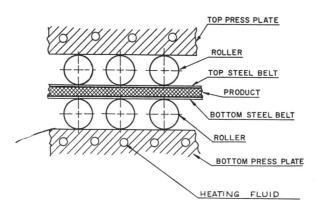

FIGURE 18–8   *Product is sandwiched between steel belts that are backed up by a traveling chain of rollers. (Source: Sandvik Process System, Inc., Totowa, NJ.)*

18-7 are detailed in Figure 18-8. The rollers are on an endless loop and travel at half the linear speed of the belt.

A large number of closely spaced moving nip pressures pass over the surface of the product through the steel belt, even though there is no relative motion between the product and the steel belt. This action tends to average out any tolerance variations, resulting in very uniform thickness control.

Original designs were for belt widths up to 1550 mm, the maximum belt width previously available. With the development of new belt grades, such as precipitation hardened stainless steel and longitudinal techniques, it is now possible to increase width, but at a somewhat lower maximum pressure due to structural deflections and machine tolerances.

As seen in Figure 18-8, heat transfer or cooling fluid is circulated through the upper and lower press platens. The heating or cooling is transmitted through the rollers to the steel belts and to the product. By controlling oil temperature and press pressure, and by using various length zones, almost any time, temperature, and pressure profile can be furnished to suit the product requirements. It should be noted that a product can be passed directly from a heating zone to a cooling zone without releasing pressure and with minimum loss of energy, since only the relatively thin steel belts are repeatedly heated and cooled.

For low pressure, up to 0.33 bar, double belt presses can be designed with a closely spaced, stationary roller support system for the steel belts. In this case, the belts can be heated or cooled in zones by direct air impingement on the back side of the belts. This approach is not practical for pressures higher than 0.33 bar due to belt deflections between rollers (and therefore high stress) and roll deflection. Maintaining thickness uniformity in the product becomes more and more difficult.

The press can be operated as a fixed pressure machine by allowing the upper platens to "float" at controlled hydraulic pressure on the product. The thickness control in this case is a result of the volume and compressibility of the material entering the press. The press can also be operated as a fixed-gap machine by pulling the top platens down against fixed stops. In this case, the thickness is held uniform; however, final density is controlled by the volume of material entering the press. In either case, a pressure release system is built in to protect against overloading and exceeding the design limits.

Some typical applications for the double belt press are as follows:

- glass-reinforced plastic laminate (PP, PET, PE, etc.)
- Ski laminates (fiberglass rovings with epoxy resin)

Table 18.5   Products Frozen on Steel Belt Conveyors

Fish fillets	Whole herring
Hamburgers	Chicken pieces
Shrimp	Food sauces
Coffee concentrate	

Source: Sandvik Process System, Inc., Totowa, NJ.

- printed circuit board (copper clad)
- fiberglass-reinforced laminate
- rubber mats
- PVC sheet form-fused granules
- synthetic leather

FREEZING CONVEYOR

The stainless steel conveyor belt forms the basis of a freezing system for a variety of foods. Ease of maintenance and excellent hygiene properties are important qualities. Most products are frozen hard and readily break loose from the belt as the belt curves around the discharge pulley. Products frozen on steel belts are given in Table 18.5. Typical freezing time for a raw fish fillet, one-half inch thick, is ten minutes. Thinner products freeze even faster. Yield from this process is virtually 100 percent, since moisture losses can usually be eliminated and product breakage is minimal because of the smooth belt surface.

The coolant used under the stainless steel belt is usually calcium chloride or propylene glycol. Upper belt coils may be arranged for ammonia or freon, and defrosting can be achieved by electric, water, or hot gas methods. Belt freezers have also been designed with refrigerated air both above and below the belt.

New products may be evaluated in laboratory tests and production equipment sized accordingly.

Index